STUDY GUIDE

Jennifer Shanoski
Merritt College

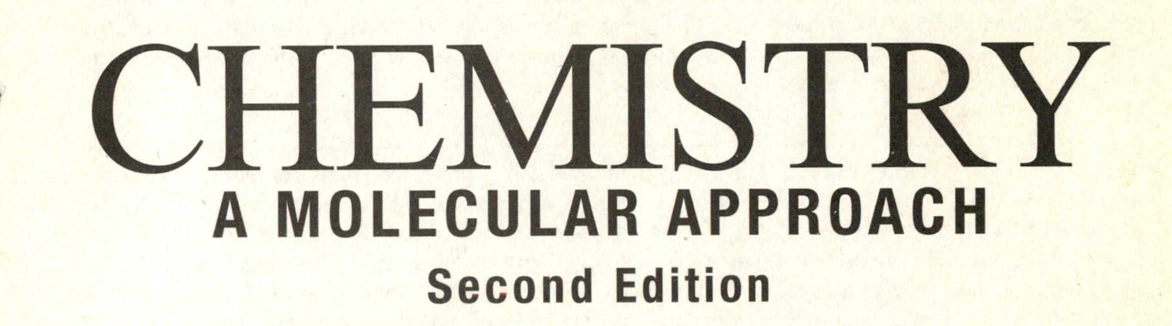

NIVALDO J. TRO

Prentice Hall
Boston Columbus Indianapolis New York San Francisco Upper Saddle River
Amsterdam Cape Town Dubai London Madrid Milan Munich Paris Montréal Toronto
Delhi Mexico City São Paulo Sydney Hong Kong Seoul Singapore Taipei Tokyo

Publisher: Dan Kaveney
Editor in Chief, Chemistry and Geosciences: Nicole Folchetti
Marketing Manager: Erin Gardner
Project Editor: Jennifer Hart
Managing Editor, Chemistry and Geosciences: Gina M. Cheselka
Project Manager: Traci Douglas
Operations Specialist: Maura Zaldivar
Supplement Cover Manager: Paul Gourhan
Supplement Cover Designer: Tina Krivoshein
Cover Illustration: Quade Paul

Pearson Prentice Hall
Upper Saddle River, NJ 07458

Printed in the United States of America

10 9 8 7 6 5 4 3 2 1

ISBN-13: 978-0-321-66788-5
ISBN-10: 0-321-66788-3

Prentice Hall
is an imprint of

www.pearsonhighered.com

Table of Contents

Preface		v
Chapter 1	Matter, Measurement, and Problem Solving	1
Chapter 2	Atoms and Elements	12
Chapter 3	Molecules, Compounds, and Chemical Equations	22
Chapter 4	Chemical Quantities and Aqueous Reactions	39
Chapter 5	Gases	51
Chapter 6	Thermochemistry	66
Chapter 7	The Quantum-Mechanical Model of the Atom	80
Chapter 8	Periodic Properties of the Elements	91
Chapter 9	Chemical Bonding I: Lewis Theory	102
Chapter 10	Chemical Bonding II: Molecular Shapes, Valance Bond Theory, and Molecular Orbital Theory	115
Chapter 11	Liquids, Solids, Intermolecular Forces	127
Chapter 12	Solutions	142
Chapter 13	Chemical Kinetics	155
Chapter 14	Chemical Equilibrium	173
Chapter 15	Acids and Bases	190
Chapter 16	Aqueous Ionic Equilibrium	206
Chapter 17	Free Energy and Thermodynamics	224
Chapter 18	Electrochemistry	238
Chapter 19	Radioactivity and Nuclear Chemistry	251
Chapter 20	Organic Chemistry	261
Chapter 21	Biochemistry	274
Chapter 22	Chemistry of the Nonmetals	282
Chapter 23	Metals and Metallurgy	288
Chapter 24	Transition Metals and Coordination Compounds	293
Answers		301

To the general chemistry student:

Chemistry is one of the subjects that college students dread. I see it on my students' faces at the beginning of a new semester: fear and anxiety about the class. In my opinion, this attitude is primarily derived from a belief that chemistry is unimportant and abstract as well as a misunderstanding of what the study of chemistry is.

The importance of chemistry is in its ubiquity – molecules, atoms, and ions are everywhere and make up everything; the study of chemistry is the study of molecules, atoms, and ions so it is the study of everything. And what does the study of chemistry entail? Mathematics, while important, is not chemistry and this is what most students do not realize upon first glance. Mathematical relationships are certainly important in a general chemistry course, but they are a means to an end. The end is an understanding of the physical principles that govern the world (and universe) that we live in.

This study guide is aimed at simplifying your exploration of general chemistry concepts so that you can find and explore the underlying physical principles, which provide meaning and context to your studies. Each chapter is divided into six sections. First, the learning objectives of each chapter are provided. These are the skills that you should have mastered upon completion of the chapter. Second, a brief summary of the chapter is given so that you have context for the topics and how they fit together. Third, an extensive outline of the chapter is given; the outline should be used as a review of your textbook reading and not in its place. In each outline, example problems are worked out with explanations for the logic and physical meaning of the steps involved and the solutions for each problem. Fourth, fill in the blank problems are provided. You should use these to test your familiarity with the main ideas and definitions of the chapter. The fifth section of each chapter contains problems for you to work out. These problems should be used as a final test of your knowledge – if you are able to complete these problems on your own, you have mastered the mechanics of the concepts presented in the chapter. Finally, and most importantly, the sixth section of each chapter is the concept questions. These questions require that you thoroughly think through the concepts presented and look at the underlying principles behind those concepts. Often, these questions require that you link together multiple topics and/or find the connections between seemingly disparate topics. If you can answer these questions thoroughly, you truly understand the material. Answers are not provided for the problems and concept questions because available answers tempt students to work backwards; you can't be sure that you understand the problem and its solution if you use the solutions. Check your work by going through your approach; a solid understanding is always better than a correct number.

The likelihood of your success, in this and other courses, is based on your plan for learning the material. I give the series of steps listed below to my own students each semester; following these steps will almost certainly lead to a passing grade and, more importantly, will generally lead you to a true understanding that you can take away with you when the course is done. You will see that this study guide is listed as an endpoint for your strategy – it is not meant to replace either you textbook nor your lectures.

1. Read the chapter before going to the lecture.
   You don't have to read for content, even skimming over the pages and looking at pictures and graphs will help. Your goal here is to get an idea of the concepts and language that you will be learning next; it is often the language of chemistry that students struggle with most and the more you expose yourself to that language, the less likely it is to be a barrier.

2. Attend lecture, take notes, participate in class, and ask questions.
   You are in class to learn and your instructor is there to help you – so use the resources that are available to you. If concepts remain unclear after lecture, go to your instructor's office hours and ask questions.

3. Read the material again after class.
   This time use a pen and some paper…create an outline of the chapter and go through the example problems. Writing concepts down and doing problems as you go will help to ensure that you retain the information. You can also make notes about concepts that are confusing or unclear; ask your instructor or your fellow students about concepts that you are struggling with.

4. Work on problems at the end of the chapter.
   The surest way to understand concepts is to practice using them in many ways. One of the most difficult parts of chemistry is the incorporation of math into physical problems – story problems. You need to practice taking apart these problems and figuring out what is important to solving them. It is a good idea to look at the physical meaning behind the problem and the answer – think about it in real world terms.

5. Form a study group.
   There is no better way to learn than to teach and there is no better way to see if you really understand something than to try to explain it. Getting a group of your peers together to discuss chemistry is one of the smartest ways that you can study.

6. Use the study guide as a review and a final test of your understanding.
   Once you have looked over the material and feel comfortable with the concepts and problems, use this study guide to help clarify difficult concepts or to test your skills. The fill in the blank questions, problems, and concept questions are all meant to be a final test – you can use these as a mock exam to see if you have truly learned the material.

I believe strongly that everyone can really learn chemistry and, once it is learned, everyone can not only excel in a general chemistry course, but also start to see the chemistry around them as interesting and exciting. Learning new material is, however, difficult and requires time and patience. Don't expect that you can study for four hours before an exam and be successful; pace yourself so that you can learn the material instead of simply trying to regurgitate facts. You will find that the slow and steady approach works best for this type of material.

I think that chemistry is a fascinating and rewarding field of study – I see it everywhere and I am always amazed at its underlying simplicity. I don't expect that you will see chemistry as simple and fascinating right from the beginning because as a student it is also challenging and frustrating. But over the course of your studies, if you look carefully and learn thoroughly, the beauty and simplicity will shine through.

I would like to thank Merritt College for supporting me while writing this book and giving me the opportunity to share my love of chemistry with my students; I am especially grateful to my fellow chemistry instructor, Ray Chamberlian, who is always encouraging and enthusiastic. I would like to thank my students for their ever-growing curiosity and desire to learn. Finally, I would like to thank my husband, Engelbert, for encouraging me to take on this project and for being so patient while I completed it.

Jennifer Shanoski, Ph.D.
Merritt College
Oakland, California

Chapter 1: Matter, Measurement, and Problem Solving

Learning Objectives:

- Understand the scientific method.
- Classify matter according to its state and composition.
- Identify physical and chemical changes as well as physical and chemical properties.
- Become familiar with the International System of Units and prefix multipliers.
- Correctly use significant figures to convey the reliability of a measurement.
- Apply dimensional analysis as a problem-solving strategy.

Chapter Summary:

Chemistry is the study of the transformations, properties, structures, and compositions of atoms and molecules. In this chapter, you will learn the basic principles that supply the scaffolding for your chemical knowledge. You will first be introduced to the scientific method so that you are able to understand where our understanding of chemistry comes from and how it changes. The various classifications of matter will then be introduced followed by an explanation of the physical and chemical changes that allow one to go from one class of matter to another. Energy, in its various forms, will then be discussed in terms of physical and chemical transformations. The second half of this chapter introduces you to the numerical side of chemistry. You will learn the importance of units and the International System of Units that will be used throughout your study of chemistry and other sciences. Measurements and the number of digits reported for them will be explained as well as scientific notation and an explanation of significant figures in the context of calculations. Finally, you will be introduced to the problem-solving technique of dimensional analysis, which will be utilized throughout this book.

Chapter Outline:

I. Atoms and Molecules
   a. Atoms are submicroscopic particles and are the fundamental building blocks of matter.
   b. Molecules are two or more atoms joined together by a chemical bond in a specific geometric arrangement.
   c. Chemistry is the science that seeks to understand the properties, structures, composition, and reactivity of atoms and molecules.

II. The Scientific Approach to Knowledge
   a. Chemistry is an empirical science, meaning that it is based on experimentation and observation.
   b. The scientific method begins with an observation that leads to a tentative explanation (hypothesis). The hypothesis is repeatedly tested by experimentation and then either accepted, revised, or discarded.
   c. Scientific laws are generalizations about nature that summarize past observations and predict future ones.
   d. Theories are models based on well-established hypotheses and validated by experimental evidence. By explaining the generalizations stated in scientific laws, theories give insight into how nature works.

e. It is important to realize that science is constantly expanding and changing because a law or theory cannot be "proven" absolutely. The "strongest" theories are those that have been consistently supported by experimental data over time, but any theory that fails to account for new observations must be modified or discarded.

III. The Classification of Matter

a. Matter is anything that occupies space and has mass.

b. There are three states of matter: solid, liquid, and gas.

i. A solid has atoms or molecules that are held close together and maintains a rigid shape. Solids can be orderly (crystalline) or lack long-range order (amorphous).

ii. A liquid has atoms or molecules that are held close together but can move relative to each other. A liquid is therefore fluid and can assume the shape of its container.

iii. A gas has atoms or molecules that are very far apart. A gas is therefore a compressible fluid. Gases assume the shape and volume of the containers that hold them.

c. Matter can be classified as a pure substance or a mixture depending on the variability of its composition.

i. A pure substance has no variability in its composition. No matter what the size of the sample is or how it is prepared, it will always be the same.

1. A pure substance can be classified as a compound or an element. A compound consists of two or more elements in fixed proportions and can be separated into elements by chemical means.

ii. A mixture can have a variable composition. A mixture can be prepared in an infinite number of ways.

1. A mixture can be classified as either homogeneous or heterogeneous. A homogeneous mixture is uniform throughout; a sample of the mixture taken from one area will be identical to a sample of the mixture taken from another area. A heterogeneous mixture has variations throughout the sample.

► In general, a homogeneous mixture forms when the constituent parts can move freely in random directions; this constant, random motion results in mixing on the molecular/atomic level. For example, gaseous water and gaseous carbon dioxide will mix to form a homogeneous mixture while solid water and solid carbon dioxide will form a heterogeneous mixture.

2. A mixture can be separated based on differences in the physical and chemical properties of its components.

► When deciding if a sample is pure or is a mixture; think about the recipe for its preparation. A pure substance has a definite recipe whereas a mixture can have an infinite number of recipes. For example, water will always be made of two hydrogen atoms and one oxygen atom so water is a pure substance; salt water can have a pinch of salt or a bucket of salt in a given amount of water so salt water is a mixture.

EXAMPLE:

Identify the following as pure substances or mixtures. If it is a substance, determine whether it is an element or a compound. If it is a mixture, determine if it is a homogeneous or heterogeneous mixture.

a. Salt and sugar that have been ground down to fine powder

This is a mixture; it contains two distinct substances: sugar and salt. This mixture is heterogeneous because no matter how small the crystals have become, there are still two distinct phases of matter.

b. Glucose

This is a pure substance with one type of molecule. Glucose is a molecule because it contains more than one type of atom.

c. Stainless steel

This is a mixture of various elements that have been combined into an alloy. Since the elements are not chemically bound to one another, they are not constituents of a molecule. This is a homogeneous mixture that has the same microscopic composition throughout.

d. Air

Air contains a number of different molecules and atoms, which means that it is a mixture. Since the mixture is a fluid that is in constant motion, the mixture is homogeneous or the same throughout.

IV. Physical and Chemical Changes and Physical and Chemical Properties

a. A physical change is an alteration of state or appearance.

   i. A physical property is a property that is displayed by a substance when the composition does not change.

b. A chemical change is a rearrangement of atoms or alteration of composition.

   i. A chemical property is a property that is displayed when the composition changes.

► Note that physical changes may occur when separating mixtures while chemical changes must occur when separating atoms in a pure substance.

EXAMPLE:

Identify the following as physical or chemical changes.

a. The distillation of salt water to produce pure water

This is a physical change in which the liquid water is undergoing a physical change to water vapor and then condensing back into water.

b. The rusting of an iron nail

This is a chemical change in which elemental iron is being oxidized to form iron oxide.

c. The burning of wood in a fireplace

This is a chemical change in which combustion is occurring to change the carbon in the wood into gaseous carbon dioxide.

d. Salt precipitating out of a solution

This is a physical change in which the aqueous salt is becoming a solid crystal, but the chemical composition remains unchanged.

V. Energy: A Fundamental Part of Physical and Chemical Change

a. Energy is the capacity to do work; work is a force applied over a distance.

b. Most physical and chemical changes are accompanied by a change in energy.

c. The total energy is always equal to the sum of kinetic energy (energy associated with motion) and the potential energy (energy associated with position or composition).

d. Thermal energy is associated with temperature; it is the average kinetic energy of the constituents of a substance. The link between temperature and kinetic energy is important to remember for future chapters.

e. The law of conservation of energy states that energy cannot be created or destroyed but can change form and flow from one object to another.

f. Systems with high potential energy tend to change in a way that lowers their potential energy.

   i. Low potential energy is more stable than high potential energy.

VI. The Units of Measurement

a. Units are standard quantities that are used to specify the meaning of a measurement.

b. In science, the metric system is used and the standard units are the International System of Units (SI). Common SI base units and their abbreviations (given in parentheses) are given below:

   i. Length is a measure of distance. The unit for length is the meter (m).

   ii. Mass is a measure of the quantity of matter in an object; mass is different from weight, which measures the gravitational pull on an object. The unit for mass is the kilogram (kg).

   iii. Time is a measure of periods between events. The unit for time is the second (s).

   iv. Temperature is a measure of the average kinetic energy of the atoms or molecules that compose matter. The unit for temperature is Kelvin (K).

      1. The Kelvin scale is often referred to as the absolute scale because the lowest possible temperature is 0 K or absolute zero. This is the temperature at which all molecular motion stops.

      2. In order to convert a temperature to the Kelvin scale, you must begin with units of °C.

         a. Conversion between °C and °F:

$$^{\circ}\mathrm{C} = \frac{(^{\circ}\mathrm{F} - 32)}{1.8}$$

         b. Conversion between °C and K:

$$\mathrm{K} = {}^{\circ}\mathrm{C} + 273.15$$

EXAMPLE:

The average temperature of the human body is 98.6°F. Convert this into °C and K.

We can convert to the Celsius scale using the above equation:

$$^{o}C = \frac{(98.6^{o}F - 32)}{1.8} \quad \rightarrow \quad 37.0^{\circ}\mathrm{C}$$

And we can now use this result to calculate the temperature in the Kelvin scale:

$$\mathrm{K} = 37^{\circ}\mathrm{C} + 273.15 \quad \rightarrow \quad 310.2\ \mathrm{K}$$

---

c. Prefix Multipliers

i. Using the metric system, it is possible to change the value of a unit by multiplying by powers of 10 and incorporating a prefix.

1. kilo (k) is 10^3 or 1000.
2. centi (c) is 10^{-2} or 0.01.
3. milli (m) is 10^{-3} or 0.001.
4. micro (μ) is 10^{-6} or 0.000001.
5. nano (n) is 10^{-9} or 0.000000001.

► You can translate the prefix multipliers directly into numbers so that 15 kilometer is really just 15 ($\times 10^3$)m which we write simply as 15×10^3 m or, in proper scientific notation as 1.5×10^4 m.

EXAMPLE:

Convert the following measurements into numbers between 1 and 10 using prefix multipliers:

a. 1908750 m

Rewriting in scientific notation: 1.90875×10^6 m. 10^6 corresponds to mega, so we have 1.90875 Mm.

b. 0.00000000000989 s

Rewriting in scientific notation: 9.89×10^{-12} s. 10^{-12} corresponds to pico, so we have 9.89 ps.

c. 6.8526×10^{-6} g

10^{-6} corresponds to micro, so we have 6.8526 μg.

d. 0.06548 m

Rewriting in scientific notation: 6.548×10^{-2} m. 10^{-2} corresponds to centi, so we have 6.548 cm.

e. 4.548×10^3 g

10^3 corresponds to kilo, so we have 4.548 kg.

---

d. Derived Units

i. Derived units are units that come from combinations of the base units; we will see many examples in the coming chapters.

1. Volume is a measure of space. Volume is an extensive property which means that it depends on the sample size or the amount of substance. The SI base unit of volume is m^3. In general, units of liter (L) are used.

$$1 \text{ L} = 1000 \text{ mL} = 1000 \text{ cm}^3 = 0.001 \text{ m}^3$$

2. Density is the ratio of mass to volume (d=m/V). Density is an intensive property, which means that it is independent of the sample size or amount of substance; density is therefore characteristic of particular materials. The common units for density are g/mL for solids and liquids and g/L for gases.

EXAMPLE:

A metal slug is found to have a mass of 13.45 g and a volume of 4.98 mL. Calculate the density of the metal and determine its possible identity using the table of known densities given below.

Metal	Density (g/mL)
Aluminum	2.70
Iron	7.87
Silver	10.5
Lead	11.4

The density is simply the mass divided by the volume: 13.45g/4.98 mL = 2.70 g/mL. We can see from the table that this is the same density as aluminum, so we can conclude that the metal slug may be aluminum.

VII. The Reliability of a Measurement

a. The number of digits in a reported measurement indicates the certainty associated with that measurement.

   i. The last digit reported is always assumed to be the uncertain digit.

b. Uncertainty comes from measurement devices, so the number of digits reported is determined by the measurement device. The rule in measurements is to report all digits given explicitly and then guess one more; the uncertain digit (the last digit) will be the one that you guessed.

c. Uncertainty is conveyed in calculations by keeping track of the number of significant figures. The rules for determining the number of significant figures (sig. figs.) are:

   i. All non-zero digits are significant.

   ii. All interior zeros are significant.

   iii. All leading zeros are not significant.

   iv. All trailing zeros after the decimal point are significant.

   v. Trailing zeros in numbers without a decimal point are ambiguous and should be avoided by using scientific notation.

d. Exact numbers are numbers that come from counting discrete objects, are defined quantities, or are integral numbers in an equation. Exact numbers have no uncertainty and should be treated as though they have unlimited significant digits.

EXAMPLE:

Determine the number of significant figures in the following numbers:

a. 0.548704

This number has 6 significant figures. The first zero is a leading zero and is not, therefore, significant. The zero between the 7 and the 4 is significant because it comes between two non-zero integers.

b. 4580000

This number is ambiguous. It could have 3, 4, 5, 6, or 7 significant figures. If we rewrite this number in scientific notation as 4.58×10^6, then it would have 3 significant figures; in that case, the four zeroes would simply be place holders. Written another way as 4.5800×10^6, it would have 5 significant figures.

c. 654089.0

This number has 7 significant figures. The zero between the 4 and the 8 is significant because it comes between two non-zero integers. The zero at the end of the number is significant because it is a trailing zero in a number with a decimal point.

d. 0.0000548

This number has 3 significant figures. All of the zeros in this number are leading zeros and serve only as place holders.

---

e. When carrying out calculations, the following rules should be followed in regards to significant figures:

  i. When multiplying or dividing, the result will have the same number of significant figures as the factor with the fewest significant figures.

  ii. When adding or subtracting, the result will have the same number of decimal places as the quantity with the fewest decimal places.

  iii. At the end of a calculation, you should round your answer. If the digit after the last significant digit is less than or equal to four, round down; if the digit after the last significant digit is greater than or equal to five, round up.

EXAMPLE:

Carry out the following operations:

a. 65.587×654.0

We first calculate the answer and then we can determine the number of significant figures:

$65.587 \times 654.0 = 42893.898$

We see that 65.587 has 5 sig. figs. and 654.0 has 4 sig. figs. Our answer should then have four sig. figs., so it is 4.289×10^4. Do not write 42890 because the trailing zero is ambiguous.

b. 648/0.02

The answer is: 648/0.02 = 32400.

648 has 2 sig. figs. and 0.02 has 1 sig. fig. so our answer should have 1 sig. fig: 3×10^4.

c. $(4.58\times10^{-6}) \times (5.89\times10^{8})$

We will take a slightly different approach here. First, let us collect the powers of ten together so that we can easily simplify that part of the equation (using exponent rules) and avoid calculator errors.

$(4.58 \times 5.89) \times (10^{-6}\times10^{8}) = (4.58 \times 5.89) \times (10^{-6+8}) = (4.58 \times 5.89) \times 10^{2}$

Now we can continue the calculation:

$26.9762\times10^{2} = 2.69762\times10^{3}$

Since both 4.58 and 5.89 have 3 sig. figs., our answer will have 3 sig. figs.: 2.70×10^{3}.

d. $5.687 - 0.000087$

We carry out the calculation: $5.687 - 0.000087 = 5.686913$. Since this is a subtraction problem, we will use the decimal place to determine the sig. figs. in our answer. 5.687 is precise to the thousandths digit while 0.000087 is precise to the millionths digit. Our answer will therefore only be reported to the thousandths digit: 5.687.

e. $5897 - 12$

The answer is: $5897 - 12 = 5885$. Since both 5897 and 12 are precise to the ones digit, our answer will be reported to the ones digit as 5885.

f. $(9.687\times10^5) + (8.568\times10^4)$

Here we again want to isolate the powers of ten, but we will make both numbers multiplied by the same power of ten in order to add or subtract and easily determine the significant figures:

$(96.87\times10^4) + (8.568\times10^4) = (96.87 + 8.568) \times 10^4$

We can now carry out the addition and determine the significant digits without worrying about the power of ten:

$(96.87 + 8.568) \times 10^4 = (105.438) \times 10^4$

Since 96.87 is precise to the hundredths digit and 8.568 is precise to the thousandths digit, our answer will be reported to the hundredths digit: 105.44×10^4. We need to convert this to proper scientific notation to get: 1.0544×10^6.

---

f. Precision is a measure of reproducibility and is a measure of how close various measurements are to one another. Accuracy is a measure of how close measurements are to the actual or true value.

   i. Random errors are those errors that have an equal probability of being higher and lower than the true value; random errors generally average out when many measurements are taken.

   ii. Systematic errors tend to cause results that are either too high or too low and do not average out with multiple measurements. Systematic errors often cause results that are precise but inaccurate.

   ► Random errors will give a less precise result while systematic errors give a less accurate result.

VIII. Solving Chemical Problems

a. Dimensional analysis uses units to guide problem solving.

b. The general strategy for solving problems in chemistry is:

   i. Identify the type of problem that you are looking at.

   ii. Make a list of information given.

   iii. Determine what you are looking for.

   iv. Develop a plan for going from given information to the answer that you want.

   v. Check your answer to be sure that it makes sense in the context of the problem and that you have the correct number of significant figures.

c. Order of magnitude estimates can be used as a fast method for checking your answer.

EXAMPLE:

Speed limits are often given with units of miles/hour. In science, speeds are often given in units of m/s. How fast are you going in m/s if you are travelling 55 mph?

First, we identify the starting point of the problem: 55 mi/hr.

Then, we identify what our ending units will be: m/s.

Next, we map out the path that will take us from the starting point to the ending point using conversion factors that we know (or can look up):

$$\frac{\text{mi.}}{\text{hr}} \xrightarrow{\text{mi. to km}} \frac{\text{km}}{\text{hr}} \xrightarrow{\text{km to m}} \frac{\text{m}}{\text{hr}} \xrightarrow{\text{hr to min}} \frac{\text{m}}{\text{min}} \xrightarrow{\text{min to s}} \frac{\text{m}}{\text{s}}$$

And now we can include the numbers and conversion factors:

$$\frac{55\text{mi}}{\text{hr}} \times \frac{1.61\text{km}}{\text{mi}} \times \frac{1000\text{m}}{\text{km}} \times \frac{1\text{hr}}{60\text{min}} \times \frac{1\text{min}}{60\text{s}} = 24.597\frac{\text{m}}{\text{s}}$$

Our final answer should have 2 sig. figs., so we report our answer as 25 m/s.

EXAMPLE:

Salt water, with a density of 1027 kg/m^3, is denser than pure deionized water. Convert the density of salt water to the more conventional units of g/mL.

First, we identify the starting point of the problem: 1027 kg/m^3.

Then, we identify what our ending units will be: g/mL.

Next, we map out our pathway:

$$\frac{\text{kg}}{\text{m}^3} \xrightarrow{\text{kg to g}} \frac{\text{g}}{\text{m}^3} \xrightarrow{\text{m to cm}} \frac{\text{g}}{\text{cm}\cdot\text{m}^2} \xrightarrow{\text{m to cm}} \frac{\text{g}}{\text{cm}^2\cdot\text{m}} \xrightarrow{\text{m to cm}} \frac{\text{g}}{\text{cm}^3} \xrightarrow{\text{cm}^3\text{ to mL}} \frac{\text{g}}{\text{mL}}$$

And now we include the numbers and conversion factors:

$$\frac{1027\,\text{kg}}{\text{m}^3} \times \frac{1000\,\text{g}}{\text{kg}} \times \frac{1\,\text{m}}{100\,\text{cm}} \times \frac{1\,\text{m}}{100\,\text{cm}} \times \frac{1\,\text{m}}{100\,\text{cm}} \times \frac{\text{cm}^3}{\text{mL}} = 1.027\frac{\text{g}}{\text{mL}}$$

The calculated answer already has the correct number of sig. figs., so we are finished!

Fill in the Blank Problems:

1. A good hypothesis makes predictions that can be confirmed or refuted by further ________________.
2. Energy associated with position or composition is ________________ energy.
3. ________________ are brief statements that summarize past observations and predict future ones.
4. The three states of matter are: ________________, ________________, and ________________.
5. A(n) ________________ mixture is one in which the composition is uniform throughout the sample.
6. A(n) ________________ change is one that changes the state or appearance of a substance, but does not alter its composition.
7. A(n) ________________ is a substance that cannot be chemically broken down into simpler substances.
8. A(n) ________________ property is one that is displayed only when the composition is changed through a chemical change.
9. The ________________ states that energy cannot be created or destroyed; it may only change form or flow from one body to another.

10. A ________________ is two or more atoms joined together via chemical bonds in a specific geometric orientation.

11. The SI unit for mass is the ________________.

12. A(n) ________________ property is one that does not depend on the size of a system.

13. The ________________ of a substance is a measure of its average kinetic energy and is measured in units of ________________.

14. A(n) ________________ solid is one in which the atoms or molecules are arranged in patterns with long-range repeating order.

15. The ________________ of a measurement is a measure of how close it is to the actual value; the ________________ is a measure of how close a set of measurements are to one another.

Problems:

1. Classify each of the following as a mixture or a substance:
   a. Gasoline
   b. Milk
   c. Nail polish remover
   d. A silver ring
2. Classify each of the following as a homogeneous or heterogeneous mixture:
   a. Freshly squeezed lemonade
   b. Oil and vinegar salad dressing
   c. Gas in a diver's tank
   d. Hydrogen peroxide solution
3. Classify each of the following as a chemical or physical change:
   a. Removing paint with a solvent
   b. Filtering a precipitate out of a solution
   c. Bread rising in the oven
   d. Ozone depletion
4. Use prefix multipliers to convert the following measurements into numbers between 1 and 10:
   a. 9800000 m
   b. 0.00000000895 s
   c. 8040 g
   d. 0.00000571 m
5. Express the following numbers in scientific notation:
   a. 65480000
   b. 0.0009800
   c. 0.005804
   d. 15290000000
6. State the number of significant figures in each of the following numbers:

a. 0.50480

b. 5400.058

c. 9735000000

d. 0.006980

7. Carry out the following mathematical manipulations and express your final answer in scientific notation.

   a. 859×0.587514

   b. $59.328 - 4.785 + 1.57625140$

   c. $5897 \times 0.589 + 4.5850$

   d. $(8.598\times10^{8}) + (1.653809\times10^{9}) / 58$

8. The density of air is 1.184 kg/m^3 at 25°C. Calculate the mass (in g) of 1 gallon of air.

9. Perform the following unit conversions:

   a. 89000 μg to pounds

   b. 8.789 km to inches

   c. 0.489 weeks to seconds

   d. 67.580 mL to dm^3

10. You measure the volume of a sample of aluminum using the method of displacement. Your measurements are 5.45 mL, 5.58 mL, 5.49 mL, and 5.51 mL. Calculate the average value and comment on your accuracy if the actual value is reported to be 5.5 mL.

11. You decide that you are going to put up a fence around your house. You go to the hardware store and find out that the fence costs $50 per yard. You then measure your yard and find that it is 40 feet wide and 150 feet long. How much will it cost to buy the fencing that you will need?

12. A light year is the distance that light will travel (in a vacuum) in one year. Given that the speed of light is 3.0×10^{8} m/s, how many miles will light travel in one year?

13. A small fish tank will hold 29 gallons of water. If the tank leaks water at a rate of 1.0 mL/min., how long (in days) will it take for all of the water to drain out of the tank if it starts full?

<u>Concept Questions:</u>

1. One of the arguments against teaching evolution in public schools is that it is just a theory. Why is this a weak argument against the inclusion of evolution in a science curriculum?

2. A student collects from her classmates the following measurements of the length of a stick: 4.57 cm, 4.67 cm, 4.64 cm, 4.58 cm, and 4.6 cm.

   a. Did all of the students use the same measuring device? Explain how you arrived at your answer.

   b. Comment on the accuracy and precision of the class results.

3. A metal slug is found to have a mass of 5.486 g and a volume of 5.897 mL. Will the metal slug float or sink when placed in a glass of room temperature water? Explain your answer.

4. Dimensional analysis is used all the time in real life, but we usually don't recognize it as such. State one example from your life where you have used dimensional analysis.

5. The chemical potential energies of wood and oxygen (the reactants in combustion) are, collectively, higher than the potential energy of carbon dioxide and water (the products in combustion). This is why heat is released when we burn wood in our fireplace. Explain, on a microscopic or molecular level, where this excess energy goes.

Chapter 2: Atoms and Elements

Learning Objectives:

- Understand the emergence of modern atomic theory and the observations that led to it.
- Calculate the average atomic mass of an element and understand how to use it in chemical calculations.
- Become familiar with the periodic table and the various groupings used within it.
- Determine the number of protons, electrons, and neutrons in a given ion or element.
- Predict the charge of an ion for a given element.
- Gain a feeling for numbers in chemistry, such as the size of atoms, the mass of atoms, and the numbers of atoms in a macroscopic sample.
- Understand the concept of mole and use Avogadro's number in calculations.
- Use the molar mass of substances in calculations.

Chapter Summary:

In this chapter, you will be introduced to the structure of atoms, the fundamental building blocks of matter. You will learn that atoms contain a small, dense nucleus—comprised of protons and neutrons—responsible for atomic mass as well as small, light electrons responsible for atomic volume. You will be introduced to the periodic table and learn how to use it to identify properties such as the numbers of protons, neutrons, and electrons in elements and ions. As you explore atomic structure, you will see the ways in which the numbers of electrons and neutrons can change and how to predict the changes. Finally, you will learn the concept of the mole and how to utilize it in calculations.

Chapter Outline:

I. Imaging and Moving Individual Atoms
   a. Scanning Tunneling Microscopy (STM) is a technique that allows atoms and molecules to be imaged and moved.
   b. Atoms are the smallest identifiable unit of an element.

II. Early Ideas about the Building Blocks of Matter
   a. Experimental evidence for the existence of atoms was not available until the early nineteenth century.

III. Modern Atomic Theory and the Laws that Led to It
   a. The law of conservation of mass states that matter is not created or destroyed in a chemical reaction. A chemical reaction results in the rearrangement of atoms.
   b. The law of definite proportions states that all samples of a given compound have the same proportion of their constituent elements no matter how the compound is prepared or where it comes from.

EXAMPLE:

A sample of methane, CH_4, is found to have 46.12 g of carbon and 3.87 g of hydrogen. Another sample is analyzed and found to have 51.25 g of carbon and 4.30 g of hydrogen. Analyze these results in the context of the law of definite proportions.

First, we will analyze the first sample. We simply take the ratio of the mass of carbon to the mass of hydrogen and simplify:

$$\frac{46.12}{3.87} = 11.9$$

So the mass ratio is 11.9:1.

The second sample will be inspected in a similar manner:

$$\frac{51.25}{4.30} = 11.9$$

Again, the mass ratio is 11.9:1.

We see from this example that methane has a 11.9:1 ratio of carbon to hydrogen despite the sample size, which is consistent with the law of definite proportions.

c. The law of multiple proportions states that when two elements combine to form two different compounds, the masses of one element that combines with 1 g of the other element can be expressed as a ratio of small whole numbers.

EXAMPLE:

Many different iron oxides are commonly encountered in nature. Wustite contains 2.62 g of iron for every 1.00 g of oxygen while magnetite contains 3.49 g of iron for every 1.00 g of oxygen. Show that this ratio is consistent with the law of multiple proportions.

We will construct a ratio of the masses of iron and find the smallest whole number ratio:

$$\frac{3.49g}{2.62g} = 1.33$$

This is not a simple whole number, so we must multiply the top and bottom by a whole number that will give us a whole number ratio:

$$1.33 \cdot \frac{3}{3} = \frac{4}{3}$$

We see that 4:3 is a simple whole number ratio consistent with the law of multiple proportions.

d. John Dalton's atomic theory is centered on the following four ideas:

   i. Each element is composed of tiny, indestructible particles called atoms.

   ii. All atoms of a given element have properties (including mass) that differentiate them from atoms of other elements.

   iii. Atoms combine in whole number ratios to form compounds.

   iv. Atoms of one element cannot change into atoms of another element via chemical means. New substances are formed in a chemical reaction by atoms rearranging the way they are bound to other atoms.

IV. The Discovery of the Electron

a. Electrostatic forces—measured in coulombs (C)—are the attractive and repulsive forces that exist between electrically charged particles.

b. When a high electrical voltage is applied between the electrodes of a cathode ray tube, cathode rays travel from the negatively charged cathode to the positively charged anode.

c. J. J. Thomson discovered that cathode rays contain electrons, which are small, negatively charged particles contained within atoms. He measured the ratio of the mass of the electron to its charge as -1.76×10^{8} C/g.

d. The charge of an electron was determined by Robert Millikan using charged droplets of oil suspended in an electric field. The charge of a single electron is -1.60×10^{-19} C. The mass of an electron is 9.10×10^{-28} g.

V. The Structure of the Atom

a. Ernest Rutherford used positively charged alpha particles aimed at gold foil to show the existence of the nucleus.

b. The nuclear theory of an atom states that:

i. Most of an atom's mass and all of its positive charge are contained in a small, dense core called the nucleus.

ii. Most of the volume of an atom is empty space; throughout this space, negatively charged electrons are dispersed.

iii. The number of electrons is equal to the number of protons so t hat charge neutrality is maintained.

c. James Chadwick showed that the nucleus contains neutrons which are massive, neutral particles in the nucleus.

VI. Subatomic Particles: Protons, Neutrons, and Electrons in Atoms

a. Protons and neutrons are both found in the nucleus of an atom and have approximately the same mass: 1 atomic mass unit (amu).

i. An atomic mass unit is defined as 1/12 the mass of a carbon-12 atom.

b. Electrons are found outside of the nucleus and have a mass of 0.00055 amu.

▶ In practice, we say that protons and neutrons each have a mass of 1 amu and electrons have zero mass–this allows for easy estimates of the atomic mass.

c. The charge of an electron in relative units is -1 and the charge of a proton in relative units is +1.

d. Elements are defined by the number of protons in the nucleus. This is called the atomic number (Z) and is the number of the element on the periodic table.

e. Each element has a unique chemical symbol that is a one- or two-letter abbreviation of its name. Note that the abbreviations are sometimes derived from non-English names; for example, the symbol for gold is Au from the Latin word aurum.

i. In a chemical symbol, the first letter is always capitalized and the second letter (if there is one) is always lowercase.

f. Not all atoms of a given element have the same number of neutrons.

i. Atoms with the same number of protons but a different number of neutrons are called isotopes.

ii. The natural abundance of an isotope is the percent of a naturally occurring sample of the atom that is a particular isotope.

EXAMPLE:

A sample of 10^{10} atoms of chlorine is collected. Chlorine has two isotopes: chlorine-35 and chlorine-37. The natural abundance of chlorine-35 is 75.77% and the natural abundance of chlorine-37 is 24.23%. How many atoms of the sample are chlorine-35 and how many are chlorine-37?

The natural abundance is the percentage of atoms in a sample that is a particular isotope, so we can simply calculate the percentage of the 10^{10} chlorine atoms that is each isotope by multiplying the total number of atoms by the percentage (as a decimal):

Cl-35: $10^{10} \times 0.7577 = 7.577 \times 10^9$ chlorine-35 atoms in the sample

Cl-37: $10^{10} \times 0.2423 = 2.423 \times 10^9$ chlorine-37 atoms in the sample

g. The mass number (A) is the sum of protons and neutrons in an atom.

h. The full symbol for an element uses the chemical symbol (X), the mass number (A), and the atomic number (Z):

$$^{A}_{Z}X$$

i. Since inclusion of the atomic number and the element symbol is redundant, the full symbol is often abbreviated in one of two ways:

$$^{A}X \text{ or } X\text{ - }Z$$

Note: the textbook does not use the ^{A}X notation, but you may see it used elsewhere.

EXAMPLE:

Give the full atomic symbol for each of the following:

a. chlorine-35

Chlorine's symbol is Cl, it has 17 protons and the mass number given is 35; the full symbol is: $^{35}_{17}Cl$

b. aluminum-27

Aluminum's symbol is Al, it has 13 protons and the mass number given is 27; the full symbol is: $^{27}_{13}Al$

c. calcium-44

Calcium's symbol is Ca, it has 20 protons and the mass number given is 44; the full symbol is: $^{44}_{20}Ca$

d. lithium-6

Lithium's symbol is Li, it has 3 protons and the mass number given is 6; the full symbol is: $^{6}_{3}Li$

j. Ions are atoms that do not have an equal number of electrons and protons.

i. Cations are positively charged ions; they have fewer electrons than protons.

ii. Anions are negatively charged ions; they have more electrons than protons.

EXAMPLE:

State the number of electrons, protons, and neutrons in each of the following examples:

a. $^{23}Na^{+}$

Sodium is number 11 on the periodic table, so it has 11 protons. The number of neutrons is given by the mass number (23 in this case) minus the number of protons, so it has 12 neutrons. The number of electrons in a neutral atom is equal to the number of protons, but in this case we have a 1+ ion, so there is one less electron than proton; thus there are 10 electrons.

b. $^{40}Ca^{2+}$

Calcium is number 20 on the periodic table, so it has 20 protons. The number of neutrons is given by the mass number (40 in this case) minus the number of protons, so it has 20 neutrons. The number of electrons in a neutral atom is equal to the number of protons, but in this case we have a 2+ ion, so there are two less electrons than protons; thus there are 18 electrons.

c. $^{54}Fe^{3+}$

Iron is number 26 on the periodic table, so it has 26 protons. The number of neutrons is given by the mass number (54 in this case) minus the number of protons, so it has 28 neutrons. The number of electrons in a neutral atom is equal to the number of protons, but in this case we have a 3+ ion, so there are three less electrons than protons; thus there are 23 electrons.

d. $^{18}O^{2-}$

Oxygen is number 8 on the periodic table, so it has 8 protons. The number of neutrons is given by the mass number (18 in this case) minus the number of protons, so it has 10 neutrons. The number of electrons in a neutral atom is equal to the number of protons, but in this case we have a 2- ion, so there are two more electrons than protons; thus there are 10 electrons.

e. $^{37}Cl^{-}$

Chlorine is number 17 on the periodic table, so it has 17 protons. The number of neutrons is given by the mass number (37 in this case) minus the number of protons, so it has 20 neutrons. The number of electrons in a neutral atom is equal to the number of protons, but in this case we have a 1- ion, so there is one more electron than protons; thus there are 18 electrons.

VII. Finding Patterns: The Periodic Law and the Periodic Table

a. The periodic law states that when elements are arranged in order of increasing mass, certain properties recur periodically.

b. The modern periodic table is arranged according to atomic number and not atomic mass.

c. The periodic table can be divided up into metals, nonmetals, and metalloids.

i. A metal is a substance that is a good conductor of heat and electricity, is usually a solid at room temperature, is malleable, is ductile, and tends to lose electrons in chemical reactions.

ii. A nonmetal has more varied properties, but is generally a poor conductor of heat and electricity and tends to gain electrons in chemical reactions.

iii. A metalloid has properties that are in between metals and nonmetals. Semiconductors are metalloids and can be conductors or insulators depending on the conditions in which they are used.

EXAMPLE:

Identify the following elements as metals, nonmetals, or metalloids.

a. Cr

Chromium is element 24 and it is found to the left of the "staircase" on the periodic table; it is therefore a metal.

b. O

Oxygen is element 8 and it is found to the right of the "staircase" on the periodic table; it is therefore a nonmetal.

c. He

Helium is element 2 and it is found to the right of the "staircase" on the periodic table; it is therefore a nonmetal.

d. Si

Silicon is element 14 and it is found on the "staircase" on the periodic table; it is therefore a metalloid.

e. Ga

Gallium is element 31 and it is found to the left of the "staircase" on the periodic table; it is therefore a metal.

---

d. Main group elements behave in a more predictable periodic manner than the transition elements.

e. Each column within the main group is called a family or group of elements.

   i. The family or group is shown as a number on the top of each column of the periodic table.

   ii. The noble gases are group 8A on the periodic table and are generally unreactive gases at room temperature.

   iii. The alkali metals are group 1A on the periodic table and react with water to form a basic solution.

   iv. The alkaline earth metals are group 2A on the periodic table and react with water to form a basic solution in a less vigorous way than the alkali metals.

   v. The halogens are group 7A on the periodic table and are reactive nonmetals.

EXAMPLE:

Identify each of the following as a noble gas, an alkali metal, an alkaline earth metal, or a halogen.

a. Ba

Barium is found in group IIA, so it is an alkali earth metal.

b. Cl

Chlorine is found in group VIIA, so it is a halogen.

c. Ar

Argon is found in group VIIIA, so it is a noble gas.

d. Li

Lithium is found in group IA, so it is an alkali metal.

e. Rb

Rubidium is found in group IA, so it is an alkali metal.

---

f. Main group metals tend to lose electrons to form cations with the same number of electrons as the nearest noble gas.

g. Main group nonmetals tend to gain electrons to form anions with the same number of electrons as the nearest noble gas.

EXAMPLE:

Identify the anion or cation that will form from each of the following elements:

a. Ca

The nearest noble gas to calcium is argon, which has 18 electrons. When calcium, with 20 protons, has 18 electrons it will form a 2+ ion and it will therefore be Ca^{2+}.

b. S

The nearest noble gas to sulfur is argon, which has 18 electrons. When sulfur, with 16 protons, has 18 electrons it will form a 2- ion and it will therefore be S^{2-}.

c. I

The nearest noble gas to iodine is xenon, which has 54 electrons. When iodine, with 53 protons, has 54 electrons it will form a 1- ion and it will therefore be I^{-}.

d. Sr

The nearest noble gas to strontium is krypton, which has 36 electrons. When strontium, with 38 protons, has 36 electrons it will form a 2+ ion and it will therefore be Sr^{2+}.

VIII. Atomic Mass: The Average Mass of an Element's Atoms

a. The atomic mass is listed on the periodic table and is the average mass of that atom's isotopes weighted by their natural abundance.

EXAMPLE:

Neon has three isotopes: neon-20, neon-21, and neon-22. Neon-20 has a mass of 19.9924 and a natural abundance of 90.60%. Neon-21 has a mass of 20.9938 and a natural abundance of 0.26%. Neon-22 has a mass of 21.9914 and a natural abundance of 9.20%. Calculate the average atomic mass of neon.

In order to calculate the average mass, we will add together the masses of each of the isotopes multiplied by their natural abundance converted to a decimal.

$$(19.9924 \times 0.9060) + (20.9938 \times 0.0026) + (21.9914 \times 0.0920) = 20.19 \text{ amu}$$

If we look on the periodic table, we see that the atomic mass is listed as 20.18 amu which is very close to our calculated answer. Note that the difference is probably a result of more or less precise values of the mass used in our calculation versus the ones used for the periodic table.

b. The masses of atoms and the percent abundances of their isotopes are determined using mass spectrometry.

IX. Molar Mass: Counting Atoms by Weighing Them

a. Avogadro's number gives the number of particles in one mole of material. There are 6.022×10^{23} particles in 1 mole of material.

b. 1 mole is defined as the number of atoms in exactly 12 g of pure carbon-12.

c. The mole is abbreviated as mol.

d. Avogadro's number is a conversion factor between moles and the number of particles.

EXAMPLE:

Calculate the number of particles in each of the following samples:

a. 3.468 moles of gold atoms

$$3.468\,\text{mol Au atoms}\times\frac{6.022\times10^{23}\,\text{Au atoms}}{1\,\text{mol Au atoms}}=2.088\times10^{24}\,\text{Au atoms}$$

b. 2.45 moles of Ca^{2+} ions

$$2.45\,\text{mol}\,Ca^{2+}\,\text{ions}\times\frac{6.022\times10^{23}\,Ca^{2+}\,\text{ions}}{1\,\text{mol}\,Ca^{2+}\,\text{ions}}=1.48\times10^{24}\,Ca^{2+}\,\text{ions}$$

c. 1.87×10^{-13} moles of CCl_4 molecules

$$1.87\times10^{-13}\,\text{mol}\,CCl_4\,\text{molecules}\times\frac{6.022\times10^{23}CCl_4\,\text{molecules}}{1\,\text{mol}\,CCl_4\,\text{molecules}}=1.13\times10^{11}\,CCl_4\,\text{molecules}$$

e. The molar mass of an element is the mass of 1 mol of its atoms and is numerically equal to the element's atomic mass in amu.

▶ You can check your work when calculating the molar mass by determining the mass number–the two should be roughly the same.

EXAMPLE:

How many gold atoms are there in 5.89 g?

In solving this problem, we follow exactly the same procedure as in the previous example but we include one additional step of going from grams to moles (using the molar mass of gold):

$$5.89\,\text{g Au atoms}\times\frac{1\,\text{mol Au Atoms}}{196.97\,\text{g Au Atoms}}\times\frac{6.022\times10^{23}\,\text{Au atoms}}{1\,\text{mol Au Atoms}}=1.80\times10^{22}\,\text{Au atoms}$$

Fill in the Blank Problems:

1. The amount of material containing 6.022×10^{23} particles is a ________________.
2. The ________________ states that all samples of a given compound have the same proportions of their constituent elements.
3. Most of the volume of an atom is empty space, throughout which are dispersed ________________.
4. Cathode ray tubes were used in the discovery of ________________.
5. The value of an element's molar mass in grams per mole is numerically equal to the element's ________________.
6. The law of conservation of mass states that in a chemical reaction, matter is never ________________.
7. Most of an atom's mass and all of its positive charge are contained in the atom's ________________.
8. Each column within the main-group elements is called a ________________ of elements.
9. Nonmetals tend to gain electrons to form ________________.
10. A (n) ________________ is defined as 1/12 the mass of a carbon-12 atom.
11. The nu mber of ________________ in an atom defines its identity.
12. The percentage of a partic ular isotope of an atom found in a natural sample of that element is called the ________________.
13. When two elements combine to form different compounds, the masses of one element that combine with 1 g of the other element can be expressed as a ________________.
14. ____ ______________ are good conductors of heat and electricity, often shiny, ductile, malleable, and tend to lose electrons when undergoing a chemical change.
15. One of the postulates of John Dalton's atomic theory is that each element is composed of tiny, indestructible particles called ________________.
16. The percent abundance of ele ments is measured using ________________.
17. ____ ______________ are positively charged ions.
18. A vogadro's number is the number of atoms contained in exactly 12 g of ________________.

Problems:

1. A sample of glucose is found to have 34.92 g of carbon, 5.87 g of hydrogen, and 46.56 g of oxygen. Another sample is found to have 0.4471 g of carbon, 0.07510 g of hydrogen, and 0.5962 g of oxygen. Show that these results are consistent with the law of definite proportions.
2. Complete the following table:

Symbol	Ion	Number of Electrons	Number of Protons	Number of Neutrons
^{40}Ca	Ca^{2+}			
	Be^{2+}			5
^{79}Se		36		
		18	17	20

3. List two metals, two metalloids, and two nonmetals.
4. Which group on the periodic table consists of the halogen family? What properties do the halogens all share?

5. Identify the anion or cation that will most likely form from each of the following:

   a. Na

   b. O

   c. Mg

   d. Al

   e. S

   f. Cl

6. Lithium has two isotopes, ^{6}Li and ^{7}Li, with natural abundances of 7% and 93% respectively. What is the average atomic mass of lithium?

7. Using the masses of the proton, neutron, and electron given in your book, calculate the mass of 1.0 mole of sodium ions and the mass of one mole of sodium atoms. Based on your answer, do you think that it is reasonable to report the mass of one mole of sodium ions as equal to the mass of one mole of sodium atoms (as is generally done)? Why or why not?

8. How many silver atoms are there in 10.5 g of silver?

9. What is the mass of 5.89×10^{30} atoms of Al?

Concept Questions:

1. Consider the alkali metals: Li, Na, and K. Using the internet, do a search on the properties of these substances and identify at least three periodic properties.

2. The law of conservation of mass states that matter is not created or destroyed in a chemical reaction. In one particular reaction, 100.0 g of water decompose to form 11.1 g of hydrogen gas according to the equation below:

$$2H_2O \rightarrow 2H_2 + O_2$$

   According to the law of conservation of mass, how much oxygen (in g) must be produced in this reaction? Explain the reasoning behind your answer.

3. The mass of a substance is the amount of matter in that substance. We define 12 amu to be the mass of an atom of carbon-12. Other elements do not have integer masses; for example, a hydrogen-1 atom has a mass of 1.007825 amu. Explain why you think that this might be.

4. Silicon has three isotopes found in nature: ^{28}Si, ^{29}Si, and ^{30}Si. Their natural abundances are 92.3%, 4.7%, and 3.0% respectively. Without doing any calculations, estimate the average molar mass of silicon to two significant figures. Explain the reasoning behind your answer.

5. The masses listed on the periodic table are averages, as discussed in your book. When we do calculations in chemistry, these are the masses that we use. Why is it reasonable to use the average mass when carrying out calculations despite the fact that we don't know the exact composition of the isotopes in a macroscopic sample?

6. Enriched uranium has a higher concentration of uranium-235 than is found in nature. One method for obtaining enriched uranium involves ionizing a raw sample of uranium (containing U-235 and U-238) and then separating the components using their mass to charge ratio. Explain how this is done using what you learned about mass spectrometry in this chapter. Which isotope will be collected first in the experiment? Explain how you arrived at your answer.

Chapter 3: Molecules, Compounds, and Chemical Equations

Learning Objectives:

- Understand the differences between ionic and covalent bonds and identify which type is present in a given compound.
- Write empirical and molecular formulas for compounds.
- Use molecular models to visualize compounds.
- Name ionic and covalent compounds using a systematic naming system.
- Calculate molar masses for compounds and utilize them in chemical problems.
- Use experimental data to determine the molecular formulas of compounds.
- Write and balance chemical equations.

Chapter Summary:

In this chapter, you will be introduced to compounds that are comprised of more than one atom bonded together. Two different bonding types will be discussed: ionic and covalent. Ionic bonding involves the transfer of electrons while covalent bonding results when electrons are shared. Both bonding types, however, result from electrostatic interactions between positive and negative charges. Once you are able to differentiate between ionic and molecular compounds, you will learn how to name them in a systematic, non-ambiguous fashion. Molar mass, the conversion factor between moles and mass for compounds, will be explained and you will be shown it can be used in chemical calculations. A further exploration of molar mass will use experimental data to determine empirical and molecular formulas. Once you are comfortable with writing formulas for compounds and understanding what they mean, you will learn how to write and balance chemical reactions. In a chemical reaction, atoms rearrange so that starting compounds are different from final compounds. Finally, in this chapter you will be introduced to organic chemistry, which will be explored further in Chapter 20.

Chapter Outline:

I. Hydrogen, Oxygen, and Water
   a. The properties of compounds are generally different from the properties of the elements that compose them.

II. Chemical Bonds
   a. A chemical bond holds atoms together and is the result of the electrostatic interactions between positive and negative charges.
   b. An ionic bond involves the transfer of electrons.
      i. When electrons are transferred between substances, a negatively charged anion (usually a non-metal) and a positively charged anion (usually a metal) are formed.
      ii. The positively and negatively charged species are attracted to each other forming a three-dimensional array, or lattice, in the solid phase.
   c. A covalent bond involves the sharing of electrons.
      i. Covalent bonds usually occur between non-metal elements and are a result of the lowering of energy that comes from the sharing of electrons between two positively charged nuclei.

ii. Covalently bonded compounds are called molecular compounds.

EXAMPLE:

Identify the type of bonding present in each of the following compounds.

a. NaCl

Sodium is a metal and chlorine is a non-metal. This is an ionic compound with ionic bonds.

b. CH_4

Carbon and hydrogen are both non-metals. This is a covalent compound with covalent bonds.

c. Al_2O_3

Aluminum is a metal and oxygen is a non-metal. This is an ionic compound with ionic bonds.

d. MgS

Magnesium is a metal and sulfur is a non-metal. This is an ionic compound with ionic bonds.

e. N_2O_4

Nitrogen and oxygen are both non-metals. This is a covalent compound with covalent bonds.

f. NO

Nitrogen and oxygen are both non-metals. This is a covalent compound with covalent bonds.

III. Representing Compounds: Chemical Formulas and Molecular Models

a. Chemical formulas give the elements and the numbers of atoms in a compound.

i. A chemical formula gives the symbols for each element present and a subscript that refers to the number of atoms of that element.

ii. The element that is most metallic (furthest left and lower on the periodic table) is generally listed first and the element that is least metallic (furthest right and higher on the periodic table) is generally listed last.

iii. The empirical formula is the simplest whole number ratio of elements in a compound; it provides a relative number of atoms of each element type.

iv. The molecular formula gives the actual number of atoms of each element in one molecule of a compound. The molecular formula is related to the empirical formula–the subscripts in the empirical formula are multiplied by a whole number in order to give the molecular formula.

EXAMPLE:

State the number of atoms of each type of element in the compounds below.

a. Na_2SO_4

There are 2 sodium atoms, 1 sulfur atom, and 4 oxygen atoms in a formula unit of sodium sulfate.

b. HCO_2H

There are 2 hydrogen atoms, 1 carbon atom, and 2 oxygen atoms in a molecule of formic acid.

c. $Mg_3(PO_4)_2$

There are 3 magnesium atoms, 2 phosphorous atoms, and 8 oxygen atoms in a formula unit of magnesium phosphate.

d. $C_6H_{12}O_6$

There are 6 carbon atoms, 12 hydrogen atoms, and 6 oxygen atoms in this fructose molecule.

EXAMPLE:

Given the molecular formulas below, provide the empirical formula for each compound.

a. Benzene, C_6H_6

Both subscripts are 6 so if we divide them both by 6 we get the empirical formula: CH

b. Dinitrogen tetroxide, N_2O_4

Both subscripts are multiples of 2 so we can divide them by 2 to get the empirical formula: NO_2

c. Hydrogen peroxide, H_2O_2

Both subscripts are multiples of 2 so we can divide them by 2 to get the empirical formula: HO

d. Glucose, $C_6H_{12}O_6$

All of the subscripts are multiples of 6 so we can divide them by 6 to get the empirical formula: CH_2O

---

v. The structural formula uses lines to represent covalent bonds and shows the connectivity of atoms in a compound.

vi. Molecular models give a visual representation of compounds.

1. Ball-and-stick models represent atoms as spheres connected by sticks representing bonds.

2. Space-filling models have the space between atoms filled in to represent a more accurate picture of how we think compounds look.

IV. An Atomic-Level View of Elements and Compounds

a. Atomic elements exist in nature as single atoms.

b. Molecular elements exist in nature as two or more atoms bound together via covalent bonds.

i. Diatomic species have two of the same element bound together.

▶ The diatomic species are H_2, N_2, O_2, F_2, Cl_2, Br_2, and I_2. These can be remembered using the periodic table – nitrogen, oxygen, and fluorine are all in the same row and then the halogens are in a single column.

ii. Polyatomic species have more than two atoms of the same element bonded together.

c. Molecular compounds are substances with two or more non-metals covalently bound together.

d. Ionic compounds are a cation (usually a metal) and an anion (usually a non-metal) bound together by ionic bonds.

i. A formula unit is the smallest electrically neutral collection of ions.

ii. Polyatomic ions are ions that contain more than one atom covalently bound together to form a charged species.

► In ionic compounds that contain polyatomic ions, both covalent and ionic bonds are present. Covalent bonds hold the polyatomic ions together and ionic bonds hold the polyatomic ion to the counterion.

V. Ionic Compounds: Formulas and Names

a. Ionic compounds consist of positive and negative ions interacting in a crystal lattice.

b. In writing formulas for ionic compounds, the following rules must be obeyed:

i. Ionic compounds always contain positive and negative ions.

ii. The sum of positive charge (from cations) and negative charge (from anions) must be zero.

► This is the key to writing formulas for ionic compounds. You will need to determine the charge of the ions (using the periodic table or, in the case of polyatomic ions, from memory) and figure out how to make the overall charge zero by combining the ions.

iii. The formula gives the smallest whole number ratio of elements.

EXAMPLE:

Give the formula for the ionic compound that forms between the two elements given:

a. Mg and O

The charge of a magnesium ion is 2+ and the charge of an oxygen ion is 2-. In order to have a charge-neutral compound, the formula for the ionic compound must be MgO.

b. Sr and Cl

The charge of a strontium ion is 2+ and the charge of a chloride ion is 1-. In order to have a charge-neutral compound, the formula for the ionic compound must be $SrCl_2$.

c. Al and S

The charge of an aluminum ion is 3+ and the charge of a sulfur ion is 2-. In order to have a charge-neutral compound, the formula for the ionic compound must be Al_2S_3.

d. Ba and N

The charge of a barium ion is 2+ and the charge of a nitrogen ion is 3-. In order to have a charge-neutral compound, the formula for the ionic compound must be Ba_3N_2.

---

c. Naming Ionic Compounds

i. Common names exist for some compounds, but require memorization.

ii. A systematic method for naming provides rules that allow any compound to be named easily:

1. Metals that have only one type of cation (all main group metals) are named using the name of the metal and the base name of the nonmetal with the suffix –ide.

EXAMPLE:

Provide names for the following ionic compounds.

a. $AgCl$

   Silver chloride

b. Na_2S

   Sodium sulfide

c. Ca_3P_2

   Calcium phosphide

d. Rb_2O

   Rubidium oxide

---

2. Metals that have multiple possible cation charges (some transition metals) are named using the name of the metal followed by the charge in parentheses and then the base name of the nonmetal with the suffix –ide.

EXAMPLE:

Provide names for the following ionic compounds.

a. $FeCl_3$

   Iron (III) chloride

b. CuO

   Copper (II) oxide

c. MnS_2

   Manganese (IV) sulfide

d. CrF_2

   Chromium (II) fluoride

---

3. Ionic compounds that contain polyatomic ions are named using the name of the metal (following by the charge if necessary) followed by the name of the polyatomic ion.

   a. Many of the polyatomic ions are oxyanions, which contain oxygen and one other atom.

   b. Once the –ate form of a polyatomic ion is memorized, all other versions of that ion can be derived from it:

      i. Add an oxygen atom to the –ate form to get the per–ate form.

      ii. Remove an oxygen atom from the –ate form to get the –ite form.

iii. Remove two oxgen atoms form the –ate form to get the hypo–ite form.

c. All forms of a given polyatomic ion have the same charge. For example, perchlorate, chlorate, chlorite, and hypochlorite ions all have a charge of -1.

d. A simple way to remember polyatomic ions is to memorize CO_3^{2-}, NO_3^-, PO_4^{3-}, SO_4^{2-}, and ClO_3^-. All other polyatomic ions will be of the form of the ion that is nearest in the same group on the periodic table. For example, the oxyanion of arsenic will be like that of phosphorous: AsO_4^{3-} is arsenate.

EXAMPLE:

Provide names for the following ionic compounds.

a. $Pb(NO_3)_2$

Lead (II) nitrate

b. $Cu(NO_2)_2$

Copper (II) nitrite

c. K_2SO_3

Potassium sulfite

d. $Mg(ClO)_2$

Magnesium hypochlorite

---

4. Hydrates are ionic compounds that have a certain number of water molecules associated with each formula unit.

a. Hydrates are named using the ionic compound name followed by (number prefix)-hydrate.

EXAMPLE:

Provide names for the following hydrates.

a. $MgSO_4 \cdot 9H_2O$

Magnesium sulfate nonahydrate

b. $BaCl_2 \cdot 2H_2O$

Barium chloride dihydrate

c. $KHCO_3 \cdot 2H_2O$

Potassium bicarbonate dihydrate

d. $CoF_2 \cdot H_2O$

Cobalt (II) fluoride monohydrate

VI. Molecular Compounds: Formulas and Names

- ▶ When naming compounds, first determine if the compound is ionic or molecular – this is the only way that you can correctly name the compound since the rules are different.

a. Non-metals can combine with each other in variable ways, the numbers of each type of element are not obvious from the group number as in ionic compounds.

b. Common names are given for many molecular compounds but this method requires extensive memorization.

c. The systematic method for naming molecular compounds is based on the number of each type of atom.

   i. The first element listed is always the most metallic (the one that is farthest left and lowest down on the periodic table).

   ii. The number of atoms of each element is denoted using a prefix.

   iii. The name is "prefix-(first element name) prefix-(second element base name)-ide."

      1. If there is only one atom of the more metallic element, the prefix mono- is dropped.

EXAMPLE:

Provide names for the following molecular compounds.

a. NO_2

Nitrogen dioxide

b. CCl_4

Carbon tetrachloride

c. CO_2

Carbon dioxide

d. P_4O_{10}

Tetraphosphorus decoxide

---

d. Acids are compounds that produce hydrogen ions (H^+) in water. Acids are named according to which type of acid it is:

   i. Binary acids contain only two elements: hydrogen and one other element.

      1. Binary acids are named using "hydro-(base name of the metal)-ic acid."

EXAMPLE:

Provide names for the following acids.

a. H_2S

Hydrosulfuric acid

b. HCl

Hydrochloric acid

c. H_2Te

Hydrotelluric acid

d. HBr

Hydrobromic acid

---

ii. Oxyacids contain one of the oxyanions and enough hydrogen atoms to make the species charge-neutral.

1. Oxyacids are named based on the relevant oxyanion.

a. If the oxyacid ends in –ate, then the acid is named "(base name of oxyanion)-ic acid"

b. If the oxyacid ends in –ite, then the acid is named "(base name of the oxyanion)-ous acid"

EXAMPLE:

Provide names for the following acids.

a. H_2CO_3

Carbonic acid

b. H_2SO_3

Sulfurous acid

c. $HClO_4$

Perchloric acid

d. H_2SeO_4

Selenic acid

VII. Formula Mass and the Mole Concept for Compounds

a. The formula mass for a compound is also called its molecular mass or molecular weight.

b. The formula mass of a molecule is equal to the sum of the molar masses of the parts of the molecule.

c. Since the mass in amu of an atom is numerically equal to the mass of 1 mole of those atoms, the same is true for molecules.

d. The molar mass is equal to the mass in grams of 1 mole of a compound.

EXAMPLE:

Calculate the molar mass of the following compounds.

a. K_3PO_4

In this case we have three potassium atoms, one phosphorous atom, and four oxygen atoms. The mass of each element is given on the periodic table. Adding up the masses gives:

$$3 \times 39.10\ \text{g/mol} + 30.97\ \text{g/mol} + 4 \times 16.00\ \text{g/mol} = 212.27\ \text{g/mol}$$

b. $Ca(OH)_2$

In this case we have one calcium atom, two oxygen atoms, and two hydrogen atoms. Adding up the masses given on the periodic table we have:

$$40.08\ \text{g/mol} + 2 \times 16.00\ \text{g/mol} + 2 \times 1.008\ \text{g/mol} = 74.10\ \text{g/mol}$$

c. $(NH_4)_2CO_3$

In this case we have two nitrogen atoms, eight hydrogen atoms, one carbon atom, and three oxygen atoms. Adding up the masses given on the periodic table we have:

$$2 \cdot 14.01\ \text{g/mol} + 8 \times 1.008\ \text{g/mol} + 12.01\ \text{g/mol} + 3 \times 16.00\ \text{g/mol} = 96.09\ \text{g/mol}$$

---

e. Molar mass can be thought of as a conversion factor between moles of a compound and grams of that compound.

EXAMPLE:

Determine the number of carbon atoms in 150.0 g of $CaCO_3$.

First we must calculate the molar mass of $CaCO_3$. We do this by finding the sum of the masses of the atoms that comprise it:

$$40.08\ \text{g/mol} + 12.01\ \text{g/mol} + 3 \times 16.00\ \text{g/mol} = 100.09\ \text{g/mol}$$

Now we can map out the path to find the answer:

$$\text{g CaCO}_3 \xrightarrow{\text{g CaCO}_3 \text{ to mol CaCO}_3} \text{mol CaCO}_3 \xrightarrow{\text{mol CaCO}_3 \text{ to mol C atoms}} \text{mol C atoms} \xrightarrow{\text{mol C atoms to C atoms}} \text{C atoms}$$

Now we will use conversion factors to solve the problem:

$$150.0\ \text{g CaCO}_3 \times \frac{1\ \text{mol CaCO}_3}{100.09\ \text{g CaCO}_3} \times \frac{1\ \text{mol C atoms}}{1\ \text{mol CaCO}_3} \times \frac{6.022 \times 10^{23}\ \text{C atoms}}{1\ \text{mol C atoms}} = 9.025 \times 10^{23}\ \text{C atoms}$$

VIII. Composition of Compounds

a. The mass percent is the percentage of a molecule's total mass that comes from one element present in that compound:

$$\text{mass percent of X} = \frac{\text{mass of X in 1 mole of compound Y}}{\text{mass of 1 mole of compound Y}} \times 100\%$$

EXAMPLE:

Calculate the mass percent of oxygen in sulfuric acid, H_2SO_4.

First we will calculate the molar mass of sulfuric acid. We do this by finding the sum of the masses of the atoms that comprise it:

$$2 \times 1.008\ \text{g/mol} + 32.07\ \text{g/mol} + 4 \times 16.00\ \text{g/mol} = 98.09\ \text{g/mol}$$

Then we can calculate the mass of oxygen that is present in sulfuric acid:

$$4 \times 16.00 \text{ g/mol} = 64.00 \text{ g/mol}$$

And now we can use the formula given to calculate the mass percent:

$$\frac{64.00 \text{ g/mol oxygen}}{98.09 \text{ g/mol sulfuric acid}} \times 100\% = 65.25\% \text{ oxygen}$$

So sulfuric acid is 65.25% oxygen by mass.

---

b. Mass percent can be viewed as a conversion factor between the element's mass and the compound's total mass.

EXAMPLE:

Using the percent mass of oxygen calculated in the previous example, determine the mass of oxygen present in a 5.43 kg sample of sulfuric acid.

This is a one-step conversion problem using percent by mass as the conversion factor:

$$5.43 \text{ kg sulfuric acid} \times \frac{65.25 \text{ g/mol oxygen}}{100.00 \text{ g/mol sulfuric acid}} = 3.54 \text{ kg oxygen}$$

We can also solve this problem by multiplying the total mass of the compound by the percentage of that mass that is oxygen to get the same result:

$$5.43 \text{ kg} \times 65.25\% = 3.54 \text{ kg oxygen}$$

---

c. In a chemical formula, we are given a ratio of atoms to molecules or moles of atoms to moles of molecules. Subscripts do not represent relative masses. For example, one mole of water molecules (H_2O) has twice as many moles of hydrogen as oxygen, but the mass of hydrogen in water is only one-eighth the mass of oxygen.

IX. Determining a Chemical Formula from Experimental Data

a. A substance can be decomposed into its elements and the masses of the elements can be measured; from this process, the empirical formula can be determined.

EXAMPLE:

Fructose is one of the sugars found in fruit. Elemental analysis of fructose gave the following mass percent composition: 40.0% carbon, 6.72% hydrogen, and 53.28% oxygen. What is the empirical formula of fructose?

First we will assume that we have 100.0 g of fructose and determine the number of moles of each element in that 100.0 g sample.

Since we have 100.0 g of fructose and 40.0% of the mass is carbon, we have 40.0 g of carbon:

$$40.0 \text{ g C} \times \frac{1 \text{ mol C}}{12.01 \text{ g C}} = 3.33 \text{ mol C}$$

Since we have 100.0 g of fructose and 6.72% of the mass is hydrogen, we have 6.72 g of hydrogen:

$$6.72\,\text{g H} \times \frac{1\,\text{mol H}}{1.008\,\text{g H}} = 6.67\,\text{mol H}$$

Since we have 100.0 g of fructose and 53.28% of the mass is oxygen, we have 53.28 g of oxygen:

$$53.28\,\text{g O} \times \frac{1\,\text{mol O}}{16.00\,\text{g O}} = 3.33\,\text{mol O}$$

We now need to find the simplest whole number ratio of each of these elements. By dividing all three by the smallest number of moles (3.33 mol in this case), we find that the ratio of C to H to O is 1:2:1. The empirical formula is therefore CH_2O.

b. In order to determine the molecular formula given the empirical formula, the molar mass of the compound must be given.

EXAMPLE:

The molar mass of fructose is 180.16 g/mol. Using the empirical formula found in the example above, determine the molecular formula of fructose.

We start out by calculating the molar mass of the empirical formula from above:

$$12.01\,\text{g/mol} + 2 \times 1.008\,\text{g/mol} + 16.00\,\text{g/mol} = 30.03\,\text{g/mol}$$

Now by dividing the molar mass by the empirical formula molar mass, we find the integer multiplier:

$$\frac{180.16\,\text{g/mol}}{30.03\,\text{g/mol}} = 5.999$$

Rounding to the nearest whole number, we see that the multiplier is 6. We multiply all of the subscripts of the empirical formula by 6 in order to find the molecular formula which is $C_6H_{12}O_6$.

c. Combustion analysis can also be used in order to determine the empirical formula of a compound.

EXAMPLE:

A certain hydrocarbon (compound containing only carbon and hydrogen) is burned in excess oxygen. Analysis of the combustion products showed that 24.38 g of water and 30.29 g of carbon dioxide were produced in the reaction. What is the empirical formula of the hydrocarbon?

We need to calculate the number of moles of carbon and the number of moles of hydrogen in the compound. Once we have this, then the problem becomes simple.

First we will determine the amount of hydrogen produced using the water. We will convert grams of water to moles of water to moles of hydrogen:

$$24.38\,\text{g H}_2\text{O} \times \frac{1\,\text{mol H}_2\text{O}}{18.02\,\text{g H}_2\text{O}} \times \frac{2\,\text{mol H}}{1\,\text{mol H}_2\text{O}} = 2.706\,\text{mol H}$$

Now we will convert grams of carbon dioxide to moles of carbon dioxide to moles of carbon:

$$30.29\,\text{g CO}_2 \times \frac{1\,\text{mol CO}_2}{44.01\,\text{g CO}_2} \times \frac{1\,\text{mol C}}{1\,\text{mol CO}_2} = 0.6883\,\text{mol C}$$

Now we will divide both numbers by the smaller of the two (0.6883 in this case) to get a whole number ratio. We see from this that the ratio of C:H is 1:4, so the empirical formula is CH_4.

X. Writing and Balancing Chemical Equations

a. In a chemical reaction, one or more substances are converted into one or more different substances.

b. We write chemical reactions in shorthand notation where the arrow represents the phrase "react(s) to form":

reactants → products

c. Chemical reactions must be balanced in order to adhere to the law of conservation of mass, which states that matter is not created or destroyed in a chemical reaction.

i. When balancing a chemical reaction, do not change the molecular identity of products or reactants, but rather change the number of each molecule. This means that coefficients may be changed while balancing a reaction, but subscripts should not.

EXAMPLE:

Balance the following reactions:

a. $H_2 + Cl_2 \rightarrow HCl$

First we will set up a table in order to balance the atoms:

Atom	# on Left Side	# on Right Side
Cl	2	1
H	2	1

We will start by putting a "2" in front of HCl so that the Cl atoms are balanced.

$H_2 + Cl_2 \rightarrow 2HCl$

We then recount everything in our table:

Atom	# on Left Side	# on Right Side
Cl	2	2
H	2	2

And we see that the reaction is now balanced.

b. $KClO_3 \rightarrow KCl + O_2$

We will again set up a table:

Atom	# on Left Side	# on Right Side
K	1	1
Cl	1	1
O	3	2

We see that the first two entries, K and Cl, are already balanced, so we look at O. In order to balance with whole numbers, we must multiply the reactant side by 2 and the product side by 3:

$2KClO_3 \rightarrow KCl + 3O_2$

Atom	# on Left Side	# on Right Side
K	2	1
Cl	2	1
O	6	6

We have now unbalanced the first two elements in the table. By multiplying by 2 on the product side to balance the K atoms we have:

$2KClO_3 \rightarrow 2KCl + 3O_2$

Atom	# on Left Side	# on Right Side
K	2	2
Cl	2	2
O	6	6

And we see that everything is balanced.

c. $CH_3OH + O_2 \rightarrow CO_2 + H_2O$

Again we will start by writing out the table:

Atom	# on Left Side	# on Right Side
C	1	1
H	4	2
O	3	3

For combustion reactions in general (the reaction with oxygen to produce water and carbon dioxide), the easiest way to balance is to consider first C, then H, and finally O. (This strategy can be remembered as alphabetical order: C, H, O.) Since carbon is already balanced in this case, we will start with hydrogen by multiplying the water coefficient by 2:

$CH_3OH + O_2 \rightarrow CO_2 + 2H_2O$

Atom	# on Left Side	# on Right Side
C	1	1
H	4	4
O	3	4

Now we need to balance the oxygen by increasing the amount on the reactant side. We have two choices: we can change the number of CH_3OH molecules or O_2 molecules. We will balance the O_2 because if we change the number of CH_3OH molecules we will have to go back and balance C and H again. In this case, in order to balance the oxygen atoms, we will need 3/2 of an O_2 molecule:

$CH_3OH + 3/2O_2 \rightarrow CO_2 + 2H_2O$

Atom	# on Left Side	# on Right Side
C	1	1
H	4	4
O	4	4

Now we see that everything balances, but the 3/2 is not conventional. We can multiply all coefficients by 2 in order to eliminate the fraction:

$2CH_3OH + 3O_2 \rightarrow 2CO_2 + 4H_2O$

Which is a correctly balanced equation.

XI. Organic Compounds

a. Organic compounds are compounds that contain mostly carbon with few other elements.
    i. Carbon forms four bonds to other atoms.
    ii. Carbon can form single, double, or triple bonds.
b. Hydrocarbons are compounds that only contain carbon and hydrogen.
    i. An alkane is a hydrocarbon that contains only single-bonded carbon atoms.
    ii. An alkene is a hydrocarbon that contains at least one carbon-carbon double bond.
    iii. An alkyne is a hydrocarbon that contains at least one carbon-carbon triple bond.
c. Functional hydrocarbons contain a hydrocarbon group and a functional group.
    i. A functional group is a characteristic atom or group of atoms.
    ii. The hydrocarbon group in a functional hydrocarbon is usually denoted by the symbol "-R."
    iii. Functional groups include alcohols, ethers, aldehydes, ketones, carboxylic acids, esters, and amines.

Fill in the Blank Problems:

1. The average mass of a molecule is the _______________.
2. A(n) _______________ reaction is a reaction in which a substance combines with oxygen to form carbon dioxide and water.
3. The _______________ is the smallest, electrically neutral collection of ions.
4. A(n) _______________ bond occurs between a metal and a non-metal.
5. The _______________ gives the number of atoms of each element in a molecule of a compound.
6. Oppositely charged species are attracted to one another via _______________ forces.
7. _______________ are compounds that contain only two elements.
8. A(n) _______________ group is a characteristic atom or group of atoms attached to a hydrocarbon.
9. The _______________ of an element is that element's percentage of a compound's total mass.
10. A(n) _______________ bond occurs between two non-metals.
11. Man y molecular elements exist as _______________ molecules.
12. Organic co mpounds that contain only carbon and hydrogen are called _______________.

Chapter 3: Molecules, Compounds, and Chemical Equations

Problems:

1. List five elements that are found in nature as diatomic molecules.
2. Provide formulas for the following compounds:
   a. Sodium bromide
   b. Magnesium sulfide
   c. Disulfur tetroxide
   d. Carbonic acid
   e. Hydroiodic acid
   f. Perchloric acid
   g. Selenium dinitride
   h. Iron (III) oxide
   i. Silver chloride
   j. Silicon tetrachloride
   k. Calcium bicarbonate
   l. Cobalt (III) phosphate monohydrate
3. Provide names for the following compounds:
   a. P_2O_5
   b. Rb_2O
   c. TiO_2
   d. BeI_2
   e. N_2O_4
   f. $Fe_2(SO_4)_3$
   g. XeF_4
   h. H_3PO_3
   i. $LiOH$
   j. CuF_2
   k. $CuSO_4 \cdot 5H_2O$
   l. $HClO_2$
4. Calculate the formula mass for the compounds given in question 3.
5. What is the mass percent of carbon in pentanol, $CH_3CH_2CH_2CH_2CH_2OH$?
6. The daily nutritional allowance for salt is about 2.5 g. How many sodium atoms are there in a day's worth of table salt, NaCl?
7. Balance the following chemical reactions:
   a. $N_2O_4 \rightarrow NO_2$
   b. $NaOH + H_2SO_4 \rightarrow H_2O + Na_2SO_4$
   c. $NaCl + BeF_2 \rightarrow NaF + BeCl$
   d. $Mg + Mn_2O_4 \rightarrow MgO + Mn$

e. $HCl + Na_2CO_3 \rightarrow NaCl + H_2O + CO_2$

f. $CH_3CH_2OH + O_2 \rightarrow CO_2 + H_2O$

g. $C_2H_6 + O_2 \rightarrow CO_2 + H_2O$

h. $Fe_2(SO_4)_3 + KSCN \rightarrow K_3Fe(SCN)_6 + K_2SO$

i. Solid copper (II) oxide and solid carbon react to form solid copper and carbon monoxide gas.

j. Aqueous cobalt (III) nitrate and aqueous ammonium sulfide react to form solid cobalt (III) sulfide and aqueous ammonium nitrate.

8. How many carbon atoms are there in a 5.0-g sample of fructose, $C_6H_{12}O_6$?

9. A hydrate of copper (II) chloride is heated until all of the water is driven off. The sample was 6.10 g before heating and was 3.41 g after heating. What is the formula of the hydrate?

10. A mixture of $CaCO_3$ and $(NH_4)_2CO_3$ is 61.9% CO_3 by mass. What is the percent of $CaCO_3$ in the mixture?

11. Sodiu m bicarbonate is a common baking ingredient. Calculate the mass percent of each element in sodium bicarbonate. Using the mass percent, determine the number of grams of oxygen present in a 5.0-g sample of sodium bicarbonate.

12. A certain hydrocarbon is placed in a sealed container with pure oxygen and burned. After the reaction is complete, all of the hydrocarbon and all of the oxygen have been consumed. Quantitative analysis of the products showed that 52.10 g of water and 84.85 g of carbon dioxide were produced. What is the empirical formula of the hydrocarbon and how many grams of oxygen were present in the container before the reaction was carried out?

13. A recent report has suggested that children are becoming more susceptible to tooth decay because they are drinking bottled water, which lacks fluoride. Studies have shown that the optimal concentration of fluoride in drinking water is 1.1 mg/L, given that the average amount of water consumed in one day is six 8-ounce glasses. How many grams of sodium fluoride will need to be consumed in order to obtain the same benefit as drinking fluoridated water?

Concept Questions:

1. What is wrong with the following statement? "When a chemical equation is balanced, the number of molecules on each side of the equation will be equal."

2. What is wrong with the following statement? "The chemical formula of ammonia indicates that the compound contains three grams of hydrogen for each gram of nitrogen."

3. As we learned in this chapter, ionic compounds exist in a crystal lattice which is a large network of anions and cations. Conversely, molecular compounds exist as discrete entities. Explain why you think this is so based on the principles of bonding that exist in ionic and covalent compounds.

4. The number of compounds that have been synthesized in the laboratory is on the order of millions. Most of these compounds are molecular instead of ionic. Suggest one reason why the number of molecular compounds is larger than the number of ionic compounds.

5. The fact that chemical reactions must be balanced is another statement of the law of conservation of mass. Explain how these are equivalent.

6. The structures of morphine (a) and codeine (b) are shown below:

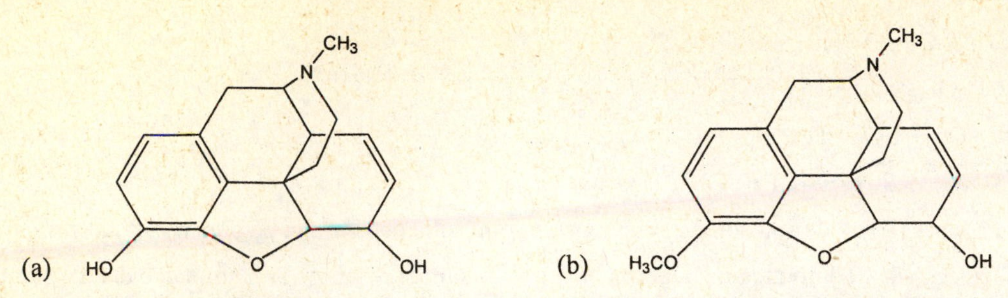

In these structures, the intersection of two lines indicates that a carbon atom is present.

These two compounds have very different properties. Identify any functional groups that you recognize in these structures. What aspect of their structure is responsible for the differences in their reactivity?

Chapter 4: Chemical Quantities and Aqueous Reactions

Learning Objectives:

- Utilize reaction stoichiometry in calculations.
- Determine theoretical and percent yields for reactions with a limiting reagent.
- Calculate solution concentrations and use concentration values as a conversion factor.
- Identify electrolyte solutions and understand their properties.
- Learn to write molecular, complete ionic, and net ionic equations for chemical reactions.
- Determine oxidation states and utilize that information in identifying oxidizing and reducing agents in redox reactions.
- Identify products in precipitation, gas evolution, acid-base, and redox reactions.

Chapter Summary:

In this chapter, you will learn how to carry out the most common type of chemistry problem: the use of reaction stoichiometry in chemical calculations. You will learn how to identify the limiting reagent in a stoichiometry problem and, from that information, determine the theoretical yield of a reaction. With the theoretical yield, you will learn how to calculate the percent yield when the actual yield is given. Electrolyte solutions will be discussed and used to predict the product(s) of a precipitation reaction. Various other reaction types including gas evolution, acid-base, and redox reactions will then be explored so that you will be able to predict product formation and carry out stoichiometry calculations on the most common reaction types. The details of these reaction types will be discussed so that you can understand the fundamental differences and similarities between them.

Chapter Outline:

I. Reaction Stoichiometry

a. The coefficients in a chemical reaction specify the relative amounts, in moles, of each substance in the reaction.

b. Stoichiometry is the numerical relationships between the amounts of each substance in a balanced chemical reaction.

i. The balanced chemical reaction can be viewed as a recipe.

c. A chemical reaction can be used to construct conversion factors between the numbers of moles of each substance:

$$H_2(g) + F_2(g) \rightarrow 2HF(g)$$

Example conversion factors for this reaction are: $\frac{1\ \text{mol}\ H_2}{1\ \text{mol}\ F_2}$, $\frac{1\ \text{mol}\ H_2}{2\ \text{mol HF}}$, and $\frac{2\ \text{mol HF}}{1\ \text{mol}\ F_2}$.

EXAMPLE:

Write a conversion factor between each species in the balanced reaction for the combustion of methane:

$$CH_4(g) + 2O_2(g) \rightarrow 2H_2O(g) + CO_2(g)$$

$$\frac{1\text{ mol } CH_4}{2\text{ mol } O_2}, \frac{1\text{ mol } CH_4}{2\text{ mol } H_2O}, \frac{1\text{ mol } CH_4}{1\text{ mol } CO_2}, \frac{1\text{ mol } O_2}{1\text{ mol } H_2O}, \frac{2\text{ mol } O_2}{1\text{ mol } CO_2}, \text{ and } \frac{2\text{ mol } H_2O}{1\text{ mol } CO_2}$$

II. Limiting Reactant, Theoretical Yield, and Percent Yield

a. The limiting reagent restricts the amount of product that can form in a chemical reaction.

   i. The limiting reactant is the reactant that is completely used up in a chemical reaction.

b. The excess reagent(s) do not limit the amount of product that forms in a chemical reaction and will be present to some degree after the reaction is complete.

▶ To determine the limiting reagent, calculate the amount of product formed using each reagent. The limiting reagent will be the species that gives the smallest amount of product and all other species will be in excess.

c. The theoretical yield is the amount of product that is formed in a reaction based on the limiting reagent.

EXAMPLE:

3.04 g of aluminum reacts with 7.42 g of iodine to form aluminum iodide. When the reaction is complete, how many grams of all species are predicted to be present in the reaction flask?

We first need to write a balanced chemical reaction:

$$2Al + 3I_2 \rightarrow 2AlI_3$$

Then we can determine the amount of product formed from each of the reactant species:

$$3.04\text{ g Al} \times \frac{1\text{ mol Al}}{26.98\text{ g Al}} \times \frac{2\text{ mol } AlI_3}{2\text{ mol Al}} \times \frac{407.68\text{ g } AlI_3}{1\text{ mol } AlI_3} = 45.94\text{ g } AlI_3 \text{ from Al}$$

$$7.42\text{ g } I_2 \times \frac{1\text{ mol } I_2}{253.8\text{ g } I_2} \times \frac{2\text{ mol } AlI_3}{3\text{ mol } I_2} \times \frac{407.68\text{ g } AlI_3}{1\text{ mol } AlI_3} = 7.946\text{ g } AlI_3 \text{ from } I_2$$

We find that I_2 is the limiting reagent since it results in a smaller amount of product, AlI_3.

Now we can calculate the amount of Al that reacted with the I_2:

$$7.946\text{ g } AlI_3 \times \frac{1\text{ mol } AlI_3}{407.68\text{ g } AlI_3} \times \frac{2\text{ mol Al}}{2\text{ mol } AlI_3} \times \frac{26.98\text{ g Al}}{1\text{ mol Al}} = 0.526\text{ g Al reacted}$$

So the mass of Al that remains after the reaction is $3.04 - 0.526 = 2.51$ g.

All of the I_2 reacts so there is none left – this is always the case for the limiting reagent.

This means that after the reaction is complete there are 2.51 g Al, 0 g of I_2, and 7.95 g of AlI_3.

---

d. The actual yield is the amount of product that is actually formed in a chemical reaction.

▶ The actual yield should always be lower than the theoretical yield – this can be a good way to check your work.

e. The percent yield is the actual yield over the theoretical yield:

$$\text{percent yield} = \frac{\text{actual yield}}{\text{theoretical yield}} \times 100\%$$

EXAMPLE:

The reaction in the example above is carried out and 7.04 g of AlI_3 are collected. What is the percent yield of the reaction?

The theoretical yield is 7.95 g as calculated above, so the percent yield is:

$$\frac{7.04}{7.95} \times 100\% = 88.6\%$$

III. Solution Concentration and Solution Stoichiometry

a. A solution is a homogeneous mixture of two or more substances.

b. The solvent is the majority component of a solution and the solute is the minor component.

c. An aqueous solution is a solution in which water is the solvent.

d. Solution concentrations can be described qualitatively as concentrated or dilute.

e. Solution concentrations can be described quantitatively using the molarity (abbreviated M) which is the moles of solute over the volume of solution.

► You will see other units for solution concentrations later on.

EXAMPLE:

Calculate the molarity of a solution that is made from dissolving 5.28 mg of sodium chloride in 48.5 mL of water. Assume that the solution volume is the same as the solvent volume.

First we need to calculate the number of moles of NaCl:

$$5.28\,\text{mg NaCl} \times \frac{1\,\text{g}}{1000\,\text{mg}} \times \frac{1\,\text{mol NaCl}}{40.0\,\text{g NaCl}} = 0.000132\,\text{mol NaCl}$$

Now we convert the volume into liters of water:

$$48.5\,\text{mL} \times \frac{1\,\text{L}}{1000\,\text{mL}} = 0.0485\,\text{L}$$

The molarity is then the number of moles of solute divided by the volume of solution:

$$\frac{0.000132\,\text{mol}}{0.0485\,\text{L}} = 0.00272\,\text{M}$$

---

f. Molarity can be used as a conversion factor.

EXAMPLE:

How many grams of sodium are in 12.5 L of a 5.82 M solution of sodium carbonate?

This is a multiple step conversion factor problem using the molarity to convert from liters to moles:

$$12.5\ \text{L} \times \frac{5.82\ \text{mol Na}_2\text{CO}_3}{1\ \text{L}} \times \frac{2\ \text{mol Na}^+}{1\ \text{mol Na}_2\text{CO}_3} = 146\ \text{g Na}^+$$

---

g. Stock solutions (concentrated solutions) can be diluted; the concentration of the dilute solution can be calculated using the concentration of the stock solution and the volumes according to $M_1V_1=M_2V_2$.

EXAMPLE:

How many liters of a 4.95 M solution of HCl are required to prepare 10.0 L of a 0.510 M solution?

This is a simple dilution problem where the unknown is the volume of the concentrated solution:

$$(4.95\ \text{M})V_1 = (0.510\ \text{M})(10.0\ \text{L})$$

$$V_1 = 1.03\ \text{L}$$

EXAMPLE:

When lead(II) nitrate and sodium chloride are combined, a lead(II) chloride precipitate forms:

$$Pb(NO_3)_2(aq) + 2NaCl(aq) \rightarrow PbCl_2(s) + 2NaNO_3(aq)$$

If 30.0 mL of a 0.541 M $Pb(NO_3)_2$ solution is added to 24.5 mL of a 1.452 M NaCl solution, how much $PbCl_2$ (in grams) is formed?

This is a conversion factor problem using molarity and the reaction stoichiometry.

We first need to calculate the grams of product using the amount of $Pb(NO_3)_2$:

$$30.0\ \text{mL} \times \frac{1\ \text{L}}{1000\ \text{mL}} \times \frac{0.541\ \text{mol Pb(NO}_3)_2}{1\ \text{L}} \times \frac{1\ \text{mol PbCl}_2}{1\ \text{mol Pb(NO}_3)_2} \times \frac{278.11\ \text{g PbCl}_2}{1\ \text{mol PbCl}_2} = 4.51\ \text{g PbCl}_2$$

Now we will do the same calculation using NaCl:

$$24.5\ \text{mL} \times \frac{1\ \text{L}}{1000\ \text{mL}} \times \frac{1.452\ \text{mol NaCl}}{1\ \text{L}} \times \frac{1\ \text{mol PbCl}_2}{2\ \text{mol NaCl}} \times \frac{278.11\ \text{g PbCl}_2}{1\ \text{mol PbCl}_2} = 4.95\ \text{g PbCl}_2$$

We can see from these calculations that $Pb(NO_3)_2$ is the limiting reagent, so 4.51 g of $PbCl_2$ will form.

IV. Types of Aqueous Solutions and Solubility

a. Water is a polar molecule meaning that one end of it is partially positive and one end is partially negative.

i. Some ionic compounds dissolve in water because the anion is attracted to the positive end of the water molecule and the cation is attracted to the negative end.

b. Substances that, when dissolved in water, form solutions that conduct electricity are called electrolytes.

   i. Strong electrolytes completely dissociate and weak electrolytes only dissociate partially when dissolved in water.

c. Non-electrolytes do not dissociate into ions when dissolved in water and their solutions, therefore, do not conduct electricity.

d. Acids are molecular compounds that dissociate into ions when dissolved in water.

   i. A strong acid completely ionizes and a weak acid only partially dissociates in water.

   ► Acids are an example of an electrolyte. Solutions of strong acids conduct electricity and solutions of weak acids weakly conduct electricity.

e. A soluble compound is, essentially, one that dissolves in water while an insoluble compound does not.

   i. The solubility rules can be simplified by realizing that exceptions always include the same six cations.

      1. Calcium, strontium, and barium are in the same group in the periodic table.

      2. Silver, mercury, and lead are one away from each other on the periodic table: start at silver, skip one to the right (cadmium), below is mercury; then skip one to the right (thallium) and end at lead.

EXAMPLE:

Predict which of the following substances will form a solution that will conduct an electrical current.

a. HNO_3

   Nitric acid is a strong acid which means that it will dissociate completely in water. A solution of nitric acid will, therefore, conduct electricity.

b. NaCl

   Sodium salts are always soluble in water and will form sodium cations and chloride ions in solution. The presence of ions in water means that the solution will conduct electricity.

c. CaS

   Calcium sulfide is not soluble in water so a solution of it will not conduct electricity.

d. $Fe(NO_3)_3$

   All salts containing the nitrate ion are soluble in water. A solution of iron(III) nitrate will conduct electricity since iron cations and nitrate anions will be present in solution.

e. $C_6H_{12}O_6$

   Glucose is a molecular compound which dissolves in water. Since glucose does not dissociate into ions, however, a solution of glucose will not conduct electricity.

f. AgI

   Silver halides are insoluble in water so a solution of it will not conduct electricity.

V. Precipitation Reactions

a. Precipitation reactions are those in which a solid (precipitate) forms upon the mixing of two solutions.

i. Only insoluble compounds form precipitates.

b. When two solutions are mixed, the anions can pair with the cations of the other species; if an insoluble species is formed, the reaction is a precipitation reaction.

VI. Representing Aqueous Reactions: Molecular, Complete Ionic, and Net Ionic Equations

a. A molecular equation gives the complete neutral formulas for all species as though they existed as molecules in solution.

b. A complete ionic equation shows the individual ions present in the solution.

▶ Any species that is in the aqueous phase can, potentially, dissociate into ions. Gases, liquids, and solids do not dissociate.

c. The net ionic equation shows only species that change during a chemical reaction and eliminates spectator ions which appear in the same form on both sides of the reaction.

▶ Note that species that appear on both sides of the equation do not change and, therefore, do not participate in the chemistry. Canceling out these spectator ions is a way to show only the chemistry that is occurring.

EXAMPLE:

When a solution of barium acetate is combined with a solution of sodium sulfate, a precipitate forms. Predict the products formed in the reaction and write the molecular, complete ionic, and net ionic equations.

We will first determine the products of the reaction by exchanging the anions of the reactants and being sure to form charge-neutral species. The phases of all species are determined using the solubility rules.

$$Ba(C_2H_5O_2)_2(aq) + Na_2SO_4(aq) \rightarrow BaSO_4(s) + NaC_2H_3O_2(aq)$$

Now we will balance the reaction to give the molecular equation:

$$Ba(C_2H_5O_2)_2(aq) + Na_2SO_4(aq) \rightarrow BaSO_4(s) + 2NaC_2H_3O_2(aq)$$

The full ionic equation shows all aqueous species separated into ions. This equation shows all species that are present in the container when these solutions are mixed.

$$Ba^{2+}(aq) + 2C_2H_5O_2^-(aq) + 2Na^+(aq) + SO_4^{2-}(aq) \rightarrow BaSO_4(s) + 2Na^+(aq) + 2C_2H_3O_2^-(aq)$$

Now we write the net ionic equation by eliminating the spectator ions (Na^+ and $C_2H_3O_2^-$):

$$Ba^{2+}(aq) + SO_4^{2-}(aq) \rightarrow BaSO_4(s)$$

VII. Acid-Base and Gas Evolution Reactions

a. Acid-base reactions are often called neutralization reactions and produce water and/or a weak electrolyte.

i. An acid is a substance that produces H_3O^+ in water.

1. H_3O^+ is a proton that is associated with a water molecule due to electrostatic interactions.

ii. A base is a substance that produces OH^- in water.

► You will see a more complete set of definitions for acids and bases later on.

b. A polyprotic acid is an acid in which more than one ionizable proton can be sequentially released. A diprotic acid is one in which two protons are sequentially released.

c. In a general acid-base reaction, the acid donates a proton to the base to produce water and a salt.

d. A titration is an analytical tool in which a known acid is added to an unknown base or a known base is added to an unknown acid; a titration problem is a stoichiometry problem involving acids and bases.

   i. The equivalence point is the point in a titration where the moles of acid and base are stoichiometrically equal.

   ii. An indicator is a dye that changes color at a particular acidity and is used to visually identify the equivalence point.

   ► Note that the point where the indicator changes color is called the endpoint of the titration. It is desirable for the endpoint to be the same as the equivalence point, but this is not always true.

EXAMPLE:

A 0.4512 M solution of nitric acid is titrated with 25.4 mL of a 0.5420 M solution of NaOH. What volume of acid was utilized in the experiment?

First we need to write a balanced chemical reaction:

$$HNO_3(aq) + NaOH(aq) \rightarrow NaNO_3(aq) + H_2O(l)$$

We can calculate the moles of sodium hydroxide used:

$$25.4\text{ mL} \times \frac{1\text{ L}}{1000\text{ mL}} \times \frac{0.5420\text{ mol NaOH}}{\text{L}} = 0.01377\text{ mol NaOH}$$

We can see from the balanced chemical reaction that we need one mole of nitric acid for every mole of sodium hydroxide. We can use this to determine the volume of acid:

$$0.01377\text{ mol NaOH} \times \frac{1\text{ mol HNO}_3}{1\text{ mol NaOH}} \times \frac{1\text{ L}}{0.4512\text{ mol}} \times \frac{1000\text{ mL}}{1\text{ L}} = 30.5\text{ mL HNO}_3$$

The two solutions have very similar molarity values, so it makes sense that the volume required for an equimolar amount of each solution is similar.

---

e. Gas evolution reactions are those where a gas evolves as a product in a reaction.

   i. Gases can form directly in a reaction or form from the decomposition of a reaction product.

      1. The formation of sulfides, carbonates, sulfites, and ammonia are all indicative of a gas evolution reaction.

EXAMPLE:

How many grams of carbon dioxide gas are produced when a stoichiometric amount of hydrochloric acid is added to 3.452 g of sodium bicarbonate?

We will first write the balanced chemical reaction:

$$NaHCO_3(aq) + HCl(aq) \rightarrow H_2O(l) + CO_2(g) + NaCl(aq)$$

In this reaction we have indicated that the carbonic acid that is produced decomposes into carbon dioxide and water.

We can use the reaction stoichiometry to calculate the theoretical yield of carbon dioxide:

$$3.452\,g \times \frac{1\,mol\,NaHCO_3}{84.01\,g} \times \frac{1\,mol\,CO_2}{1\,mol\,NaHCO_3} \times \frac{44.01\,g\,CO_2}{1\,mol\,CO_2} = 1.808\,g\,CO_2$$

VIII. Oxidation-Reduction Reactions

a. An oxidation-reduction reaction (also called a redox reaction) involves the transfer of electrons between species in a chemical reaction.

  i. Many redox reactions involve oxygen.

  ii. Oxidation is the loss of electrons and reduction is the gain of electrons.

    1. A mnemonic device for redox reactions is "leo the lion goes ger" where "leo" stands for "loss of electrons is oxidation" and "ger" stands for "gain of electrons is reduction."

b. The oxidation state (or oxidation number) is the charge of an atom if all of the electrons in its bonds are assigned to the more electronegative atom sharing them.

c. The rules for assigning oxidation state are listed below and go in order of priority (meaning that the first item takes priority over any item below it):

  i. Free elements all have an oxidation state of zero.

  ii. The sum of the oxidation states of a species must add up to the overall charge.

  iii. Group IA elements have an oxidation state of +1.

  iv. Group IIA elements have an oxidation state of +2.

  v. Fluorine has an oxidation state of -1.

  vi. Hydrogen has an oxidation state of +1 when bonded to a nonmetal and an oxidation state of -1 when bonded to a metal.

  vii. Oxygen has an oxidation state of -2.

  viii. Group VIIA elements have an oxidation state of -1.

  ix. Group VIA elements have an oxidation state of -2.

  x. Group VA elements have an oxidation state of -3.

EXAMPLE:

What is the oxidation state of the underlined atom in each of the following compounds?

a. $H\underline{F}$

According to rule v. from above, fluorine has an oxidation state of -1.

b. $H\underline{N}O_3$

According to rule vi. from above, hydrogen has an oxidation state of +1. According to rule vii., oxygen has an oxidation state of -2. Since the sum of the oxidation states must equal the overall charge of the compound (zero in this case), nitrogen has an oxidation state of 0–(+1+3(-2))=+5.

c. $\underline{Na}_2SO_4$

Sodium is a Group IA element, which means that it has an oxidation state of +1 according to rule iii.

d. $C_2\underline{H}_6$

According to rule vi., hydrogen has an oxidation state of +1.

---

d. Redox reactions always involve a change in the oxidation numbers of the elements in the reaction.

   i. A species is oxidized when its oxidation number increases.

   ii. A species is reduced when its oxidation number decreases.

EXAMPLE:

Determine the species that is oxidized and the species that is reduced in the oxidation of magnesium:

$$2Mg(s) + O_2(g) \rightarrow 2MgO(s)$$

Using the rules for oxidation states given above, we can determine the oxidation states of all species.

On the reactant side:

Mg: 0

O: 0

On the product side:

Mg: +2

O: -2

And we see that magnesium is oxidized (increase in oxidation number) and oxygen is reduced (decrease in oxidation number).

---

e. The reducing agent is the species that is oxidized; it is the agent of reduction.

f. The oxidizing agent is the species that is reduced; it is the agent of oxidation.

EXAMPLE:

Which species is the oxidizing agent and which species is the reducing agent in the reaction of magnesium and oxygen given in the example above?

Since magnesium is oxidized, it is the reducing agent, and since oxygen is reduced, it is the oxidizing agent.

---

g. Combustion reactions are a specific example of a redox reaction.

i. Combustion reactions of compounds with only carbon and hydrogen produce carbon dioxide and water.

EXAMPLE:

Write the combustion reaction for methane gas, CH_4, and determine which species is oxidized and which species is reduced.

The combustion reaction of a hydrocarbon results in the production of carbon dioxide and water:

$$CH_4(g) + O_2(g) \rightarrow CO_2(g) + 2H_2O(g)$$

Using the rules for oxidation states, we can find the oxidation states of all species:

On the reactant side:

C: -4

H: +1

O: 0

On the product side:

C: +4

H: +1

O: -2

And we see that carbon is oxidized while oxygen is reduced. Notice that oxygen is the oxidizing agent which is often the case, thus the name "oxidizing agent."

Fill in the Blank Problems:

1. The species that limits the amount of product that is formed in a chemical reaction is the ________________.
2. ________________ reactions are those in which electrons are transferred from one species to another.
3. When a species is dissolved in water but does not form a solution that conducts electricity, it is called a(n) ________________.
4. Acids that contain more than one ionizable proton are called ________________.
5. The ________________ is a numerical measure of the relationship between the actual yield and the theoretical yield.
6. We can qualitatively describe the concentration of a solution as ________________ or ________________.
7. A reaction in which two solutions combine to form a solid is called a(n) ________________.
8. The numerical relationships between chemical amounts in a balanced reaction are called reaction ________________.
9. The oxidation number of the oxidizing agent ________________ in a redox reaction.
10. The equivalence point i n a titration is visualized using a(n) ________________.
11. The su m of oxidation states of all atoms in an ion is equal to ________________.
12. ____ ________________ are species that are not involved in the chemistry of a solution phase reaction.

13. A(n) ________________ is a species that dissociates completely in solution to form ions.

14. A(n) ________________ is a solution stored in concentrated form.

Problems:

1. Urea (CH_4N_2O) is a common fertilizer that can be synthesized by reacting ammonia with carbon dioxide.
   a. What is the balanced chemical reaction for the synthesis of urea given that water is the byproduct of the reaction?
   b. 136.4 kg of ammonia and 211.4 kg of carbon dioxide are combined. How much (in kg) urea will be produced in this reaction?
   c. How many grams of ammonia and carbon dioxide are left over upon completion of the reaction?
   d. What is the percent yield of the reaction if 168.4 kg of urea are collected?
2. When sodium carbonate and hydrochloric acid react, the production of a gas is observed.
   a. Write the balanced molecular equation for the reaction of sodium carbonate and hydrochloric acid.
   b. Write the net ionic equation for the reaction.
   c. What volume of a 1.54 M HCl solution will be required to neutralize 60.5 g of sodium carbonate?
   d. How much gas is produced (in grams) when 45.7 mL of a 1.54 M HCl solution is combined with 52.4 mL of a 1.38 M Na_2CO_3 solution?
3. 10.0 g of NaOH are dissolved to prepare 150.0 mL of solution.
   a. What is the concentration of the initial NaOH solution?
   b. The NaOH stock solution is then diluted by taking 10.0 mL of the original solution and adding 90.0 mL to it. What is the concentration of the newly prepared solution?
   c. The diluted solution (from part b) is used to titrate 53.0 mL of a H_2SO_4 solution. Write the molecular, complete ionic, and net ionic equations for this reaction.
   d. If 32.4 mL of the NaOH solution are required to completely neutralize the acid, what is the concentration of the H_2SO_4 solution?
4. Consider a solution made from combining 10.0 mL of a 1.0 M $Mg(C_2H_3O_2)_2$ solution and 40.0 mL of a 2.0 M Na_2CO_3 solution.
   a. When these solutions are combined, but before any reaction occurs, what will the concentration of sodium carbonate be?
   b. What chemical reaction would occur between these two solutions?
   c. What will the concentration of sodium ions be before the reaction?
   d. What will the concentration of sodium ions be after the reaction?
   e. What mass of solid can be collected after the reaction occurs?
   f. Imagine that we could test the conductivity of the solution (using a light bulb) before the reaction took place (but after the solutions had been combined) and again after the reaction took place. Would the bulb be brighter before or after the reaction occurs? Explain your answer.
5. The reaction of hydrogen and oxygen gas to produce water is utilized in hydrogen fuel cells.
   a. Write the balanced chemical reaction.

b. About 23.2% of air (by mass) is oxygen. What mass of air will be required to react with 10.0 g of hydrogen gas in order for the hydrogen to react completely?

c. If 10.0 g of oxygen gas and 10.0 g of hydrogen gas are allowed to react in a sealed container, what mass of each species will be present after the reaction is complete?

d. This is an oxidation-reduction reaction. What are the oxidation states of all species in this reaction?

e. What is the oxidizing agent in this reaction?

Concept Questions:

1. What is wrong with the following statement? "When a chemical equation is balanced, the number of molecules on each side of the equation will be equal."
2. A solution is a homogeneous mixture of two or more substances. Explain why this means that the concentration of the solute can be expressed as moles of solute per liter of solution even when the sample size is on the milliliter scale.
3. There are situations when the qualitative description of a solution as dilute or concentrated is just as valuable as a quantitative statement of molarity. Describe a situation where this would be appropriate.
4. When solid NaCl and solid $AgNO_3$ are crushed and combined, no reaction takes place. When the two solids are dissolved in water and mixed, however, a precipitate forms. Explain why there is such a dramatic difference between these two scenarios.
5. The percent yield of a reaction relates the theoretical amount of product formed to the actual amount.
   a. Suggest at least two reasons why the theoretical and actual yields differ.
   b. The percent yield should never be greater than 100%; explain why this is.
   c. The percent yield sometimes IS greater than 100%; describe one experimental error that could account for this.
6. The net ionic equation provides a complete description of the chemistry of a reaction while the complete ionic equation shows the actual nature of the reaction. Explain this statement.
7. The solubility of ionic substances is explained using the attraction of polar water molecules to the charged species that comprise ionic substances. Some ionic substances, however, do not dissolve in water; for example sulfates that do not contain Group IA cations are insoluble. Suggest one possible reason for this.

Chapter 5: Gases

Learning Objectives:

- Understand the meaning and origin of gas pressure.
- Utilize the simple gas laws and their combined form, the ideal gas law, in chemical problems.
- Determine the partial pressure of a component in a mixture of gases.
- Understand kinetic molecular theory and use it to derive the ideal gas law and Dalton's law of partial pressures.
- Predict the manner in which temperature and mass affect the speed of gas molecules.
- Compare mean free paths, diffusion times, and effusion times for different gases.
- Understand how and why real gases deviate from ideal behavior.
- Identify the main components responsible for air pollution and ozone depletion.

Chapter Summary:

In this chapter, you will be introduced to the properties and behaviors of gases. You will begin by exploring the simple gas laws, each of which relate two of the four variables that determine gas behaviors: temperature, pressure, volume, and sample size (number of moles). You will then see how the simple gas laws are combined to give the ideal gas law, which can be used in stoichiometry, density, molar mass, and molar volume calculations. Once you are comfortable working with gases in calculations, you will explore mixtures of gases and will learn how to calculate partial pressures of components in mixtures. Kinetic molecular theory will then be explored to provide a physical basis for the ideal gas law. Using the kinetic molecular theory, you will learn to compare molecular velocities and their relationship to pressure; an extension of this theme will introduce the concepts of mean-free-path, diffusion, and effusion. With a firm grasp on ideal gases in hand, real gases and their deviations from ideal behavior will be explored and quantized in the van der Waals equation. Finally, you will look at the chemistry of the atmosphere with an emphasis on air pollution and ozone depletion.

Chapter Outline:

I. Water from Wells: Atmospheric Pressure at Work

   a. Pressure is the force exerted per area by gas molecules as they strike a surface.

II. Pressure: The Result of Molecular Collisions

   a. Pressure can be measured using a barometer, an evacuated tube inverted in a pool of mercury. The pressure from the atmosphere pushes the mercury up the tube until the weight of the mercury column is equal and opposite to the force from the atmosphere.

   b. Common units of pressure are mmHg, torr, atm, psi, and the SI unit of pressure, Pa.

EXAMPLE:

The barometric pressure is often given in millibars. The barometric pressure in Northern California today is 1018 millibars. Given that 1 atm is equal to 1013.2 millibars, determine the barometric pressure of Northern California in the following units:

a. mmHg

$$1018\,\text{millibars} \times \frac{1\,\text{atm}}{1013.2\,\text{millibars}} \times \frac{760\,\text{mmHg}}{1\,\text{atm}} = 763.6\,\text{mmHg}$$

b. torr

$$1018\,\text{millibars} \times \frac{1\,\text{atm}}{1013.2\,\text{millibars}} \times \frac{760\,\text{torr}}{1\,\text{atm}} = 763.6\,\text{torr}$$

c. atm

$$1018\,\text{millibars} \times \frac{1\,\text{atm}}{1013.2\,\text{millibars}} = 1.005\,\text{atm}$$

d. psi

$$1018\,\text{millibars} \times \frac{1\,\text{atm}}{1013.2\,\text{millibars}} \times \frac{14.7\,\text{psi}}{1\,\text{atm}} = 14.8\,\text{torr}$$

e. kPa

$$1018\,\text{millibars} \times \frac{1\,\text{atm}}{1013.2\,\text{millibars}} \times \frac{101325\,\text{Pa}}{1\,\text{atm}} \times \frac{1\,\text{kPa}}{1000\,\text{Pa}} = 101.8\,\text{kPa}$$

---

c. A manometer is used to measure pressures of samples in a laboratory. It consists of a "u-tube" that contains mercury and is open to the atmosphere on one end and the sample vessel on the other. The difference in mercury height between the two sides of the "u-tube" is equal to the difference in pressure between the vessel and the atmosphere.

III. The Simple Gas Laws: Boyle's Law, Charles's Law, and Avogadro's Law

a. The properties of gases can be described completely using the temperature, pressure, volume, and sample size (moles). We can look at the relationships between any two of these variables by holding the other two constant.

b. Boyle's law: volume and pressure are inversely related when the temperature and number of moles are held constant.

$$V \propto \frac{1}{P}$$

We can compare two different sets of conditions (that have the same temperature and sample size) with the following equation:

$$V_1P_1 = V_2P_2$$

EXAMPLE:

A 5.0 L plastic container is opened to the atmosphere and then sealed. The container is crushed so that the volume is 3.2 L. What is the pressure (in atm) in the crushed container?

We first need to realize that the initial pressure in the container will be equal to atmospheric pressure. Then we can use Boyle's law to solve for the final pressure:

$$V_1P_1 = V_2P_2$$

$$(5.0\,\text{L}) \times (1\,\text{atm}) = (3.2\,\text{L}) \times P_2$$

$$P_2 = 1.56\,\text{atm}$$

Our answer is reasonable since the pressure is expected to increase when the volume decreases.

---

c. Charles's law: volume and temperature are directly related when the number of moles and pressure are held constant:

$$V \propto T$$

We can compare two different sets of conditions with an equation:

$$\frac{V_1}{T_1} = \frac{V_2}{T_2}$$

In order for these relationships to hold, the temperature must be expressed in kelvins.

EXAMPLE:

A sample of gas at 2.73 dm^3 at 21.0°C is warmed to 100.0°C. What is the volume of the sample?

We will use Charles's law to solve this problem, but we first need to convert the temperatures to kelvins.

$$T_1 = 21.0 + 273.15 = 294.2 \text{ K}$$

$$T_2 = 100.0 + 273.15 = 373.2 \text{ K}$$

$$\frac{2.73 \text{ dm}^3}{294.2 \text{ K}} = \frac{V_2}{373.2 \text{ K}}$$

$$V_2 = 3.46 \text{ dm}^3$$

Our answer is reasonable since we expect the volume to increase when the temperature increases.

---

d. Avogadro's law: number of moles and volume are directly related when the temperature and pressure are held constant:

$$V \propto n$$

We can compare two different sets of conditions with an equation:

$$\frac{V_1}{n_1} = \frac{V_2}{n_2}$$

EXAMPLE:

10.0 mol of H_2 and 5.0 mol of O_2 react to completion according to the equation:

$$2H_2(g) + O_2(g) \rightarrow 2H_2O(g)$$

If the reactants are placed in a sealed 12.0 L container, what will the volume of the container be after the reaction takes place?

The number of moles before the reaction takes place is 15.0 mol and the number of moles after the reaction takes place is 10.0 moles (using the reaction stoichiometry). So we can utilize Avogadro's law to solve for V_2:

$$\frac{12.0 \text{ L}}{15.0 \text{ mol}} = \frac{V_2}{10.0 \text{ mol}}$$

$$V_2 = 8.0 \text{ L}$$

Our answer is reasonable since a decrease in the number of moles should result in a decreased volume.

IV. The Ideal Gas Law

a. We can combine the simple gas laws into a single expression:

$$V \propto \frac{nT}{P}$$

This can be rearranged slightly to eliminate the fractional form:

$$PV \propto nT$$

And the proportionality can be converted into an equality with the introduction of a constant, R:

$$PV = n\mathrm{R}T$$

$$\mathrm{R} = 0.08206\frac{\mathrm{L \cdot atm}}{\mathrm{mol \cdot K}}$$

This is called the ideal gas law and R is called the gas constant or ideal gas constant.

EXAMPLE:

What is the volume of a container that holds 46.3 g of $CO_2(g)$ at 283 K and 1.4 atm?

In order to use the ideal gas law, we need to convert the mass of CO_2 into the number of moles:

$$46.3\,\mathrm{g} \times \frac{1\,\mathrm{mol}}{44.01\,\mathrm{g}} = 1.05\,\mathrm{mol\,CO_2}$$

Rearranging the ideal gas law and solving for volume gives:

$$V = \frac{nRT}{P} = \frac{(1.05\,\mathrm{mol}) \times (0.08206\frac{\mathrm{L \cdot atm}}{\mathrm{mol \cdot K}}) \times (283\,\mathrm{K})}{1.4\,\mathrm{atm}}$$

$$V = 17\,\mathrm{L}$$

EXAMPLE:

What is the temperature of 0.587 moles of a gas that is held in a sealed 4.73 L container at 34.3 psi?

In order to use the ideal gas equation, we need to convert the units of pressure to the same pressure units in the gas constant (atm):

$$34.3\,\mathrm{psi} \times \frac{1\,\mathrm{atm}}{14.7\,\mathrm{psi}} = 2.33\,\mathrm{atm}$$

Now we can rearrange the ideal gas law to solve for temperature:

$$T = \frac{PV}{nR} = \frac{(2.33\,\mathrm{atm}) \times (4.73\,\mathrm{L})}{(0.587\,\mathrm{mol}) \times \left(0.08206\frac{\mathrm{L \cdot atm}}{\mathrm{mol \cdot K}}\right)}$$

$$T = 229\,\mathrm{K}$$

V. Applications of the Ideal Gas Law: Molar Volume, Density, and the Molar Mass of a Gas

a. The molar volume of a gas at standard temperature and pressure (STP) is the volume that an ideal gas occupies at 0°C and 1 atm.

i. Standard temperature and pressure is also known as standard conditions.

ii. We can calculate the molar volume using the ideal gas law:

$$V = \frac{nRT}{P}$$

iii. The molar volume of a gas at STP is 22.4 L.

$$V = \frac{(1\ \text{mol}) \times (0.08206 \frac{\text{L} \cdot \text{atm}}{\text{mol} \cdot \text{K}}) \times (273.15\ \text{K})}{1\ \text{atm}} = 22.4\text{L}$$

b. The density of one mole of a gas at standard conditions can be calculated by dividing the molar mass by the molar volume:

$$\text{density} = \frac{\text{molar mass}}{\text{molar volume}}$$

i. Density can also be calculated using the ideal gas law and the relationship between number of moles, mass (m), and molar mass (MM):

$$V = \frac{\frac{m}{\text{MM}} RT}{P}$$

$$\text{d} = \frac{m}{V} = \frac{\text{MM}P}{RT}$$

EXAMPLE:

1.58 atm of ammonia is held in a closed container at 456 K. What is the density of ammonia in the container?

The molar mass of ammonia is 17.04 g/mol; using the density form of the ideal gas law gives:

$$d = \frac{m}{V} = \frac{(17.04\ \text{g/mol}) \times (1.58\ \text{atm})}{(0.08206 \frac{\text{L} \cdot \text{atm}}{\text{mol} \cdot \text{K}}) \times (456\ \text{K})}$$

$$d = \frac{m}{V} = 0.719\ \text{g/L}$$

---

c. The molar mass of a gas can be calculated using a rearranged form of the above expression for density:

$$\text{MM} = \frac{mRT}{PV}$$

EXAMPLE:

A 5.6 L holds 39 g of an unknown gas at a temperature of 100°C and a pressure of 4.8 atm. Is the identity of the gas O_2, CO, CO_2, or CH_4?

We will use the information provided and compare the experimental molar mass to the molar masses of the given species.

$$\text{MM} = \frac{mRT}{PV}$$

$$\text{MM} = \frac{(39\text{ g}) \times (0.08206 \frac{\text{L} \cdot \text{atm}}{\text{K} \cdot \text{mol}}) \times (373.2\text{ K})}{(4.8\text{ atm}) \times (5.6\text{ L})}$$

$$\text{MM} = 44\text{ g/mol}$$

This is the same molar mass as CO_2, so we can conclude that the unknown gas is CO_2.

VI. Mixtures of Gases and Partial Pressure

a. The pressure exerted by a single component of a mixture of gases is the partial pressure of that component.

b. The partial pressure of a gas, at a given temperature and volume, is determined by the number of moles of that gas:

$$P_a = \frac{n_a RT}{V}$$

c. Dalton's law of partial pressures states that the total pressure exerted by a mixture of gases is equal to the sum of the partial pressures of each component of the mixture:

$$P_{total} = P_a + P_b + P_c + \ldots$$

d. The partial pressure of a gas is related to the mole fraction (X_a) of the gas:

$$P_a = X_a P_{total}$$

$$X_a = \frac{n_a}{n_{total}}$$

EXAMPLE:

A mixture of neon, argon, and krypton gases has a total mass of 10.45 g and a total pressure of 1.25 atm. If there are 0.278 mol of neon and 0.0589 mol of argon in the mixture, what is the partial pressure of krypton?

We first need to find out how many moles of krypton are present in the mixture. In order to do this, we need to calculate the grams of neon and grams of argon:

$$0.278\text{ mol Ne} \times \frac{20.18\text{ g Ne}}{1\text{ mol Ne}} = 5.61\text{ g Ne}$$

$$0.0589\text{ mol Ar} \times \frac{39.95\text{ g Ar}}{1\text{ mol Ar}} = 2.35\text{ g Ar}$$

So the remaining mass must be Kr:

$$10.45\text{ g gas} - 5.61\text{ g Ne} - 2.35\text{ g Ar} = 2.49\text{ g Kr}$$

We can now calculate the moles of Kr:

$$2.49 \text{ g Kr} \times \frac{1 \text{ mol Kr}}{83.80 \text{ g Kr}} = 0.0297 \text{ mol Kr}$$

The mole fraction of Kr is:

$$\frac{0.0297 \text{ mol Kr}}{0.0297 \text{ mol Kr} + 0.278 \text{ mol Ne} + 0.0589 \text{ mol Ar}} = 0.0811$$

The partial pressure of Kr is the mole fraction of Kr times the total pressure:

$$0.081 \times 1.25 \text{ atm} = 0.101 \text{ atm}$$

EXAMPLE:

An equimolar amount of CO, CO_2, and O_2 are placed in a sealed 1.24 L container at 250 K. The total pressure in the container is 5.4 kPa. What is the mass of O_2 in the container?

Since there is an equimolar amount of each gas, the mole fraction of O_2 in the container is 1/3. Since the partial pressure of O_2 is the mole fraction times the total pressure, the partial pressure of O_2 is 1/3 of the total pressure or 1.8 kPa.

We can calculate the number of moles of O_2 using the ideal gas equation (with pressure converted to atm):

$$n = \frac{(1.8 \text{ kPa} \times \frac{1 \text{ atm}}{101.325 \text{ kPa}}) \times (1.24 \text{ L})}{(0.08206 \frac{\text{L} \cdot \text{atm}}{\text{mol} \cdot \text{K}}) \times (250 \text{ K})}$$

$$n = 0.00107 \text{ mol}$$

Now we can convert the number of moles to grams to solve the problem:

$$0.00107 \text{ mol} \times \frac{32.00 \text{ g}}{1 \text{ mol}} = 0.034 \text{ g } O_2$$

---

e. The vapor pressure of a liquid is the pressure that the gaseous phase of the liquid exerts in a closed container; the vapor pressure is temperature-dependent.

   i. When gases are collected over a liquid, the vapor pressure of the liquid must be taken into account.

EXAMPLE:

Potassium chlorate decomposes to potassium chloride and oxygen gas according to the equation:

$$2KClO_3(s) \rightarrow 2KCl(s) + 3O_2(g)$$

When potassium chlorate is allowed to decompose at 50.0°C, the oxygen gas is collected over water in a 2.45-L container. If the total pressure in the container is 0.875 atm, what mass of the $KClO_3$ decomposed?

Since the gas is collected over water, we first need to find the partial pressure of water at 50.0°C. In table 5.4 we see that the vapor pressure of water at 50.0°C is 92.6 mmHg. We can find the partial pressure of oxygen gas by subtracting the vapor pressure of water from the total pressure:

$$92.6\ \text{mmHg} \times \frac{1\ \text{atm}}{760\ \text{mmHg}} = 0.1218\ \text{atm}$$

$$0.875 - 0.122 = 0.753\ \text{atm}$$

We can now use the ideal gas law to find the number of moles of O_2 that were produced:

$$n = \frac{(0.753\ \text{atm}) \times (2.45\ \text{L})}{(0.08206 \frac{\text{L} \cdot \text{atm}}{\text{mol} \cdot \text{K}}) \times (323.15\ \text{K})}$$

$$n = 0.06957\ \text{mol}$$

We can use the reaction stoichiometry and the molar mass of potassium chlorate to find the mass:

$$0.06957\ \text{mol}\,O_2 \times \frac{2\ \text{mol}\ KClO_3}{3\ \text{mol}\ O_2} \times \frac{122.55\ \text{g}}{1\ \text{mol}} = 5.68\ \text{g}\ KClO_3$$

VII. Kinetic Molecular Theory

a. Kinetic molecular theory assumes that gases are in constant random motion and that:

   i. The size of the gas particles is negligible as compared to the empty space in a container.

   ii. The kinetic energy of the particles is proportional to the temperature in kelvins.

   iii. The collisions between gas particles and the walls of a container are elastic.

      1. An elastic collision is one in which no energy is lost, although energy can be exchanged between colliding particles.

b. Kinetic molecular theory can be used to understand the simple gas laws.

   i. Boyle's law states that pressure and volume are indirectly proportional; when the pressure increases, the number of collisions that occur with the walls of the container will increase. If the temperature and number of particles remain constant, then the volume must decrease in order for this to occur.

   ii. Charles's law states that the volume and temperature are directly related; when the volume increases, the number of collisions with the walls of the container can only remain constant (constant pressure and number of moles) if the particles are moving faster. Particles move faster when the temperature increases.

   iii. Avogadro's law states that the volume and number of moles are directly proportional; when the volume increases the number of collisions with the walls of the container can only remain constant (constant pressure and temperature) if the number of moles of the gas increases.

   iv. Dalton's law states that the total pressure is the sum of the partial pressures of the components of the gas. Kinetic molecular theory states that pressure is a result of collisions between the gas particles and containers of the wall; because gases do not interact and have no size differences, the identity of the gas particles is not relevant to their behavior in this context.

► Instead of memorizing the simple gas laws and/or the ideal gas law, focus on understanding kinetic molecular theory. Using the ideas of KMT, you can easily derive the simple gas laws and the ideal gas law.

c. The ideal gas law can be derived directly using the postulates of kinetic molecular theory.

   i. The pressure of a sample of gas(es) is equal to the total force divided by the area:

$$P = \frac{F_{total}}{A}$$

   ii. The total force is the sum of the force from collisions between gas particles and the wall multiplied by the number of collisions that occur.

      1. The force from collisions is equal to the mass of the particles times their acceleration:

$$F_{collisions} = ma$$

      2. The acceleration of the gas particles is equal to their change in velocity over the time period:

$$F_{collisions} = m\frac{\Delta v}{\Delta t}$$

      3. The change in velocity that results from an elastic collision with the wall will be twice the velocity. This is because the velocity changes direction:

$$F_{collisions} = m\frac{2v}{\Delta t}$$

      4. The total number of collisions will be proportional to the number of particles that are within a distance in order to hit the wall ($v\Delta t$), the area of the wall (A), and the density of the particles (moles per volume):

$$\text{number of collisions} \propto v\Delta t \times A \times \frac{n}{V}$$

      5. The total force is then:

$$F_{total} \propto m\frac{2v}{\Delta t} \times v\Delta t \times A \times \frac{n}{V}$$

$$F_{total} \propto 2mv^2 \times A \times \frac{n}{V}$$

   iii. The mass time velocity squared (kinetic energy) is proportional to the temperature, the pressure is equal to the total force over area, and the constants can be removed because it is a proportionality:

$$P \propto \frac{T \times n}{V}$$

      1. We can introduce a constant and rearrange to get the ideal gas law:

$$PV = nRT$$

d. Since the temperature determines the average kinetic energy of a sample of gases, gases with a lower mass will travel faster at a given temperature:

$$KE = \frac{1}{2}mv^2$$

i. The average kinetic energy of one mole of an ideal gas will be related to the average molecular speed:

$$KE_{ave} = \frac{1}{2} N_A m\bar{u}^2$$

Where $\bar{u}$ is the average speed and N_A is Avogadro's number.

ii. Kinetic molecular theory states that the kinetic energy is proportional to the temperature:

$$KE_{ave} = \frac{3}{2} RT$$

Where 3/2 R is the proportionality constant.

iii. We can set the two equations for the average kinetic energy equal to one another:

$$\frac{3}{2} RT = \frac{1}{2} N_A m\bar{u}^2$$

iv. Solving for the speed gives:

$$\bar{u}^2 = \frac{3RT}{N_A m}$$

Since $N_A m$ is equal to the molar mass (MM):

$$\sqrt{\bar{u}^2} = \sqrt{\frac{3RT}{\text{MM}}}$$

v. The expression $\sqrt{\bar{u}^2}$ is called the root-mean-squared speed (u_{rms}) and is a measure of the average molecular speed of a gas:

$$u_{rms} = \sqrt{\frac{3RT}{\text{MM}}}$$

e. When the temperature of a gas increases, the distribution of molecular speeds gets broader and has a higher average value.

f. When the mass of a gas sample increases, the distribution of molecular speeds gets narrower and has a lower average value.

EXAMPLE:

The root-mean-squared speed of a diatomic gas is measured at 543 K and found to be 21.99 m/s. What is the identity of the gas?

We can rearrange the expression above to solve for the molar mass:

$$\text{MM} = \frac{3RT}{\bar{u}^2} = \frac{3 \times (8.3145 \frac{\text{J}}{\text{mol} \cdot \text{K}}) \times (543\ \text{K})}{(21.99\ \text{m/s})^2}$$

$$\text{MM} = 28.0\ \text{g/mol}$$

Since we are told that the gas is diatomic, it must be N_2 gas.

VIII. Mean Free Path, Diffusion, and Effusion of Gases

a. The mean free path of a gas is the average distance that a gas particle can travel before colliding with another gas particle.

b. Diffusion is the process by which molecules spread out.

c. Effusion is the process of a gas escaping a container with a small hole in it.

i. Graham's law of effusion relates the rate of molecular effusion to the molar mass of the gas:

$$\frac{rate_A}{rate_B} = \sqrt{\frac{\text{MM}_B}{\text{MM}_A}}$$

EXAMPLE:

How much longer will methane gas take to effuse out of a container than ammonia gas?

We can calculate the rate at which ammonia will effuse as compared to methane:

$$\frac{rate_{CH_4}}{rate_{NH_3}} = \sqrt{\frac{17.0\text{ g/mol}}{16.0\text{ g/mol}}}$$

CH_4 effuses out of the container 1.03 times faster than NH_3 does, so it will take 1.03 times as long for ammonia to effuse out of the container.

IX. Real Gases: The Effects of Size and Intermolecular Forces

a. Real gases behave ideally when the size of the gas particle is negligible compared to the distance between the gas particles and when forces between particles are insignificant.

i. Ideal behavior is best approximated at low pressures and high temperatures.

b. The effect of finite volume of gas particles becomes important at high pressures. A correction to the ideal gas law for volume is:

$$V = \frac{nRT}{P} + nb$$

The value of b depends on the identity of the gas.

c. Intermolecular forces are attractions and repulsions that exist between atoms or molecules; at low pressures particles are too far away from each other to exert large forces and at high temperatures particles are moving too fast to be affected by these forces.

i. Intermolecular forces become important at high pressures and low temperatures. The correction to the ideal gas law for intermolecular forces is:

$$P = \frac{nRT}{V} - a\left(\frac{n}{V}\right)^2$$

The value of "a" depends on the identity of the gas, because different gases have different intermolecular forces.

d. We can combine the corrections for volume and intermolecular forces into a single expression called the van der Waals equation:

$$\left[P + a\left(\frac{n}{V}\right)^2\right] \times (V - nb) = nRT$$

e. Different molecules and atoms will have different deviations from ideal behavior:

i. Negative deviations (nT/PV < R) occur when intermolecular forces are great.

ii. Positive deviations (nT/PV > R) occur when particle volume has a large effect.

EXAMPLE:

The van der Waals constants for CO are a=1.485 and b=0.03985. What will the temperature of a sample of CO gas be when 5.4 moles are placed in a 10.4 L container at 1.5 atm?

We can use the van der Waals equation and solve for temperature:

$$T = \frac{\left[P + a\left(\frac{n}{V}\right)^2\right] \times (V - nb)}{nR}$$

$$T = \frac{\left[1.5\,\text{atm} + (1.485\,\frac{\text{L}^2\cdot\text{atm}}{\text{mol}^2})\left(\frac{5.4\,\text{mol}}{10.4\,\text{L}}\right)^2\right] \times \left(10.4\,\text{L} - (5.4\,\text{mol})\left(0.03985\,\frac{\text{L}}{\text{mol}}\right)\right)}{(5.4\,\text{mol})(0.08206\,\frac{\text{L}\cdot\text{atm}}{\text{mol}\cdot\text{K}})}$$

$$T = 44\,\text{K}$$

X. Chemistry of the Atmosphere: Air Pollution and Ozone Depletion

a. The troposphere is the region of the atmosphere that is closest to the earth and is most affected by air pollution.

i. Pollutants in the troposphere include sulfur oxides (SO_x), carbon monoxide (CO), nitrogen oxides (NO_x), and ozone (O_3).

1. Ozone in the troposphere is called "bad ozone" because it is a respiratory irritant. It is one of the components of photochemical smog.

b. The stratosphere is above the troposphere and absorbs UV light.

i. Ozone in the stratosphere is called "good ozone" because it absorbs UV light, which is harmful to humans and other organisms.

ii. Chlorine, found in chlorofluorohydrocarbons, acts as a catalyst to destroy ozone.

Fill in the Blank Problems:

1. ________________ states that the total pressure of a gas sample is equal to the sum of the partial pressures of its components.
2. The average distance that a particle can travel in between collisions is called its ________________.
3. The average kinetic energy of a particle is proportional to its ________________ in units of ________________.

4. _______________ is the force exerted per unit area by gas particles as they strike the walls of a container.
5. Pressure and volume are _______________ related.
6. _______________ law relates the temperature of a gas to its volume.
7. A positive deviation from ideal behavior occurs primarily because of the particle's _______________.
8. In a gas at a given temperature, lighter particles travel _______________ than heavier particles.
9. A barometer is a device that measures pressure in units of _______________.
10. The SI unit of pressure is the _______________.
11. The molar vol ume of an ideal gas at STP is always _______________.
12. The partial pressure of a liqui d in equilibrium with its gas phase is its _______________.
13. The constant of proportionalit y in the ideal gas law is equal to _______________.

Problems:

1. The density of salt water is 1.025 g/mL. Calculate the conversion factor between mmHg and millimeters of salt water.
2. The pressure on the top of Mt. Everest is about 30. kPa.
   a. What is the pressure in units of atm, psi, torr, and mmHg?
   b. If the percentage of oxygen in air on top of Mt. Everest is about 21.0%, what is the partial pressure of oxygen on top of Mt. Everest?
   c. If the content of oxygen in the air on top of Mt. Everest is only 2/3 of that at sea level, what is the partial pressure of oxygen at sea level?
   d. A sample of air from on top of Mt. Everest is collected in a 1.0-L jar and brought back down. If the sample was collected on a "warm" sunny day, the temperature of the sample would be about -15°F on the mountain. If the temperature is 50°F in Oakland when the jar is brought home, what will be the new pressure of the gas in the jar?
3. A 12.0 g sample of nitrogen gas is placed into a sealed 5.0 L container at 25°C. What will the pressure be inside the container?
4. The gas sample in problem 3 is heated up to 100°C. What will the pressure be?
5. A cylinder of helium gas is advertised to fill 50 9" balloons.
   a. If a 9" balloon can hold a volume of 0.27 cubic feet and the temperature of the gas is 25°C, what is the mass of helium contained in the full cylinder? You can assume that the pressure inside a balloon is equal to 1.01 atm.
   b. What is the density of helium in one balloon?
   c. What would the density of a balloon filled with nitrogen be if all other factors remained constant?
   d. Air leaks out of the balloon so that the pressure decreases by a rate of 0.0012 atm per day. How long will it take (in hours) for the balloon to be half of its original size?
6. A mixture of gases is prepared containing 5.0 g of He, 2.0 g of Ne, and 5.0 g of Ar at STP.
   a. What is the partial pressure (in atm) of helium in the mixture?
   b. What volume will the mixture occupy?
7. When 10.0 g of hydrogen gas and 10.0 grams of chlorine gas react together, they form hydrogen chloride gas.

a. What will the partial pressure of hydrogen gas be at the end of the reaction if it is carried out in a sealed 10.0-gallon container at 100°C?

b. What will the total pressure in the container be?

8. Calcium phosphide is reacted with an excess of water to produce calcium hydroxide and phosphine:

$$Ca_3P_2(s) + 6H_2O(l) \rightarrow 3Ca(OH)_2(aq) + 2PH_3(g)$$

What will the total pressure of the container be when 3.0 grams of Ca_3P_2 are reacted in a closed 5.5-mL container at 20°C?

9. What volume of carbon dioxide will be produced when 2.0 g of methane is combusted in a sealed container with a stoichiometric amount of oxygen at STP?

10. Pennies are c urrently made by coating zinc with copper. The mass of a penny is determined to be 2.482 g, and then the penny is scratched to expose the zinc. The scratched penny is placed in a solution of HCl, where it decomposes in a single displacement reaction.

a. Write the balanced chemical reaction that occurs between zinc and HCl.

b. The product gas is collected over water at 25°C. The total pressure after the reaction is 791 mmHg. What is the partial pressure of the product gas?

c. The gas collected occupies a volume of 0.899 L. What is the percent zinc in the penny if all of the zinc reacts with the acid?

11. State the t hree main postulates of the kinetic molecular theory.

12. What is t he average molecular speed of a collection of nitrogen molecules at 100°C?

13. Ho w much more time will it take for a sample of nitrogen gas to effuse out of a container as compared to a sample of hydrogen gas?

14. Metha ne, CH_4, is held in a 4.0-L container at standard conditions.

a. What mass of methane is in the container if its behavior is ideal?

b. Use the values of a and b listed in Table 5.5 of your textbook to determine the mass of methane in the container if its behavior is better described using the van der Waals equation.

c. Do you think that methane is appropriately described as an ideal gas at STP based on your calculations in parts a and b? Explain your answer.

Concept Questions:

1. You car has tires that each contain gas at 25–35 psi.

a. Without doing any calculations, is the pressure in your tires greater than, less than, or equal to the atmospheric pressure? How can you tell? (Hint: what happens when you put a tire gauge on your valve stem?)

b. If the tire pressure is different from atmospheric pressure, why don't your tires collapse or explode as you are driving down the street?

c. Most people do not check their tire pressure often enough and drive around with tires that are underinflated. Explain why this is very dangerous.

2. Equal molar amounts of gas A and gas B are combined in a 1.0-L container at room temperature. Gas B has a molar mass that is twice that of gas A. Answer the following questions about this system with a brief explanation of your reasoning.

a. Which gas, A or B, has a higher kinetic energy?

b. Which gas, A or B, has a higher partial pressure?

c. Which gas, A or B, has a higher root-mean-squared speed?

d. Which gas, A or B, contributes more to the average density of the mixture?

3. Discuss the relationship between pressure and temperature using the postulates from the kinetic theory of gases.

4. Compare and contrast the deviation from ideal behavior for carbon dioxide and carbon tetrachloride using the values of the van der Waals constants listed in Table 5.5 of your textbook.

Chapter 6: Thermochemistry

Learning Objectives:

- Understand the various forms of energy and energy transfer.
- Understand the First Law of Thermodynamics and its application to chemical problems.
- Use calorimetry to determine the quantity of heat transferred for various reactions at a variety of experimental conditions.
- Calculate work done by or on a system.
- Understand the relationship between energy and enthalpy.
- Calculate enthalpy changes for reactions using experimental data and/or Hess's law.
- Identify standard-state conditions and the zero of enthalpy.

Chapter Summary:

This chapter introduces you to the subject of thermochemistry, which is the study of energy in chemical reactions. You will begin this exploration by gaining an understanding of the forms of energy and the manners in which it can be transferred. The first law of thermodynamics will then be explained to provide context for understanding energy transfer. Calorimetry will be explained as a method for determining the heat transferred to and from a substance as well as the heat transfers that occur in chemical reactions. Pressure-volume work will then be explained so that you can calculate the energy change of any system using heat and work. The relationship between internal energy and enthalpy will be introduced and the conditions under which they are equal will be explored. Experimental and theoretical methods for determining the enthalpy of a chemical reaction will be explained alongside their physical meanings. Finally, we will define the zero of enthalpy and the standard state so that reaction enthalpies can be calculated from tabulated values.

Chapter Outline:

I. Chemical Handwarmers
   a. Thermochemistry is the study of the relationships between chemistry and energy.

II. The Nature of Energy: Key Definitions
   a. Energy is the capacity to do work.
      i. Energy is possessed by an object.
   b. Energy is transferred or exchanged via heat or work.
      i. Work is the result of a force acting through a distance.
      ii. Heat is the flow of energy that results from temperature differences.
   c. Kinetic energy is the energy associated with motion.
      i. Thermal energy is associated with temperature; it is a type of kinetic energy – it results from the motion of atoms or molecules.
   d. Potential energy is the energy associated with position or composition.
      i. Chemical energy (or chemical potential) is associated with the relative positions of electrons and nuclei; it is a type of potential energy.

e. The law of conservation of energy states that energy cannot be created or destroyed, although it can be transferred between objects and change form.

f. A thermodynamic system is divided into the system and the surroundings.

  i. The system is the part of the universe that is of interest; in a chemical reaction, the system is the reactants and products of the reaction.

  ii. The surroundings are everything in the universe other than the system.

    1. There are certain situations where the universe does not need to be considered. As we will see, there are times when a portion of the universe can be isolated for study.

g. The joule (J) is the SI unit of energy. Calories (cal), nutritional calories (Cal), and kilowatt-hours (kWh) are also commonly used.

  i. $1\,\text{J} = 1\dfrac{\text{kg} \cdot \text{m}^2}{\text{s}^2}$

  ▶ We can remember the units of energy by remembering that KE = $\frac{1}{2}mv^2$

  ii. 1 cal = 4.184 J

    1. The calorie is defined as the amount of energy required to raise the temperature of 1 g of a substance by 1°C.

  iii. 1 Cal = 1000 cal

  iv. $1\,\text{kWh} = 3.60 \times 10^6\,\text{J}$

EXAMPLE:

A typical candy bar has about 250 Cal in it. How long could you light up a 100.-watt light bulb using the energy from a candy bar? (1 watt is defined as 1 joule per second)

We need to convert the energy from the candy bar into joules to find the amount of energy available:

$$250\,\text{Cal} \times \frac{1000\,\text{cal}}{1\,\text{Cal}} \times \frac{4.184\,\text{J}}{1\,\text{cal}} = 1.046 \times 10^6\,\text{J}$$

Now we can convert to time using the conversion factor given and the fact that a 100.-watt light bulb will use 100 joules per second:

$$1.046 \times 10^6\,\text{J} \times \frac{1\,\text{s}}{100\,\text{J}} = 1.046 \times 10^4\,\text{s}$$

Our answer should have two significant figures, so the energy from a candy bar can light up a 100.-watt light bulb for 1.0×10^4 s.

III. The First Law of Thermodynamics: There Is No Free Lunch

a. The first law of thermodynamics states that the total energy of the universe is constant.

  i. This is another statement of the law of conservation of energy.

b. The internal energy of any system is the sum of the kinetic and potential energy of all of the particles in the system.

  i. The internal energy of a system is a state function.

1. A state function is any function that depends only on the state of a system; a state function does not depend on how the system was prepared.

2. The change in a state function is defined as the difference between the final and initial states:

$$\Delta E = E_{final} - E_{initial}$$

ii. For a chemical reaction, the change in energy is defined by the internal energy of the products and the reactants:

$$\Delta E_{reaction} = E_{products} - E_{reactants}$$

iii. The sum of the energies of the system and the surroundings is the energy of the universe. Since the total energy change of the universe is zero, the energy change of the system (ΔE_{sys}) is equal and opposite to the energy change of the surroundings (ΔE_{surr}):

$$\Delta E_{sys} = -\Delta E_{surr}$$

EXAMPLE:

What is the energy change of a system if the surroundings absorb 843 kJ of energy during the change?

The energy change of the system is equal and opposite to the energy change of the surroundings. Since the surroundings are absorbing 843 kJ of energy, ΔE_{surr} is equal to +843 kJ of energy. This means that the system has lost 843 kJ of energy so ΔE_{sys} = -843 kJ.

---

iv. The system of a chemical reaction is defined as the products and reactants.

1. If $\Delta E_{sys} > 0$, the products must have a higher internal energy than the reactants do because energy is absorbed from the surroundings.

2. If $\Delta E_{sys} < 0$, the reactants must have a higher internal energy than the products do because energy is released to the surroundings.

v. Energy can be exchanged through heat (q) and work (w):

$$\Delta E = q + w$$

vi. Unlike energy, heat and work are path functions (as opposed to state functions) and their values, therefore, depend on how a change is carried out.

EXAMPLE:

When a 5.0-g ball is dropped from a building, work is done as the ball moves over the distance that it falls. If the building is 100. m tall, what is the energy change that results from the work and where does the energy go when the ball stops moving?

The work done is the force times distance. The force is the acceleration due to gravity (9.81 m/s^2) times the mass of the ball (in SI units of kg):

$$w = 5.00\,g \times \frac{1\,kg}{1000\,g} \times 9.81\frac{m}{s^2} \times 100\,m = 0.04905\frac{kg \cdot m^2}{s^2} = 0.04905\,J$$

This is the work that is done by the ball. Since the ball is doing work, it is losing energy, so the sign of the energy change is negative. Our answer should have two significant figures, so the final answer is that the energy change is -0.049 J.

The energy must go somewhere after the work is done because $\Delta E_{univ}=0$ for any process. The energy is being transferred to the molecules of the ground as the ball hits it. The molecules will gain kinetic energy, which we do not see but could feel as heat.

IV. Quantifying Heat and Work

a. Temperature is a measure of thermal energy; thermal energy is transferred via heat.

   i. Thermal energy always flows from high to low temperature.

b. Thermal equilibrium is the condition at which no further heat transfer takes place.

c. The heat capacity (C) of a system is the quantity of heat required to raise its temperature by 1°C.

$$q = C \times \Delta T$$

   i. Heat capacity has units of J/°C and is an extensive property, meaning that the heat transfer depends on the quantity of material used.

   ii. The specific heat capacity (C_s) is the quantity of heat required to raise the temperature of 1 g of a substance by 1°C. The specific heat capacity is an intensive property: it depends only on the identity of the substance, not on the size of the sample.

$$q = C_s \times m \times \Delta T$$

   iii. The molar heat capacity (C_m) is the quantity of heat required to raise the temperature of 1 mole of a substance by 1°C. Molar heat capacity is an intensive property.

EXAMPLE:

How much heat is transferred to the air when 1.54 L of water at 125°C is allowed to sit in a room that is at 25°C?

First we need to convert the volume of water to a mass. The density of water at 100°C is 0.958 g/mL.

$$1.54\,\text{L} \times \frac{1000\,\text{mL}}{\text{L}} \times \frac{0.958\,\text{g}}{\text{mL}} = 1475.32\,\text{g}$$

Recalling that the definition of a calorie is the amount of energy required to change the temperature of one gram of water by 1°C, we know that the specific heat capacity of water must be 4.184 J/g°C. We also need to consider the final temperature of the water. Because the room is open to the universe, the temperature change of the air will be negligible, so we can assume that the final temperature of the water will be the same as the initial temperature of the room. We can then calculate the heat lost by the water:

$$q = 4.184\frac{\text{J}}{\text{g}\cdot{}^{\circ}\text{C}} \times 1475.32\text{g} \times (125^{\circ}\text{C} - 25^{\circ}\text{C}) = 6.173 \times 10^{5}\,\text{J}$$

We need three significant figures in our answer, so we report that 6.17×10^5 J was transferred to the air in the room when the water cools.

---

iv. When heat is transferred from the system to the surroundings, energy is conserved and:

$$q_1 = -q_2$$

$$m_1 \times s_1 \times \Delta T_1 = -m_2 \times s_2 \times \Delta T_2$$

EXAMPLE:

A 5.4-g sample of an unknown metal is heated to 100.0 °C and is placed in a beaker containing 142 g of water at 24.2 °C. The final temperature of the water is 25.1 °C. What is the heat capacity of the metal?

Heat is being transferred from the metal to the water so:

$$q_{water} = -q_{metal}$$

$$m_{metal} \times s_{metal} \times \Delta T_{metal} = -m_{water} \times s_{water} \times \Delta T_{water}$$

Using the information given in the problem we have:

$$5.4g \times s_{metal} \times (25.1^\circ C - 100.0^\circ C) = -142g \times 4.184 \frac{J}{g^\circ C} \times (25.1^\circ C - 24.2^\circ C)$$

Rearranging to solve for s_{metal} gives:

$$s_{metal} \times = \frac{-142g \times 4.184 \frac{J}{g^\circ C} \times (25.1^\circ C - 24.2^\circ C)}{5.4g \times (25.1^\circ C - 100.0^\circ C)}$$

And we find that the specific heat capacity of the metal is 1.3 J/g °C.

---

d. Pressure-volume work occurs when an applied force is the result of a volume change against a constant pressure.

i. Work is force times a distance (D), and pressure (P) is force over area (A). Combining these:

$$w = P \times A \times D$$

ii. We can change the A×D term to ΔV since the area over a distance will be equal to the change in volume of a reaction vessel:

$$w = -P \times \Delta V$$

The negative sign is included by the convention of defining work of the system; when the system does work on the surroundings (pushes against the atmosphere, meaning $\Delta V > 0$) the system loses energy and so the work is negative.

EXAMPLE:

A large-displacement motorcycle can have cylinders that are as large as 0.500 L. What is the energy change due to the work done by a motorcycle piston when it travels its maximum distance at sea level?

The change in volume will be +0.500 L since the maximum distance will be completely closed (0.000 L) to completely open (0.500 L). The pressure will be 1.0 atm since the change in volume is carried out at sea level against the atmosphere. The work is therefore:

$$w = -1\text{ atm} \times 0.500\text{ L} = -0.500\text{ atm} \cdot \text{L}$$

The work is negative because the motorcycle is doing work on the surroundings. Now, we need to convert to more conventional units:

$$-0.500\ \text{atm}\cdot\text{L}\times\frac{101.3\ \text{J}}{\text{atm}\cdot\text{L}}=-50.65\ \text{J}$$

Our answer should have three significant figures and should be negative, since the problem suggests that the motorcycle engine is doing work. The change in energy from the work is therefore -50.7 J.

V. Measuring the Change in Energy for Chemical Reactions: Constant-Volume Calorimetry

a. In order to calculate the energy change of a system, we need to calculate the heat and the work.

   i. The heat is calculated by measuring the temperature change.

   ii. The work is calculated by measuring the volume change.

b. When a reaction is carried out at a constant volume, the change in volume is zero and therefore the pressure-volume work is zero.

   i. The energy change will therefore result only from heat transfer:

$$\Delta E_{rxn} = q_v$$

c. We can measure the constant-volume heat transfer using a bomb calorimeter that maintains a constant volume.

   i. If the calorimeter is designed so that no heat can escape it, then the heat lost/gained by the reaction will be gained/lost by the calorimeter:

$$q_{cal} = -\ q_{rxn}$$

   ii. By measuring the change in temperature of the calorimeter, we can determine the constant volume heat transfer and the energy change of the reaction:

$$\Delta E_{rxn} = q_{rxn} = C_{cal} \times \Delta T$$

EXAMPLE:

Bomb calorimeters are often used to measure the heat released in a combustion reaction as mentioned in the text. 1.76 g of methane is burned in a bomb calorimeter with excess oxygen at 25°C. What is the final temperature of the calorimeter and its contents given that one mole of methane gas releases 882 kJ of heat when 1.0 mol of it is combusted? The heat capacity of the calorimeter is 4.319 kJ/°C.

We first need to find out how much heat is released when 1.76 g of methane is combusted; this is a conversion factor problem:

$$1.76\ \text{g}\times\frac{1\ \text{mol}\ CH_4}{16.042\ \text{g}}\times\frac{882\ \text{kJ}}{1\ \text{mol}\ CH_4}=96.7660\ \text{kJ}$$

In a calorimetry experiment, the heat of the reaction is equal to the heat capacity of the calorimetry times the temperature change. In this case, the heat of the reaction is negative since heat is lost by the reaction:

$$q_{rxn} = C_{cal} \times \Delta T$$

$$96.7660\ \text{kJ} = 4.319\frac{\text{kJ}}{^{\circ}\text{C}}\times\Delta T$$

$$\Delta T = 22.405^{\circ}\text{C}$$

Since the change in temperature is the final temperature minus the initial temperature, the final temperature will be:

$$T_f - 25.0°C = 22.405°C$$

$$T_f = 47.405°C$$

Our final answer is reported to the tenths place, so the final temperature of the calorimeter is 47.4°C.

VI. Enthalpy: The Heat Evolved in a Chemical Reaction at Constant Pressure

a. Enthalpy (H) is defined as the energy plus the pressure times volume:

$$H = E + PV$$

b. At constant pressure, the change in enthalpy is:

$$\Delta H = \Delta E + P\Delta V$$

i. The change in energy is equal to the heat at constant pressure (q_p) and the work:

$$\Delta H = (q_p + w) + P\Delta V$$

$$\Delta H = (q_p + w) + -w$$

$$\Delta H = q_p$$

ii. The change in energy is a measure of all of the energy exchanged while the change in enthalpy is only a measure of the heat exchanged at constant pressure.

c. At constant volume, the change in enthalpy is equal to the change in energy.

d. When the change in enthalpy of a reaction is positive, heat is flowing into the system; this is an endothermic reaction.

i. In an endothermic reaction, the potential energy of the reactants is lower than the potential energy of the products.

e. When the change in enthalpy of a reaction is negative, heat is flowing out of the system; this is an exothermic reaction.

i. In an exothermic reaction, the potential energy of the reactants is higher than the potential energy of the products.

f. The change in enthalpy of a reaction is called the heat of a reaction and is an extensive property.

EXAMPLE:

State whether each of the following examples is endothermic or exothermic:

a. Water freezing

Water freezes when heat is removed (the converse, ice melting, requires heat) from the system meaning that the potential energy of the reactants is higher than the potential energy of the reactants. This is therefore an exothermic reaction.

b. Wood burning

When wood burns, heat is released from the system (which is why a fireplace keeps us warm). Heat is released from a system when the potential energy of the products is lower than the potential energy of the reactants, which is the case for an exothermic reaction.

c. Alcohol evaporating from your skin

When alcohol evaporates from our skin, it feels cold. This is an indication that heat has been absorbed in the process meaning that it is an endothermic reaction.

d. Combustion of propane

When propane is burned in oxygen (combusted) heat is released (and we can use it to cook our food). Heat is released from the system because the products are more stable (at a lower potential energy) than the reactants meaning that the reaction is exothermic.

e. Dry ice subliming

Sublimation occurs when a solid goes directly to a gas. Gas molecules have greater kinetic energy than solid molecules (gas molecules move a lot while the motion of molecules in the solid phase is restricted). Since energy must be provided to the system in order for this to occur, it is an endothermic reaction.

VII. Constant-Pressure Calorimetry: Measuring ΔH_{rxn}

a. The heat of a reaction can be measured with a coffee-cup calorimeter:

$$q_{soln} = m_{soln} \times C_{s,soln} \times \Delta T$$

b. The solution will absorb the heat released or will provide the heat absorbed by the reaction.

$$q_{rxn} = -q_{soln} = \Delta H_{rxn}$$

EXAMPLE:

A common experiment that is carried out in a coffee-cup calorimeter is the dissolution of a salt in water. When 5.45 g of ammonium nitrate is dissolved in 154.2 g of water, the temperature of the water changes from 25.5°C to 25.0°C. What is the enthalpy change for the dissolution of ammonium nitrate in water in kJ/mol? Assume that the specific heat of the solution is equal to the heat capacity of the water.

We will use the expression for the q_{soln} given above with the information provided in the problem:

$$q_{soln} = (5.45 \text{ g} + 154.2 \text{ g}) \times 4.184 \text{ J/g°C} \times (25.0\text{ °C} - 25.5\text{°C})$$

$$q_{soln} = -333.99 \text{ J}$$

We know that the enthalpy change of the reaction is equal and opposite to the heat of the solution. ΔH is therefore 333.99 J.

This is the heat that is absorbed from the solution when 5.45 grams of ammonium nitrate are dissolved in the water. We can normalize the calculated value of ΔH to find ΔH_{rxn} by dividing the calculated value by the number of moles of ammonium nitrate used.

$$5.45 \text{ g } NH_4NO_3 \times \frac{1 \text{ mol } NH_4NO_3}{80.04 \text{ g } NH_4NO_3} = 0.06809 \text{ mol } NH_4NO_3$$

$$\frac{333.99 \text{ J}}{0.06809 \text{ mol } NH_4NO_3} \times \frac{1 \text{ kJ}}{1000 \text{ J}} = 4.905 \text{ kJ/mol}$$

Our answer should have two significant figures (resulting from the temperature change term), so the ΔH_{rxn} for the dissolution of ammonium nitrate in water is 4.9 kJ/mol.

VIII. Relationships Involving ΔH_{rxn}

a. If a chemical reaction is multiplied by a factor, then the ΔH_{rxn} value must be multiplied by that factor; this reflects the fact that ΔH is extensive.

b. If a chemical reaction is reversed, the ΔH_{rxn} value changes sign.

c. If a reaction can be expressed as the sum of a series of steps, ΔH_{rxn} for the overall reaction is the sum of the heats of reaction for each step.

i. This is called Hess's law and is valid because ΔH_{rxn} is a state function.

EXAMPLE:

Calculate the enthalpy change for the combustion of propane, C_3H_8, using the reactions given below:

1. $3C(s) + 4H_2(g) \rightarrow C_3H_8(g)$ $\Delta H = -103.8$ kJ
2. $2H_2(g) + O_2(g) \rightarrow 2H_2O(g)$ $\Delta H = -484.0$ kJ
3. $C(s) + O_2(g) \rightarrow CO_2(g)$ $\Delta H = -393.5$ kJ

We first need to write the balanced chemical reaction for the combustion of propane:

$C_3H_8(g) + 5O_2(g) \rightarrow 3CO_2(g) + 4H_2O(g)$

Now we can manipulate the given reactions so that they add up to give the overall reaction. We will start by figuring out whether any of the reactions need to be reversed. We will do this by looking for species in the given reactions that also appear in the overall reaction.

The first reaction has propane as a product; in our reaction we want propane to be a reactant, so we will flip the reaction and reverse the sign of the enthalpy change. The second reaction has oxygen as a reactant and water as a product; in our reaction this is also what we want, so we will leave reaction two alone. The third reaction has oxygen as a reactant and carbon dioxide as a product; again the positions of these species are the same in our overall reaction so we will leave them alone. We now have:

1. $C_3H_8(g) \rightarrow 3C(s) + 4H_2(g)$ $\Delta H = +103.8$ kJ
2. $2H_2(g) + O_2(g) \rightarrow 2H_2O(g)$ $\Delta H = -484.0$ kJ
3. $C(s) + O_2(g) \rightarrow CO_2(g)$ $\Delta H = -393.5$ kJ

Now we need to multiply each reaction by a factor so that the stoichiometry of the overall reaction is obtained and all species that are not in the overall reaction cancel out. The first reaction has one propane molecule, which is the same amount of propane in our reaction, so we will not change this reaction. In the second reaction we see that two molecules of water are formed; we need four molecules of water so we will multiply the entire reaction by two (note that we are not worried about oxygen yet because it appears as a reactant in both reactions two and three). The third reaction forms one mole of carbon dioxide while the overall reaction produces three moles of carbon dioxide; we will multiply the third reaction by three.

1. $C_3H_8(g) \rightarrow 3C(s) + 4H_2(g)$ $\Delta H = +103.8$ kJ
2. $2\times\{2H_2(g) + O_2(g) \rightarrow 2H_2O(g)\}$ $\Delta H = 2\times(-484.0$ kJ)
   $4H_2(g) + 2O_2(g) \rightarrow 4H_2O(g))$ $\Delta H = -968.0$ kJ
3. $3\times\{C(s) + O_2(g) \rightarrow CO_2(g)\}$ $\Delta H = 3\times(-393.5$ kJ)
   $3C(s) + 3O_2(g) \rightarrow 3CO_2(g)$ $\Delta H = -1180.5$ kJ

Now we will add up the reactions and the corresponding enthalpy changes:

$C_3H_8(g) + 4H_2(g) + 2O_2(g) + 3C(s) + 3O_2(g) \rightarrow 3C(s) + 4H_2(g) + 4H_2O(g) + 3CO_2(g)$ $\Delta H = +1774.7$ kJ

Now we will collect like species and cancel out any substances that appear on both sides of the equation:

$$C_3H_8(g) + 5O_2(g) \rightarrow 4H_2O(g) + 3CO_2(g) \qquad \Delta H = +1774.7 \text{ kJ}$$

The reaction is the same as the overall reaction that we want, so the enthalpy change is +1774.7 kJ.

IX. Enthalpies of Reaction from Standard Heats of Formation

a. The standard state is defined depending on the phase of the substance.
   i. The standard state of a gas is 1 atm.
   ii. The standard state for a liquid or solid is the pure substance in its most stable form at a pressure of 1 atm and a given temperature (usually 25°C).
   iii. The standard state for a substance in solution is a concentration of exactly 1 M.

b. The standard enthalpy change (ΔH^o) is the change in enthalpy when all reactants and products are in their standard states.

c. The standard enthalpy of formation (ΔH_f^o) is defined using the zero of enthalpy.
   i. The zero of enthalpy is the ΔH_f^o of a pure element in its standard state.
   ii. The ΔH_f^o for a pure compound is the change in enthalpy when one mole is produced from its constituent elements in their standard states.

d. The $\Delta H°$ for a reaction can be determined by breaking the reactants into their constituent elements (- ΔH_f^o(reactants)) and then forming products from the elements (ΔH_f^o (products)):

$$\Delta H^o_{rxn} = \sum n_p \Delta H_f^o(\text{products}) - \sum n_r \Delta H_f^o(\text{reactants})$$

where the ΔH_f^o values are weighted by the stoichiometry of the reaction.

EXAMPLE:

Using the enthalpies of formation in your book, calculate the standard enthalpy of formation for the dissolution of silver nitrate in a 1 M solution of sodium chloride to form silver chloride and sodium nitrate. Will the solution get hotter or colder when the silver nitrate is added?

First we need to write the balanced chemical reaction for the process:

$$AgNO_3(s) + NaCl(aq) \rightarrow AgCl(s) + NaNO_3(aq)$$

We can write an expression for the standard enthalpy change of the reaction using the enthalpies of formation of the reactants and products:

$$\Delta H^o_{rxn} = \left[\Delta H_f^o(AgCl(s)) + \Delta H_f^o(NaNO_3(s))\right] - \left[\Delta H_f^o(AgNO_3(s)) + \Delta H_f^o(NaCl(aq))\right]$$

The values are given in Appendix II in your book:

$$\Delta H^o_{rxn} = \left[-127.0 \text{ kJ/mol} + -447.5 \text{ kJ/mol}\right] - \left[-124.4 \text{ kJ/mol} + -407.2 \text{ kJ/mol}\right]$$

$$\Delta H^o_{rxn} = -42.9 \text{ kJ/mol}$$

Our answer has the correct number of significant figures since this is an addition/subtraction problem.

Since the change in enthalpy is negative, the system loses energy in the course of the reaction meaning that the solution will absorb energy and get warmer.

Fill in the Blank Problems:

1. ________________ is energy associated with motion while ________________ is energy associated with position.
2. The SI units of energy are the ________________.
3. A(n) ________________ is defined as the amount of energy required to raise the temperature of 1 g of water by exactly 1°C.
4. The First Law of Thermodynamics states that the total energy of the universe is ________________.
5. A(n) ________________ function is one that depends only on the initial and final values, but not on the manner in which the system was prepared.
6. When heat is absorbed by a system, the sign of ΔH is ________________.
7. A positive value of work is obtained when work is done ________________ the system.
8. When two objects are at the same temperature, they are said to be at ________________ equilibrium.
9. The measure of a substance's intrinsic capacity to absorb heat is called its ________________.
10. When a closed s ystem expands, the sign of the work is ________________.
11. At co nstant pressure, the enthalpy is equal to the ________________.
12. A(n) ________________ reaction is one that absorbs heats from its surroundings.
13. Accordin g to ________________, the heat of an overall reaction is equal to the sum of the heats of reaction for each step.
14. The standard state of a gas i s defined as the pure gas at ________________.
15. The zero of enthalpy is defined as the enthalpy of formation for ________________.

Problems:

1. Determine the sign of heat, work, and change in energy for the systems described.
   a. A candle is burned in an open room.
   b. A car rolls down a hill.
   c. A piano is lowered down the side of a building.
   d. Water is heated on a stovetop.
2. When ammonia is oxidized it produces nitrogen monoxide and heat:

$$4NH_3(g) + 5O_2(g) \rightarrow 4NO(g) + 6H_2O(g) \qquad \Delta H^\circ = -906 \text{ kJ}$$

   a. Is the reaction endothermic or exothermic? Explain your answer.
   b. If the reaction is carried out in a bomb calorimeter, will the temperature of the calorimeter increase or decrease? Explain your answer.
   c. Is the internal energy of the products or reactants higher? Explain your answer.
3. Butane is a fuel for lighters. When 1.0 mol of butane burns at a constant pressure, it produces 2658 kJ of heat and does 3 kJ of work.

a. What is the enthalpy change of this reaction?

b. What is the internal energy change of this reaction?

c. Explain what kind of work is done in this process.

d. In this reaction, is it reasonable to use the values of energy and enthalpy interchangeably? Explain your answer.

4. One method for rating fuels with respect to global warming is to determine the heat that they release relative to the amount of CO_2 produced (kJ/mol CO_2). The greater the heat produced relative to the amount of CO_2 released, the better the fuel.

   a. Using the standard enthalpies of formation, calculate the enthalpy change for the combustion of solid carbon, gaseous methane (CH_4), and liquid octane (C_8H_{18}). Assume that each substance reacts with a stoichiometric amount of oxygen to produce carbon dioxide gas and water vapor.

   b. Calculate the amount of heat produced per mol of CO_2 released for each fuel.

   c. Which gas is most environmentally friendly based on your results in part b?

5. Calculate the molar heat capacity (in cal/J°C) given the specific heat capacities for the following substances:

   a. aluminum, C_s=0.897 J/g°C

   b. copper, C_s=0.385 J/g°C

   c. gold, C_s=0.129 J/g°C

   d. magnesium, C_s=1.02 J/g°C

   e. zinc, C_s=0.387 J/g°C

   f. iron, C_s=0.450 J/g°C

6. A typical experiment used to measure the specific heat capacity of a metal is to heat the metal in a boiling water bath for a period of time and then place the hot metal sample into a coffee cup calorimeter. The temperature change of the water is measured and the specific heat capacity of the metal is calculated.

   a. Calculate the specific heat capacity of an unknown 5.423-g metal sample if the temperature of 30.68 g of water in the calorimeter changes by 2.72°C to a final temperature of 28.22°C.

   b. The metal is one of the substances listed in Problem 4. Which metal is it?

   c. In this type of experiment, it is important that the metal sample is placed in a clean dry test tube that is then submerged in a boiling water bath until thermal equilibrium is established. Explain how your results would be different if you instead placed the metal sample directly into the boiling water.

7. Hydrogen and oxygen gas are converted to water in a hydrogen fuel cell.

   a. Given that hydrogen fuel cells are being made to replace combustion engines, what do you expect the sign of the ΔH_{rxn} to be?

   b. Calculate the ΔH_{rxn} for this reaction using the standard enthalpies of formation.

   c. How many moles of oxygen and hydrogen would you need to convert to water in order to accelerate a large family car (~3500 lbs) from 0. to 60. mph over a ten-second time period?

   d. How many moles of oxygen and hydrogen would you need to convert to water in order to move the same car from part c from 0. to 60. mph over a 30-second time period?

   e. How many moles of oxygen and hydrogen would you need to convert to water in order to move a small SUV (~4200 lbs) from 0. to 60. mph over a ten-second time period?

f. According to the values that you calculated in parts c and d, which do you think has a greater effect on vehicle efficiency: driving habits or vehicle mass?

8. Hydrogen and iodide gas react together to form hydrogen iodide gas in a bomb calorimeter in an endothermic reaction.

   a. Using the enthalpies of formation, determine the enthalpy of the reaction.

   b. Equimolar amounts of the reactants are placed in a 5.0-L bomb calorimeter. The initial pressure inside the calorimeter is 2.5 atm. What is the change in internal energy of the reactants and products when the reaction is carried out in the bomb?

   c. Calculate the temperature change of the calorimeter when the reactant gases are initially at 25.4°C. The heat capacity of the calorimeter is determined to be 4.230 kJ/°C.

9. Calculate the heat of combustion of methane using the equations below:

   i. $2O(g) \rightarrow O_2(g)$ $\Delta H = -249$ kJ/mol

   ii. $H_2O(l) \rightarrow H_2O(g)$ $\Delta H = +44$ kJ/mol

   iii. $2H(g) \rightarrow H_2(g)$ $\Delta H = -803$ kJ/mol

   iv. $C(s) + 2O(g) \rightarrow CO_2(g)$ $\Delta H = -643$ kJ/mol

   v. $C(s) + O_2(g) \rightarrow CO_2(g)$ $\Delta H = -394$ kJ/mol

   vi. $C(s) + 2H_2(g) \rightarrow CH_4(g)$ $\Delta H = -75$ kJ/mol

   vii. $2H_2(g) + O_2(g) \rightarrow 2H_2O(g)$ $\Delta H = -484$ kJ/mol

10. Use the equations from problem 9 to determine the heat of the reaction between methane and oxygen to produce liquid water instead of gaseous water.

11. Dra w the potential energy diagrams for the processes carried out in Problems 9 and 10.

Concept Questions:

1. Most chefs prefer gas stove tops because they allow greater control over the heat that is delivered to food. Gas stoves, however, are the least efficient type of stove. Electric stoves are about 15% more efficient. Even better are induction cooktops, which utilize electromagnetic radiation to heat up the cooking vessel itself instead of transferring heat to the cooking vessel. Explain, using the principles used in this chapter, why induction cooktops are more efficient than electric cooktops and why electric cooktops are more efficient than gas cooktops.

2. Using the specific heat capacity values of various metals in given in Problem 5 above, which substance would be most ideal for use as a cooking pan? Explain your answer.

3. Molar heat capacities are related to the microscopic complexity of the substance: the more complexity, the higher the heat capacity.

   a. Considering the values that you calculated in Problem 5 above, what general statement can you make about the microscopic complexities of metals?

   b. Specific heat capacity is often used to confirm the identity of a metal. As you saw in Problem 5 above, molar heat capacity is not as useful. Explain how these two statements can be rectified.

   c. The molar heat capacity of liquid water is approximately 18.0 cal/mol°C. What conclusion can you draw about the molecular complexity of liquid water?

4. When an ice cube is placed in water, the water molecules from the ice change temperature more than the water molecules of the liquid.

   a. Which has a higher molar heat capacity, water or ice?

b. State which water molecules (those from ice or those from liquid) will have a larger change in their molecular motions.

c. We will learn later that heat transferred to a substance results in an increase in the random motion of the components of the substance. We will also learn that random motion increases for all processes. Explain why this means that when the ice cube is placed in the glass of liquid water, the liquid will never freeze.

5. The enthalpy of a reaction is also called the heat of a reaction. This seems a bit troubling since the enthalpy of a reaction is actually the heat of a reaction at constant pressure. No one, however, makes this distinction because of the conditions in which most reactions are carried out in a laboratory. Explain.

Chapter 7: The Quantum-Mechanical Model of the Atom

Learning Objectives:

- Calculate the frequency, wavelength, and energy of light.
- Understand the wave nature of light using the two-slit experiment.
- Understand the particle nature of light using the photoelectric effect.
- Understand the two-slit experiment for electron beams.
- Use the de Broglie relation in calculations and understand its implications about matter properties.
- Use Heisenberg's uncertainty principle as an example of complementary properties and as an explanation for the wave-particle duality of matter.
- Identify the three quantum numbers and understand their physical meanings.
- Use quantum mechanical ideas to explain atomic spectra and to calculate energy level differences.
- Understand the nature of and sketch the probability distribution of a quantum mechanical orbital.

Chapter Summary:

In this chapter, you will explore the subject of quantum mechanics which provides a model for atomic structure and behavior. You will begin by looking at electromagnetic radiation (light) and the electromagnetic spectrum. Properties of light such as frequency, wavelength, and amplitude will be explained and then you will learn the dependence of these properties on one another. The wave nature of light will be discussed in terms of interference and then the particle properties of light will be explored using the photoelectric effect. The connection between light and matter will be introduced using atomic spectroscopy and the first theory to explain emission spectra, the Bohr model, will be discussed. In order to explore a more correct description of electron behavior, you will then look at the wave-particle duality of electrons. Your understanding of electron wave properties will be used to introduce quantum mechanical orbitals, which are specified using three quantum numbers and provide a probability distribution map for electrons in atoms. You will then learn how to calculate the energy associated with an orbital using spectroscopic data. Finally, we will explore each of the different orbital shapes in order to understand the physical characteristics of the atoms that result from them.

Chapter Outline:

I. Quantum Mechanics: A Theory That Explains the Behavior of the Absolutely Small
   a. Behaviors of very small things are affected by our observations of them.
   b. The quantum mechanical model explains how electrons exist in atoms and how these electrons determine the chemical and physical properties of elements.

II. The Nature of Light
   a. Some properties of light are best described using its wave characteristics.
      i. Light is electromagnetic radiation that consists of electric and magnetic fields oscillating perpendicular to one another while propagating through space.
      ii. The speed of light in a vacuum is a constant, $c = 3.00\times10^8$ m/s.
      iii. Electromagnetic waves are characterized by their amplitude, wavelength, and frequency.
         1. The amplitude is the intensity of light.

2. The wavelength (λ) is the distance between two crests of a light wave.
3. The frequency (ν) is the number of cycles that pass a point in a given period of time. The units of frequency are s^{-1} or Hz, where 1 Hz=1 s^{-1}.
4. The wavelength and frequency are related through the speed of light:

$$\nu = \frac{c}{\lambda}$$

EXAMPLE:

Calculate the frequency of red light with a wavelength of 700. nm.

We will first convert the wavelength of light from nm to m:

$$700.\,\text{nm} \times \frac{1\,\text{m}}{10^9\,\text{nm}} = 7.00 \times 10^{-7}\ \text{m}$$

We can now use this to calculate the frequency:

$$\nu = \frac{c}{\lambda} = \frac{3.00 \times 10^8\,\text{m/s}}{7.00 \times 10^{-7}\,\text{m}} = 4.29 \times 10^{14}\,\text{s}^{-1}$$

---

5. The electromagnetic spectrum includes all wavelengths of light.
6. The regions of the electromagnetic spectrum have different names.
   a. In order of increasing wavelength or decreasing frequency: gamma rays, x-rays, ultraviolet light, visible light, infrared light, microwaves, and radio waves.

iv. Interference occurs when waves interact with one another.
   1. Constructive interference occurs when waves travel in phase with one another. When the crests of waves overlap, their amplitudes are additive.
   2. Destructive interference occurs when waves travel out of phase with respect to one another.

v. Diffraction is the phenomenon of a wave bending around an obstacle or slit.

vi. The diffraction of light through two closely spaced slits results in an interference pattern that represents regions of constructive and destructive interference.

b. Some properties of light are best described using its particle characteristics.

i. The photoelectric effect occurs when electrons are emitted from metals upon their exposure to light.
   1. Electrons are emitted from a metal when light of sufficient energy is shined upon its surface.
      a. The minimum energy of light required to emit electrons from the surface of a metal is the threshold energy.
      b. When the intensity of the light is increased, more electrons are emitted and when the intensity is decreased, fewer electrons are emitted.

2. If light with an energy lower than the threshold frequency is shined on a metal, electrons will not be emitted no matter the light intensity.

ii. Light comes in packets (called photons or quantum) that have a particular energy that is quantized:

$$E = h\nu = \frac{hc}{\lambda}$$

where h is Planck's constant and is equal to 6.626×10^{-34} J·s.

EXAMPLE:

What is the energy of red light with a wavelength of 700. nm?

We calculated the frequency of this light in the last problem, so we can easily calculate the energy:

$$E = (6.626\times10^{-34}\ \text{J}\cdot\text{s})\times(4.29\times10^{14}\ \text{s}^{-1}) = 2.84\times10^{-19}\ \text{J}$$

Note that visible light will always have energy that is on the order of 10^{-19} as in this problem.

---

iii. The threshold frequency is related to the binding energy of the electron, ϕ:

$$\phi = h\nu$$

iv. When energy greater than the binding energy is incident on the surface of a metal, electrons are emitted with a kinetic energy (KE) that is equal to the difference between the energy of the incident light and the binding energy:

$$KE = h\nu - \phi$$

► The idea here is that electrons need a certain energy in order to "escape" the atom. Any additional energy makes them move faster (kinetic energy).

EXAMPLE:

What is the threshold frequency of a metal that ejects electrons with a kinetic energy of 1.58×10^{-19} J when 400. nm light is shined upon its surface?

We will first calculate the binding energy for the metal using the kinetic energy and the energy of incident light. In order to do this, we will express the previous equation in terms of wavelength:

$$KE = \frac{hc}{\lambda} - \phi$$

Converting the wavelength from nm to m:

$$400\ \text{nm} \times \frac{1\ \text{m}}{10^{9}\ \text{nm}} = 4.00\times10^{-7}\ \text{m}$$

Using this to calculate the binding energy:

$$1.58\times10^{-19}\ \text{J} = \frac{(6.626\times10^{-34}\ \text{J}\cdot\text{s})(3.00\times10^{8}\ \text{m/s})}{4.00\times10^{-7}\ \text{m}} - \phi$$

$$\phi = 3.39\times10^{-19}\ \text{J}$$

Now we will convert the energy into frequency:

$$(3.39\times10^{-19}\,\text{J}) = (6.626\times10^{-34}\,\text{J·s})\,\nu$$

$$\nu = 5.12\times10^{14}\,\text{s}^{-1}$$

III. Atomic Spectroscopy and the Bohr Model

a. After energy is added to an atom it is often released in the form of light.

b. Atomic spectroscopy is the study of the electromagnetic radiation that is absorbed and/or emitted by atoms.

c. Line spectra are emission spectra of atoms with the emitted wavelengths of light separated from each other.

    i. Line spectra are unique for different elements and are always the same for a given element.

d. Continuous spectra have no sudden interruptions in the intensity of light as a function of the wavelength.

e. The Bohr model was the first model introduced to explain line spectra.

    i. In the Bohr model, electrons were described as orbiting the nucleus at particular distances; these distances were quantized (at particular energy values) so that electrons could only jump from one orbit to another but could not be in between.

        1. Energy is absorbed when an electron goes from a lower energy orbit to a higher energy orbit.

        2. Energy is emitted when an electron goes from a higher energy orbit to a lower energy orbit.

f. The Bohr model was replaced by a model that considers electrons as waves.

► It is important to realize that electrons do NOT travel in orbits – this was the first promising model set forth, but was replaced with a more thorough and correct description of electron behavior: the quantum mechanical description.

IV. The Wave Nature of Matter: The de Broglie Wavelength, the Uncertainty Principle, and Indeterminacy

a. Electron beams that travel through two closely spaced slits exhibit a diffraction pattern.

    i. The diffraction pattern occurs even when electrons are sent through the slits one at a time.

    ii. The diffraction pattern occurs from single electrons interfering with themselves.

    iii. The wave nature of an electron is an inherent property of an individual electron.

b. The de Broglie relation mathematically relates the wave properties (wavelength: λ) to the particle properties (momentum: mv) of matter.

$$\lambda = \frac{h}{mv}$$

EXAMPLE:

Calculate the speed of an electron whose wavelength is 8.54 μm.

We will first rearrange the de Broglie relation to solve for velocity (which is equal in magnitude to the speed):

$$v = \frac{h}{m\lambda}$$

Now we can convert the wavelength into SI units of m:

$$8.54\ \mu m \times \frac{1\ m}{10^6\ \mu m} = 8.54 \times 10^{-6}\ m$$

The mass of an electron is given in the back cover of your book, allowing us to solve for the velocity:

$$v = \frac{(6.626 \times 10^{-34}\ J \cdot s)}{(9.109 \times 10^{-31}\ kg)(8.54 \times 10^{-6}\ m)} = 85.2\ m/s$$

---

c. Complementary properties are those that exclude one another; the more certainly you know one of the properties, the less certain you are of the value of the other.

   i. If we try to measure the position of an electron in a two-slit experiment, no interference pattern will be observed.

   ii. It is not possible to measure the wave properties (interference) and particle properties (position) simultaneously because they are complementary.

   iii. The Heisenberg principle is a statement of the uncertainty associated with complementary properties:

$$\Delta x \times m\Delta v \geq \frac{h}{4\pi}$$

   Where Δx is the uncertainty in the position, m is the mass, Δv is the uncertainty in the velocity, and h is Planck's constant.

d. The wave nature of electrons makes them non-deterministic, which means that their trajectories are unknowable.

   i. In quantum mechanics, trajectories are replaced with probability distribution maps that show where electrons are likely to be found.

V. Quantum Mechanics and the Atom

a. Energy and position are complimentary properties.

b. Electron positions are described using orbitals, which are probability distribution maps.

c. The energy of an electron's wave function is given by the Schrödinger equation:

$$\hat{H}\Psi = E\Psi$$

where $\hat{H}$ is the Hamiltonian operator, which represents the total energy dependencies, E is the energy, and ψ is the wave function that describes the wave nature of the electron.

d. Orbitals are specified by three quantum numbers.

   i. The principal quantum number, n, is also called the principal level or shell.

1. The principle level determines the energy of the orbital for hydrogen:

$$E = -2.18\times10^{-18}\,J\left(\frac{1}{n^2}\right) \qquad n = 1, 2, 3, 4, \ldots$$

EXAMPLE:

What is the energy of the n=2 level for the hydrogen atom?

$$E = -2.18\times10^{-18}\,J\left(\frac{1}{2^2}\right) = -5.45\times10^{-19}\,J$$

---

2. The energy is negative because an electron close to the nucleus is at a lower energy than when it is far away and zero energy is defined as infinite separation between electrons and protons.
3. As the principal quantum number increases, the spacing between subsequent levels decreases.

ii. The angular momentum quantum number, l, is also called the azimuthal quantum number, the sublevel, or the subshell.

1. The value of l determines the shape of the orbital.
2. l can take on values of 0, 1, 2, 3...(n-1). For n=1, l must equal zero. For n=2, l can equal either zero or one.
3. Letters are often used to replace the number of l.
   a. $l = 0$ is an s orbital.
   b. $l = 1$ is a p orbital.
   c. $l = 2$ is a d orbital.
   d. $l = 3$ is an f orbital.

iii. The magnetic quantum number, m_l, gives the orientation of the orbital in space.

1. The magnetic quantum number has values of: -l, -(l-1)...(l-1), l.

EXAMPLE:

What are the three quantum numbers for an electron in the 4d orbital?

The principle quantum number is 4 and the angular momentum quantum number is 2 (d orbital). The magnetic quantum number an be -2, -1, 0, +1, or +2; we cannot determine which without more information.

---

e. The wavelengths of light that appear in atomic spectra correspond to transitions between quantum orbitals. The energy of the light is equal to the energy of the emitted/absorbed photon:

$$\Delta E_{atom} = -E_{photon}$$

i. The addition of energy to an atom/ion results in excitation: electrons move from a lower energy orbital to a higher energy orbital.

ii. When an electron moves from a higher energy orbital to a lower energy orbital, energy is released as the electron relaxes.

EXAMPLE:

What wavelength of light is emitted when an electron moves from the n=3 shell to the n=2 shell?

First, we will calculate the energy associated with each level:

$$E_2 = -2.18\times10^{-18}\ \text{J}\left(\frac{1}{2^2}\right) = -5.45\times10^{-19}\ \text{J}$$

$$E_3 = -2.18\times10^{-18}\ \text{J}\left(\frac{1}{3^2}\right) = -2.42\times10^{-19}\ \text{J}$$

The difference in energy is:

$$E_2\text{-}E_3 = (-5.45\times10^{-19}\ \text{J}) - (-2.42\times10^{-19}\ \text{J}) = (-5.45 + 2.42)\times10^{-19}\ \text{J} = -3.03\times10^{-19}\ \text{J}$$

The energy change associated with this transition is equal in magnitude to the energy of the emitted photon. The wavelength of emitted light is then:

$$E = \frac{hc}{\lambda}$$

$$\lambda = \frac{(6.626\times10^{-34}\ \text{J}\cdot\text{s})(3.00\times10^{8}\ \text{m})}{3.03\times10^{-19}\ \text{J}} = 6.56\times10^{-7}\ \text{m}$$

VI. The Shapes of Atomic Orbitals

a. The shape of an atomic orbital is determined primarily by the angular momentum quantum number.

i. The shape of an atomic orbital is actually a probability distribution map.

1. The orbital shapes are drawn to represent the region in which the probability of finding an electron is 90%.

ii. The square of the wave function (ψ^2) is the probability (per unit volume) of finding an electron at a given position in space.

b. We plot the probability of finding an electron at a given distance using the radial distribution function as a function of the distance from the nucleus.

i. The radial distribution function is the total probability of finding an electron in a thin spherical shell at a distance, r, from the nucleus.

ii. The probability of finding an electron at the nucleus is zero because the size of a shell on the nucleus, effectively a mathematical point compared to the volume of the atom, is zero.

c. A node is a point where the wave function (ψ)—and therefore the probability of finding an electron (ψ^2)—is equal to zero.

i. There are (n-1) nodes in each shell.

ii. There are l angular nodes in each subshell.

d. s orbitals have an angular momentum quantum number of zero (l=0).

i. s orbitals are spherically symmetric.

ii. s orbitals have no angular nodes and have (n-1) radial nodes.

e. p orbitals have an angular momentum quantum number of one (l=1).

i. There are three p orbitals in each shell (where n>2), one each with m_l = -1, 0, or +1.

ii. p orbitals have two lobes and one angular node.

f. d orbitals have an angular momentum quantum number of two (l=2).

i. There are five d orbitals in each shell (where n>3) with m_l = -2, -1, 0, +1, or +2.

ii. Four of the d orbitals have four lobes, and the fifth d orbital has two lobes with a ring where an electron is most likely to be found.

iii. d orbitals have two angular nodes.

g. f orbitals have an angular momentum quantum number of three (l=3).

i. There are seven f orbitals in each shell (where n>4) with m_l = -3, -2, -1, 0, +1, +2, or +3.

h. The phase of a wave is the sign of the wave's amplitude – positive or negative.

<u>Fill in the Blank Problems:</u>

1. The phenomenon of light bending around an obstacle or slit is called ________________.
2. The ________________ determines the orientation of an orbital.
3. ________________ is a type of energy embodied in oscillating electric and magnetic fields.
4. The ________________ of light is the number of cycles that pass through a stationary point in a given period of time.
5. A(n) ________________ is a point where the wave function goes through zero.
6. An increase in the intensity of light that causes electrons to be emitted causes ________________ electrons to be emitted.
7. The ________________ is the minimum frequency of light that must be incident on a metal surface in order to emit electrons.
8. The highest energy region of the electromagnetic spectrum is ________________.
9. ________________ occurs when two waves travel in phase with one another.
10. Energ y is emitted from an atom when an electron ________________.
11. The ________________ of light is the distance in space between any two analogous points of a wave.
12. The ________________ explains how electrons exist in atoms and how these electrons determine the chemical and physical properties of elements.
13. ____ ______________ relates the uncertainty in an electron's position to the uncertainty in its velocity.
14. The series of wavelengths that are emitted from an excited species is called a(n) ________________.
15. The more certainly you know a given property, the less certain you know its ________________ property.

16. There are ________________ p orbitals with magnetic quantum numbers of ________________.

17. The speed of light is a con stant in a ________________.

18. The __ ________________ determines the overall size and approximate energy of an orbital.

19. The path of an electron is _______ __________; it can only be described statistically.

20. The interference pattern that occurs when electrons are passed through two closely spaced slits is a result of the electron's ________________.

21. Th e wave function squared can be plotted in three dimensions and represents the ________________.

22. A n orbital is specified by three interrelated ________________.

23. The __ ________________ primarily determines the shape of the orbital.

24. The ________________ is the observation that many metals emit electrons when light is shined upon them.

25. A s the energies of orbitals increases, the distance between subsequent levels ________________.

26. The energ y of an emitted photon is equal to ________________ of the atomic orbitals.

Problems:

1. Sketch the radial distribution plots for the 2s and 3s orbitals.
   a. On your plots, label the axes appropriately.
   b. Identify the radial nodes on your graphs.
   c. Use your plots to determine if an electron in the 3s orbital can be closer to the nucleus than an electron in the 2s orbital. Explain.
2. The threshold energy of chromium is 7.00×10^{-19} J.
   a. What wavelength of light will cause an electron to be emitted with zero kinetic energy?
   b. What is the threshold frequency for chromium metal?
   c. What do you expect to observe when 500 nm light is incident on the metal surface?
   d. What to you expect to observe when the intensity of the light at the threshold frequency is increased tenfold?
   e. What will the kinetic energy of the electron be when light that has a frequency twice that of the threshold frequency is incident on chromium?
   f. When an electron is ejected from a metal, it is ejected from the highest occupied orbital. In chromium, the n=4 level is the highest populated level. Will the transition from n=1 to n=4 require more or less energy than the threshold energy?
3. How many different electrons can be in the n=4 level? What will the four quantum numbers of these electrons be?
4. All matter has wave and particle characteristics, but we don't often consider wave properties of large things.
   a. Calculate the de Broglie wavelength of a 1-lb ball moving at a speed of 90.0 mph.
   b. Based on your answer in part a, explain why the wave properties are unimportant in describing the behavior of the ball.
   c. Recently, scientists have measured the de Broglie wavelengths of large carbon molecules called fullerenes. Current detection limits do not allow for the detection of wavelengths for larger

molecules. Explain how these experiments represent the current transition between quantum and classical behavior.

d. What do you expect to happen when the two-slit experiment is carried out using baseballs? Explain the reasoning behind your answer.

e. Zero-point energy is the minimum energy that a quantum mechanical system can have (the energy can't actually be zero). What will the zero-point energy of a baseball be?

5. A blue advertising sign emits light with a wavelength of 465 nm.

   a. What is the frequency of the light?

   b. What is the energy of 1.00 mol of photons of this light?

   c. When the power of the light is reduced, what about the light has changed?

6. Photons of light can be split using special materials. When a single photon is split into two, the total energy must be conserved. Derive an expression that will allow you to determine the energy of the two photons emitted when ultraviolet light of 200 nm is split into two equivalent photons.

Concept Questions:

1. For enthalpy, we defined the zero to be the enthalpy of formation for elements in their most stable states at standard condition. What is the zero of energy for an electron in an atom?

2. The uncertainty principle states that the position and velocity (or momentum) of a particle cannot be determined exactly at the same time.

   a. Why, despite uncertainty, is it possible for astronomers to track the position of satellites to a very high precision?

   b. The uncertainty principle necessarily discredits the possibility that electrons orbit the nucleus. Explain why it is impossible for an electron to be in an orbit if uncertainty is true.

3. In this chapter, radial distribution plots were drawn for you; in these plots, all angles were considered at a given radius. We could also draw angular distribution plots for the various orbitals by plotting the probability of finding an electron at a given angle for all radii. In doing this, we would need to consider two different angles to define all of three-dimensional space: the angle defined by motion away from the z-axis in the x-z plane and the angle defined by motion away from the x-axis in the x-y plane. Draw such orbitals for the 1s, 2s, and 2p orbitals. Hint: consider the number of angular nodes present for each orbital in order to situate your graphs.

4. Radio waves are permitted, by law, to be "everywhere" around us.

   a. How can you tell that radio waves are everywhere?

   b. Why are radio waves the chosen frequency range for so many common applications?

5. The two-slit experiment results in an interference pattern that depends on the distance between the slits and the wavelength of the incident light/electrons.

   a. How will the interference pattern change as the slits are moved apart from one another?

   b. How will the interference pattern change as the wavelength of light is increased?

   c. In this chapter, the experiment which attempted to identify the positions of electrons in the two-slit experiment was discussed. We can also gain a better understanding of the position by decreasing the size of the slits. Use the uncertainty principle to explain how making the slits narrower will affect the interference pattern.

6. In this chapter, you saw that the spacing between energy levels gets smaller as the energy of the levels increases. The levels become infinitely close together when the value of n approaches infinity. Explain how this relates to the idea that kinetic energy is not quantized.

7. Radioactive material is often detected using a Geiger counter, which detects the electromagnetic radiation that is emitted upon radioactive decay. It is often possible to hear single clicks, each click representing a single decay process. Explain how this experiment reinforces the idea that light has particle properties.

8. Electron microscopes allow one to image very small things. You can search for electron microscope images on the internet to see some amazing pictures. Based on the idea that one can only use waves that are comparable to the size of the particle to be imaged, do you expect electrons to have longer or shorter wavelengths than visible light (used in an ordinary microscope)?

Chapter 8: Periodic Properties of the Elements

Learning Objectives:

- Write electron configurations and orbital diagrams for multi-electron atoms.
- Understand electron spin and how it relates to the magnetic properties of elements.
- Understand energy-level splitting for multi-electron atoms.
- Identify the number of valence electrons in an atom and understand the correlation between the number of valence electrons and periodic properties.
- Determine relative atomic radii of atoms and ions.
- Predict relative ionization energies of atoms based on their positions on the periodic table.
- Understand the relationship between electron affinity and electron configuration.
- Identify common chemical reactions and their correlation to electron configurations.

Chapter Summary:

In this chapter, you will learn about the predictive properties of the quantum mechanical model discussed in Chapter 7. Specifically, you will learn the correlation between the electron wave functions (orbitals) in atoms and the periodic nature of the elements. We will begin by learning how to represent the energetic positions of electrons using electron configurations and orbital diagrams. In order to construct these representations, you will be introduced to electron spin and the splitting of sublevels in multi-electron atoms. Once you are able to write electron configurations, you will learn about the unique properties of the electrons in the outermost shell called valence electrons. You will look at the periodic table using these configurations and identify many important periodic properties including atomic radius, ionization energy, electron affinity, and metallic character. Finally, we will look at the reactivities of the elements from Groups 1A, 7A, and 8A in order to correlate electron configurations to chemical behavior.

Chapter Outline:

I. Nerve Signal Transmission
   a. Periodic properties are those properties that can be predicted based on an element's position on the periodic table.

II. The Development of the Periodic Table
   a. Mendeleev's periodic table arranges elements by their mass.
      i. Unknown elements were predicted to exist before they were discovered.
   b. The modern periodic table arranges elements by their atomic number.
   c. This chapter focuses on the connection between a theory (quantum mechanics) and a law (the periodic law).

III. Electron Configurations: How Electrons Occupy Orbitals
   a. Electron configurations show the orbitals occupied for an element.
      i. The ground-state configuration is the lowest energy configuration.
      ii. In an electron configuration, the positions of the electrons are written by stating the value of n, the symbol of *l*, and the number of electrons in the orbital as a superscript.

b. Orbital diagrams symbolize electrons as arrows in boxes that represent orbitals.

   i. The direction of the arrow represents the electron's spin.

      1. Electrons all have the same amount of spin, but it is quantized with spin up or down as demonstrated by the Stern-Gerlach experiment.

      2. The electron spin is a fourth quantum number, m_s, and can equal +½ (spin up) or -½ (spin down).

   ii. The Pauli exclusion principle states that no two electrons in an atom can have the same four quantum numbers.

      1. Each orbital can hold two electrons: each with opposite spin.

c. In hydrogen, the energy of an orbital depends only on the value of the principal quantum number, n.

   i. The energies of orbitals in a given shell (n) are the same; they are degenerate.

d. In multi-electron atoms, the energy depends on the principal quantum number, n, and the orbital angular momentum quantum number, *l*.

   i. The lower the value of *l*, the lower the energy of a subshell:

$$E(s) < E(p) < E(d) < E(f)$$

   ii. This splitting is the result of electron-electron repulsions.

   iii. The presence of multiple electrons in an atoms results in shielding of the nucleus by some of the electrons. Electrons closer to the nucleus block the outer electrons from experiencing the full charge of the nucleus.

      1. The effect of shielding is quantified by the effective nuclear charge, Z_{eff}.

      2. The energy of s orbitals is lowest because s orbitals have a higher probability of an electron being closest to the nucleus; s orbitals have the greatest penetration meaning that they shield outer electrons most.

e. The aufbau (or building up) principle states that electrons fill the orbitals from lowest to highest energy.

f. Hund's rule states that electrons fill orbitals singly with parallel spins before pairing up.

   i. When an atom has unpaired electrons it is called paramagnetic because it displays magnetic properties.

   ii. An atom with all of its electrons paired is called diamagnetic.

g. We can abbreviate electron diagrams using the noble gas abbreviation for the inner electron configuration. We write the symbol for the noble gas in square brackets, followed by the outer electron configuration.

EXAMPLE:

Write electron configurations and orbital diagrams for the following elements.

a. Si

Silicon is in the third period and Group 6A. We will fill up all of the n=1 and n=2 orbitals, then fill the 3s orbital and finally have four electrons in the 3p orbital:

$$1s^2 2s^2 2p^6 3s^2 3p^4$$

1s 2s 2p 3s 3p

Notice that the two unpaired electrons in the 3p orbital are in the same direction – they have parallel spin according to Hund's rule.

b. Sr

Strontium is in the fifth period and Group 2A. We will fill up all of the n=1, n=2, n=3, and n=4 s and p orbitals as well as the n=3 d orbitals:

$$1s^2 2s^2 2p^6 3s^2 3p^6 4s^2 3d^{10} 4p^6 5s^2$$

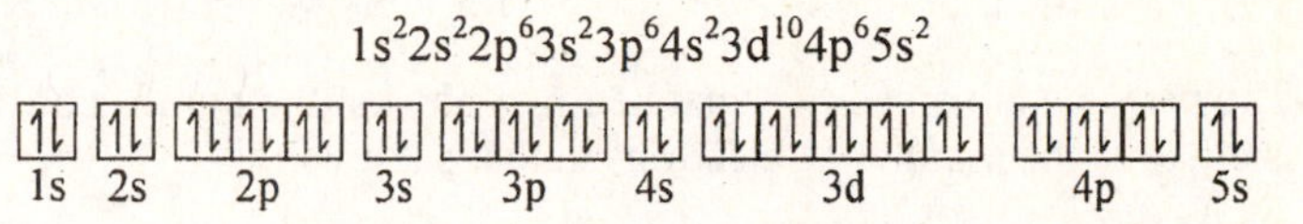

The orbitals are listed according to increasing energy and all of the orbitals are filled.

c. Ti

Titanium is a transition metal with valence electrons in the 3d orbital. We will fill the n=1, n=2, and n=3 s and p orbitals, then fill the 4s orbital and partially fill the 3d orbital:

$$1s^2 2s^2 2p^6 3s^2 3p^6 4s^2 3d^2$$

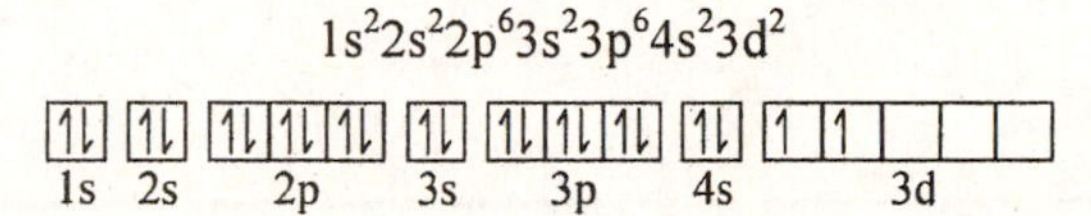

We see from the orbital diagram that titanium has two unpaired electrons in the 3d orbital and that these electrons have parallel spin.

d. P

Phosphorus has filled 1s, 2s, 2p, and 3s orbitals; it has a partially filled 3p orbital:

$$1s^2 2s^2 2p^6 3s^2 3p^3$$

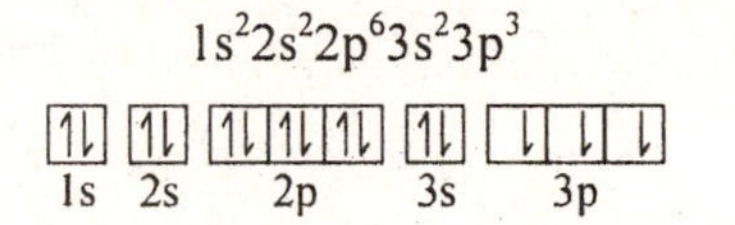

Again we see that the partially filled 3p orbital has unpaired electrons with parallel spin. In this orbital diagram, the unpaired electrons have been drawn with their spins directed down to emphasize the fact that the spin direction does not matter; only the fact that the spins are in the same direction matters.

IV. Electron Configurations, Valence Electrons, and the Periodic Table

a. The electrons in the outermost shell of an atom are called the valence electrons; down a column, the number of valence electrons remains constant.

i. Valence electrons for main group elements are those in the outermost principal quantum level (or outermost shell).

ii. Valence electrons for transition metals are those in the outermost shell plus the outermost d orbital electrons.

b. All other electrons in an atom are called core electrons.

EXAMPLE:

Determine the number of valence electrons for each of the elements in the previous example.

a. We can see from the orbital diagram that silicon has six valence electrons which is consistent with it being in Group 6A.

b. From the orbital diagram, we see that strontium has two electrons in the outermost shell (n=5) and therefore has two valence electrons.

c. Because titanium is a transition metal, we include the 3d electrons in the valence electron count. Titanium therefore has 4 valence electrons.

d. Phosphorus has five valence electrons since there are five electrons in the n=3 shell.

---

c. The periodic table can be separated into blocks; the number of the columns in each block is equal to the number of electrons in that sublevel.

d. For main group elements, the number of the group is equal to the number of valence electrons.

e. The row number is equal to the principle quantum level for the main group elements.

f. For transition metals, the d orbital will be in the n = (row-1) level and there are anomalies introduced in orbital filling because ns and (n-1)d orbitals are very close in energy.

EXAMPLE:

Use the periodic table to write electron configurations for the following elements using noble gas abbreviations.

a. F

Fluorine is in the second row of the periodic table and is located in the p block. Since fluorine is in Group 7A, it will have seven valence electrons. Helium is the noble gas that comes before fluorine and will be the noble gas core. The electron configuration is: $[He]2s^2 2p^5$.

b. As

Arsenic is in the fourth row of the periodic table and is located in the p block. Since arsenic is in Group 5A, it has five electrons in the n=4 shell. Argon is the noble gas core for arsenic. Since the 3d orbital is not part of the argon core, we will need to write the full 3d shell in the electron configuration. The electron configuration is: $[Ar]4s^2 3d^{10} 4p^3$.

c. Ca

Calcium is in the s-block of the fourth row of the periodic table and has the argon core since argon is the noble gas that immediately precedes it on the periodic table. Since calcium is in Group 2A, it has two valence electrons and an electron configuration of: $[Ar]4s^2$.

d. Zr

Zirconium is a transition metal in the fifth row of the periodic table. Krypton is the noble gas that immediately precedes it on the periodic table and the d orbital electrons will be in the (row-1)d or 4d orbital. The electron configuration of zirconium is: $[Kr]5s^2 4d^2$.

e. Mo

Molybdenum is in the fifth row of the periodic table and will have an electron configuration that is very similar to that of zirconium. We could expect the electron configuration of zirconium to be $[Kr]5s^2 4d^4$, but because of the close energy spacing of the 5s and 4d orbitals, the electron configuration is actually $[Kr]5s^1 4d^5$, which results in two half-filled shells.

V. The Explanatory Power of the Quantum Mechanical Model

a. Chemical properties are largely determined by the number of valence electrons that an atom or ion has.

i. Elements with full shells are the least reactive.

ii. Elements with almost filled shells are most reactive.

b. The formation of ions is often predictable because of the electron configurations.

EXAMPLE:

Use the electron configurations of the following elements to predict the charge of the ion that will form for each.

a. S

Sulfur has an electron configuration of $[Ne]3s^2 3p^4$. We can see that if sulfur gained two more electrons, it would have an electron configuration equivalent to that of Ar: $[Ne]3s^2 3p^6$. Therefore, sulfur will gain two electrons and form a 2- ion: S^{2-}.

b. K

Potassium has an electron configuration of $[Ar]4s^1$. If potassium loses one electron to become a 1+ ion, it will have the same electron configuration as argon. Potassium forms the K^+ ion.

c. Be

Beryllium has an electron configuration of $[He]2s^2$. By losing two electrons, beryllium will have the same electron configuration as helium and form a 2+ ion: Be^{2+}.

VI. Periodic Trends in the Size of Atoms and Effective Nuclear Charge

a. The non-bonding atomic radius or van der Waals radius is the distance between the centers (nuclei) of adjacent nonbonding atoms.

b. The bonding atomic radius or covalent radius for non-metals is ½ the distance between two nuclei that are bonded together.

c. The bonding atomic radius for metals is ½ the distance between two atoms next to each other in a crystal lattice.

d. The atomic radius is the average radius value based on many measurements of different types.

e. The atomic radius increases down the periodic table and decreases across a row.

i. The radius increases down a column because the electrons are at higher energy and are further away from the nucleus on average.

ii. The radius decreases across a row because the effective nuclear charge, Z_{eff}, increases across a row.

$$Z_{eff} = Z - S$$

Where Z is the nuclear charge (number of protons) and S is the shielding.

1. The further away the electron is from the nucleus, the more electrons there are present to shield the charge of the protons in the nucleus.

a. Core electrons effectively shield the electrons in the outermost shell.

b. Outer shell electrons do not efficiently shield each other, but do to some extent.

f. The radii of transition metals increase down a column, but remain approximately constant across a row.

i. Electrons in d-orbitals all experience approximately the same nuclear charge.

EXAMPLE:

List the following atoms in order of increasing atomic radius: Ca, F, Ba, Se, and Cl.

The element that is furthest down and to the left on the periodic table will have the largest radius, so Ba is the largest of these. Calcium is the element that is the next lowest and to the left, so it is the second largest. The element that is furthest to the right and the top of the periodic table will have the smallest radius, so F is the smallest of these. Chlorine is also very far to the right and is just under chlorine, so it is the next smallest. Finally, we see that selenium is below and to the left of chlorine making it larger while it is in the same row and to the right of calcium making it smaller. The order is then: F<Cl<Se<Ca<Ba.

VII. Ions: Electron Configurations, Magnetic Properties, Ionic Radii, and Ionization Energies

a. Electron configurations of ions are the same as their neutral counterparts with the appropriate number of electrons added (anions) or subtracted (cations).

i. For main group elements, the electrons are removed in the reverse order that they are added.

ii. For transition metals, the electrons are not removed in the reverse order that they are added.

1. s and d orbitals are very close in energy.

2. As d orbitals are filled, they are stabilized with respect to the s orbitals.

a. ns electrons are removed before (n-1)d electrons.

EXAMPLE:

Write electron configurations for the following ions:

a. Br^-

Bromine has an electron configuration of $[Ar]4s^2 3d^{10} 4p^5$ when it is neutral. The 1- ion has one additional electron so the configuration becomes $[Ar]4s^2 3d^{10} 4p^6$.

b. Cd^{2+}

Cadmium has an electron configuration of $[Kr]5s^2 4d^{10}$ when it is neutral. The 2+ ion has lost two electrons which are removed from the 5s orbital, so the electron configuration becomes $[Kr]4d^{10}$.

c. O^{2-}

A neutral oxygen atom has an electron configuration of $[He]2s^2 2p^4$. Oxygen becomes a 2- ion upon the addition of two electrons to give an electron configuration of $[He]2s^2 2p^6$.

---

b. The size of an ion relative to its neutral counterpart depends on the charge.

i. For cations, the size of the ion is much smaller than the neutral element due to the increase in Z_{eff}.

ii. For anions, the size of the ion is much larger than the neutral element because the value of Z_{eff} has not changed, while the number of electrons has.

EXAMPLE:

Rank the following in order of increasing ionic radius: Ca^{2+}, Ba^{2+}, Cs^{+}, and F^{-}.

We will first focus on the cations listed. Ca^{2+} will be the smallest since it is highest up on the periodic table and has a 2+ charge. Barium and cesium are in the same row of the periodic table, and a neutral barium atom will have a smaller radius than cesium's neutral atomic radius. This difference will be exaggerated by the charge difference of their ions: since barium is a 2+ ion, its radius will decrease more, with respect to the neutral atom, than cesium with a 1+ charge. The order for increasing radius is therefore $Ca^{2+} < Ba^{2+} < Cs^{+}$. We now need to consider where F^{-} fits into this. Although anions increase in size relative to their neutral counterparts, F^{-} will still be the smallest of the species listed because it has its valence electrons in the n=2 shell whereas the smallest of the cations, Ca^{2+}, has a full n=3 shell. The ions are thus ordered according to increasing radius: $F^{-} < Ca^{2+} < Ba^{2+} < Cs^{+}$.

---

c. The ionization energy is the energy required to remove an electron from an atom in the gas phase:

$$X + \text{energy} \rightarrow X^{+} + e^{-}$$

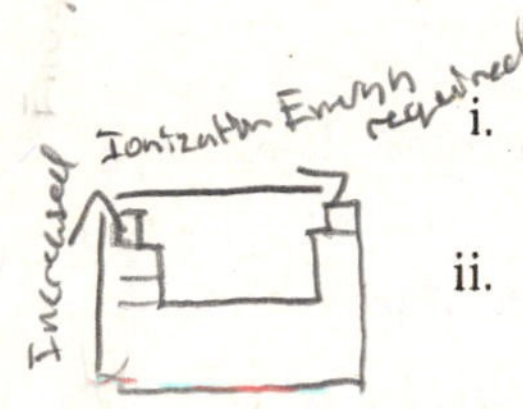

i. Ionization energies are always positive. Energy is always required in order to remove an electron from a neutral atom.

ii. Ionization energy decreases down the periodic table.

1. As the distance between the electron and the nucleus increases, the energy of the electron increases and the amount of energy required to remove it decreases.

iii. Ionization energy generally increases across a row on the periodic table.

1. As the effective nuclear charge increases, more energy is required to remove an electron.
2. The ionization energy decreases slightly when the ion formed has a full or half-full shell.

iv. The first ionization energy is less than the second ionization energy which is less than the third ionization energy and so on.

1. Successive ionization energies increase due to the increase in Z_{eff}.
2. Large jumps in the ionization energy value occur when ionizing a full or half-full shell.

EXAMPLE:

List the following neutral species in order of increasing ionization energy: K, Mg, P, S, and O.

In general, we expect the ionization energy to increase across a row and up a column, so we tentatively order these species as: K<Mg<P<S<O. We need to recognize, however, that phosphorus has a half-filled shell and will therefore have a slightly higher ionization energy than sulfur which is next to it on the periodic table. The order is, therefore, K<Mg<S<P<O.

VIII. Electron Affinities and Metallic Character

a. The electron affinity is the energy change that results from gaining an electron:

$$A + e^- \rightarrow A^- + \text{(energy)}$$

Note that energy is usually, but not always, released.

  i. Electron affinity is not as regular of a periodic trend down a column.

  ii. Across a row the trend generally increases as expected.

    1. The electron affinity is most exothermic for Group 7A elements.

b. Metallic character decreases across a row and increases down a column.

EXAMPLE:

Rank the following in order of increasing electron affinity: N, S, F, Al, Mg.

The more metallic an element is, the lower its electron affinity generally is and the closer the element is to having a full shell (the more valence electrons it has), the higher its electron affinity is. So we have: $Mg < N < Al < S < F$. Note that the EA of N is smaller than that of Al because N has a half-filled shell.

IX. Some Examples of Periodic Chemical Behavior: The Alkali Metals, the Halogens, and the Noble Gases

a. Alkali metals are in Group 1A and have an electron configuration of ns^1.

  i. Alkali metals are the most active metals on the periodic table – they are excellent reducing agents.

  ii. The radius and density increase down the column.

  iii. The ionization energy and melting point decrease down the column.

  iv. Alkali metals react with halogens to form MX and water to form MOH.

b. Halogens are in Group 7A and have an electron configuration of ns^2np^5.

  i. Halogens are the most active non-metals on the periodic table – they are excellent oxidizing agents.

  ii. The radius, melting point, boiling point, and density all increase down the periodic table.

  iii. Halogens react with metals to form metal halides, react with hydrogen to form HX, and react with each other to form interhalogen compounds.

c. Noble gases are in Group 8A and have an electron configuration of ns^2np^6.

  i. Noble gases are stable (unreactive) due to their full shells.

  ii. The radius, ionization energy, boiling point, and density increase down the column.

  iii. Krypton and xenon can be made to react with F_2 under extreme conditions to form: KrF_2, XeF_2, XeF_4, and XeF_6.

Fill in the Blank Problems:

1. The lowest energy state of an atom is its ________________.
2. ________________ are charged species that are much smaller than their corresponding atoms.
3. An element with unpaired electrons is ________________.
4. Metallic character ________________ across the periodic table and ________________ down the periodic table.
5. The principal quantum number of the d orbital being filled for a transition metal is equal to the ________________.
6. The second ionization energy is ________________ than the first ionization energy.
7. Electron affinity is the energy change associated with ________________.
8. The orientation of an electron's spin is quantized: it is either ________________ or ________________.
9. The Pauli exclusion principle states that no two electrons of an atom can have the same four ________________.
10. The effecti ve nuclear charge differs from the actual nuclear charge due to ________________.
11. Hu nd's rule states that when filling degenerate orbitals, electrons fill them singly first with parallel ________________.
12. Al kali metals have an electron configuration of ________________.
13. A(n) ________________ is one that is predictable based on an element's position on the periodic table.
14. Noble gases are very stable and unreactive; under extreme conditions ________________ and ________________ can be made to react with fluorine.
15. The atomic radius is the ________________ based on measurements of a large number of elements and compounds.
16. The ________________ is the energy required to remove an electron from an atom or ion in the gaseous state.
17. ____ ______________ are the electrons in the outermost shell and are important in chemical bonding.
18. The van der Waals radius represent s the radius of an atom when it is ________________.
19. The atomic radiu s ________________ across a row and ________________ down a column.
20. Dege nerate orbitals have the same ________________.

Problems:

1. Consider the following atoms: B, O, F, He, Na, K, Rb.
   a. Write the electron configurations and orbital diagrams for each of these species.
   b. How many valence electrons will each of the species listed have?
   c. Which of these species, if any, is paramagnetic?
   d. Rank these elements in order of increasing effective nuclear charge.
   e. Rank these elements in order of increasing atomic radius.
   f. Rank these elements in order of increasing ionization energy.
   g. Rank these elements in terms of their reactivity with $F_2(g)$.
2. Consider the following atoms and ion: P, Ca, Ga, Ca^{2+}, and Cl.

a. Write the electron configurations and orbital diagrams (using noble gas abbreviations) for each of these species.

b. How many valence electrons will each of the species listed have?

c. Which of these species, if any, is paramagnetic?

d. Estimate the effective nuclear charge for each of these species.

e. Rank these species in order of increasing atomic radius.

f. Rank these species in order of decreasing ionization energy.

g. Rank these species in order of increasing metallic character.

h. In 1–2 sentences, explain why the ionization energy follows the particular periodic trend that it does. Relate the ionization energy trend to the size of the atom/ion.

i. Which of these species do you expect to have the largest ionization energy? Explain why in terms of the explanation that you provided in part h.

3. Consider the following set of isoelectronic species: S^{2-}, Cl^-, Ar, K^+, Ca^{2+}.

   a. Write the electron configurations for all species.

   b. Rank these species according to increasing atomic radius.

   c. Rank these species according to increasing electron affinity.

   d. Rank these species in order of increasing ionization energy.

   e. How do you expect the chemical reactivity of these species to compare? Explain your answer.

4. The ground-state configuration is the lowest energy configuration of an element.

   a. Write the ground-state configuration for Zn.

   b. Will Zn be paramagnetic or diamagnetic? Explain.

   c. Based on the ground-state configuration for Zn, what will the charge of a zinc ion be when it reacts with chlorine to form zinc chloride?

   d. What are the four quantum numbers for the highest energy electron in the ground state configuration of Zn?

   e. Write an excited-state configuration of Zn.

   f. Will the ionization energy of the ground state be higher or lower than the ionization energy of the excited state that you wrote in part e? Explain.

5. The first ionization energy of lithium is 520 kJ/mol. What is the electron affinity for a lithium ion?

6. Using the size of their atomic radii, explain why you think that some noble gases, Kr and Xe, will react to form compounds while others will not.

Concept Questions:

1. Sketch the radial distribution plots for the 2s and 3s orbitals of hydrogen.

   a. Using the radial distribution plots that you have drawn, explain the periodic trend for atomic radius as you move down the periodic table.

   b. Using the radial distribution plots that you have drawn, explain the periodic trend for ionization energy as you move down the periodic table.

2. In Chapter 2, we saw that many transition metals can have variable charges in ionic compounds. Based on what you have learned in this chapter, explain why you think that this is.

3. The ionization energy for any species is always positive; i.e. energy is always required to remove an electron from a neutral species. Sodium metal, however, reacts violently with chlorine gas to form sodium chloride.

   a. Explain how this is possible and why the reaction is exothermic.

   b. Do you expect the reaction of magnesium with chlorine to be more or less exothermic? Explain.

4. Figure 8.14 shows the first ionization energy versus the atomic number for all elements through xenon.

   a. The differences in IE for subsequent elements in the first two rows of the periodic table are much larger than for subsequent rows. Explain why you think this might be.

   b. Notice that there is a jump in the region for the period 4 transition elements. Explain why this jump occurs for the IE of Mn.

   c. The decrease in IE that occurs between Group 5A and 6A disappears for period 5. Explain why you think this might be.

5. In Chapter 7, we saw the line spectrum of hydrogen.

   a. Using what you have learned about orbital energies in this chapter, qualitatively describe how the line spectra of multi-electron atoms will be different and how they will be the same.

   b. Experimentally, we observe that the spacing of energy levels gets smaller as the atomic number increases. Explain this observation using the ideal of effective nuclear charge and atomic radius.

6. According to Hund's rule, electrons are placed in orbitals singly with parallel spin before they are paired up.

   a. Recall that a spinning electron creates a magnetic field. Explain why it is more energetically favorable for the spins to be parallel than for them to be randomly oriented.

   b. The energy associated with pairing electrons in an orbital is often called the pairing energy. Explain two reasons for the existence of pairing energy.

   c. In this chapter, you learned that some transition metals fill orbitals in a seemingly anomalous way. Based on the fact that s and d orbitals are close in energy, use the pairing energy to explain how this seemingly anomalous behavior actually follows the aufbau principle.

Chapter 9: Chemical Bonding I: Lewis Theory

Learning Objectives:

- Understand the energy associated with chemical bonds.
- Identify ionic, covalent, and metallic bonds and understand the properties associated with each type of compound.
- Draw Lewis structures and use them in conjunction with formal charge to predict molecular structure.
- Use the Born-Haber cycle to determine lattice energies of ionic compounds.
- Use electronegativity values and trends to predict bond polarity.
- Understand and draw resonance structures.
- Calculate enthalpy of reaction estimates using average bond energies.

Chapter Summary:

In this chapter, you will learn how to use Lewis theory to predict compound formation. You will first be introduced to the energy associated with bond formation in order to understand why bonds form. You will next learn how to draw Lewis dot structures to represent the valence electrons of elements. Using Lewis dot structures, you will predict the formation of ionic compounds using the octet rule. You will learn about the energetic considerations associated with ion formation and how to calculate the lattice energy associated with this process. The properties of ionic substances will then be discussed in the context of their bonding. Covalent bonding will then be explored using Lewis structures. Electronegativity will be explored and the polarity of covalent bonds that result from electronegativity differences will be explained. A further expansion of covalent bonding will be discussed using resonance structures, formal charge considerations, and exceptions to the octet rule. Bond energies will be explained and the connection between reaction enthalpies and average bond energies will be explored. Finally, you will learn about metallic bonding and how the type of bonding leads to particular properties of metals.

Chapter Outline:

I. Bonding Models and AIDS Drugs
   a. Bonding theories are models that predict how atoms bond together to form molecules.
      i. Lewis theory is an example of a bonding theory.

II. Types of Chemical Bonds
   a. Chemical bonds form because the potential energy of the charged particles is lower in a compound than in an atom.
      i. The energy is described using Coulomb's law:

$$E = \frac{1}{4\pi\varepsilon_0}\frac{q_1 q_2}{r}$$

where ε_0 is a constant, q_1 is the charge of one particle, q_2 is the charge of another particle, and r is the distance between the two particles.

      ii. The potential energy is negative for oppositely charged species and is positive for like charged species.

1. In bonding, there is a competition between the repulsions of like charged particles and the attraction of oppositely charged particles.

iii. The further away particles are, the lower is the magnitude of their potential energy.

b. Ionic bonds form between metals and nonmetals because of electron transfer.

c. Covalent bonds form between nonmetals and result from electron pair sharing.

d. Metallic bonds form between metals and result because electrons are delocalized over the metal.

III. Representing Valence Electrons with Dots

a. In Lewis structures, valence electrons are represented as dots around an atom's symbol.

i. Dots are placed on each of the four sides of an element. There are a maximum of two dots per side and dots remain unpaired if possible.

ii. An octet is eight electrons and represents a full outer shell.

1. Noble gases have full octets except for helium which forms a duet (two paired valence electrons).

b. Bonding occurs between atoms in order to form stable electron configurations.

i. Bonds occur so that atoms can have full electron shells.

IV. Ionic Bonding: Lewis Structures and Lattice Energies

a. In ionic compounds, electrons are transferred from a metal to a nonmetal.

EXAMPLE:

Draw Lewis structures in order to predict the ionic compound that forms from the following metals and nonmetals.

a. K and F

Potassium has one valence electron and fluorine has seven valence electrons:

K· :F̤̈·

In order for both to have a full valence shell, potassium must lose one electron and fluorine will gain one electron to form K^+ and F^-:

K^+ [:F̤̈:]$^-$

b. Mg and Cl

Magnesium has two valence electrons and chlorine has seven valence electrons:

·Mg· :C̈l̤·

In order for both species to have a full valence shell, two chlorine atoms need to be associated with each magnesium atom:

[:C̈l̤:]$^-$ Mg^+ [:C̈l̤:]$^-$

c. Li and O

Lithium has one valence electron and oxygen has two valence electrons:

Li· ·Ö·

In order for both species to have a full valence shell, two lithium atoms need to be associated with each oxygen atom:

$Li^+ [:\ddot{O}:]^- Li^+$

---

b. The lattice energy is the energy associated with forming a crystal lattice from ions in the gas phase.

   i. The lattice energy decreases with increasing cation and anion size because the distance between the ions increases (r in Coulomb's law).

   ii. The lattice energy increases with increasing ion charge because of an increase in attraction (q_1 and q_2 in Coulomb's law).

c. The Born-Haber cycle is a hypothetical series of steps that together describe the formation of an ionic compound from its constituent elements:

$$Na(s) + \frac{1}{2}Cl_2(g) \rightarrow NaCl(s)$$

   i. The first step is the formation of gaseous Na from the solid (sublimation energy).

   ii. The second step is the formation of chlorine atoms from molecular chlorine (bond energy).

   iii. The third step is the ionization of sodium to form Na^+ ions (ionization energy).

   iv. The fourth step is the ionization of chlorine to form Cl^- ions (electron affinity).

   v. The fifth step is the formation of the ionic solid (lattice energy).

$$\Delta H_f^\circ = \Delta H_1^\circ + \Delta H_2^\circ + \Delta H_3^\circ + \Delta H_4^\circ + \Delta H_5^\circ$$

EXAMPLE:

Determine the lattice energy for the formation of Li_2O using Appendix II in your textbook. The ionization energy of lithium is 520.2 kJ/mol and the energy required for oxygen to gain two electrons is +650. kJ/mol.

ΔH_1 is equal to the ΔH_f° of Li(g). From Appendix II, ΔH_f° (Li(g)) = +159.3 kJ/mol. Note that we need two lithium ions because of the formula of lithium oxide, so double the enthalpy: ΔH_1° = +318.6 kJ/mol.

ΔH_2 is equal to the ΔH_f° of O(g). From Appendix II, ΔH_f° (O(g)) = +249.2 kJ/mol.

ΔH_3 is equal to the ionization energy of lithium, ΔH_{IE}° = +520.2 kJ/mol. Again, we need to multiply this by two in order to account for the formation of two lithium ions: ΔH_3° = 1040.4 kJ/mol.

ΔH_4 is the energy required for oxygen to gain two electrons (the sum of the first and second electron affinity values for oxygen), ΔH_4°=+650. kJ/mol.

ΔH_f° is the enthalpy of formation for Li_2O. From Appendix II, ΔH_f° = -597.9 kJ/mol.

Rearranging the expression for the Born-Haber cycle to solve for the lattice energy gives:

$$\Delta H_5^\circ = \Delta H_f^\circ - (\Delta H_1^\circ + \Delta H_2^\circ + \Delta H_3^\circ + \Delta H_4^\circ)$$

Using the values found above, we have:

$$\Delta H_5^\circ = (597.9 - (318.6 + 249.2 + 1040.4 + 650.))\ \text{kJ/mol}$$

$$\Delta H_5^\circ = -1660.3\ \text{kJ/mol}$$

According to significant figure rules, we should only report our answer to the ones digit. The lattice energy is -1660. kJ/mol.

---

d. Ionic substances have high melting points due to the extended nature of the lattice.

e. Ionic solids do not conduct electricity because the electrons are localized on a particular ion. When melted or dissolved in water, ionic substance dissociate into ions and can move in response to forces creating electrical current. Solutions of ionic substances do, therefore, conduct electricity.

V. Covalent Bonding: Lewis Structures

a. Bonding pairs are the electrons shared between two atoms.

  i. Bonding pairs are represented as dashes in order to emphasize the fact that they represent a pair of shared electrons that form a bond.

  ii. Double bonds have two electron pairs shared between two atoms.

  iii. Triple bonds have three electron pairs shared between two atoms.

b. Lone pairs are electron pairs that are associated only with a single atom.

c. Covalent bonds are directional (as opposed to ionic bonds), which means that individual molecules form.

  i. Interactions between molecules are much weaker than the bonds themselves, which means that molecular substances have much lower melting and boiling points than ionic compounds do.

VI. Electronegativity and Bond Polarity

a. Electrons are not usually shared equally; the unequal sharing of electrons results in polarity.

b. Electronegativity (EN) is the ability of an atom to attract electrons to itself.

  i. Fluorine is the most EN atom on the periodic table and is arbitrarily assigned an electronegativity value of 4.0 and all other elements are given values in reference to fluorine.

  ii. Electronegativity increases across and decreases down the periodic table.

c. The greater the electronegativity difference between two atoms is, the more polar is the bond.

  i. A small ΔEN (0.0–0.4) results in a covalent bond.

  ii. An intermediate ΔEN (0.4–2.0) results in a polar covalent bond.

  iii. A large ΔEN (>2.0) results in an ionic bond.

d. The dipole moment (μ) of a bond quantifies the bond polarity. The units of the dipole moment are Debye (D).

$$\mu = q \cdot r$$

where q is the charge of an electron, and r is the bond distance.

e. The percent ionic character is the ratio of the actual dipole moment to the dipole moment that would exist if the electrons were transferred to the more electromagnetic atom (as in an ionic compound).

i. If the percent ionic character is greater than 50%, the bond is considered ionic.

EXAMPLE:

Use the electronegativity values in Figure 9.10 of your textbook to predict whether the bond that forms between the following atoms will be ionic, covalent, or polar covalent.

a. Li and O

Lithium has an EN of 1.0 and oxygen has an EN of 3.5. The ΔEN for this bond is 2.5; this value is greater than 2.0 so the bond is ionic.

b. N and H

Nitrogen has an EN of 3.0 and hydrogen has an EN of 2.1. The ΔEN for this bond is 0.9; this value is between 0.4 and 2.0 so the bond is polar covalent.

c. S and Br

Sulfur has an EN of 2.5 and bromine has an EN of 2.8. The ΔEN for this bond is 0.3; this value is less than 0.4 so the bond is covalent.

d. Al and F

Aluminum has an EN of 1.5 and fluorine has an EN of 4.0. The ΔEN for this bond is 2.5; this value is greater than 2.0 so the bond is ionic.

VII. Lewis Structures of Molecular Compounds and Polyatomic Ions

a. Lewis structures can easily be written by following four steps.

i. Connect atoms together with bonds.

1. Hydrogen atoms are always terminal – they only connect to one other atom and are never the central atom.

2. The less EN atoms are used as central atoms and the more EN atoms are terminal atoms.

► In general, the first non-hydrogen atom in the formula is the central atom.

ii. Calculate the total number of valence electrons in the molecule (adding an electron for each negative charge and subtracting an electron for each positive charge).

iii. Distribute remaining electrons as lone pairs around as many atoms as possible.

iv. Make double and/or triple bonds for electron deficient (less than eight electrons) atoms by sharing electron pairs from the least EN adjacent atom.

EXAMPLE:

Draw Lewis structures for each of the following.

a. CH_4

1. Connect carbon to four hydrogen atoms with bonding pairs:

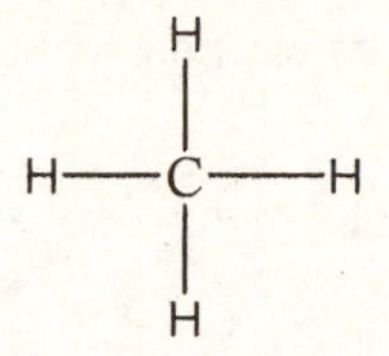

2. The total number of valence electrons is 4(from carbon) and 4×1(from hydrogen) = 8.
3. All electrons are used in the four bonds of methane.
4. All atoms have noble gas configurations: an octet around carbon and duets around each hydrogen.

b. HCOH

1. Connect carbon to the oxygen atom and the two hydrogen atoms:

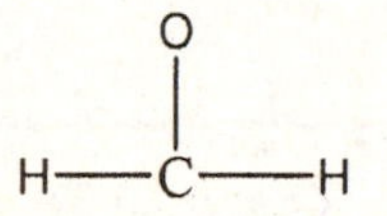

2. The number of valence electrons is 4(from carbon)+6(from oxygen)+2×1(from hydrogen)=12.
3. Six of the electrons have been used in forming the bonds. The remaining 6 electrons will be distributed around the carbon and oxygen atoms as lone pairs.

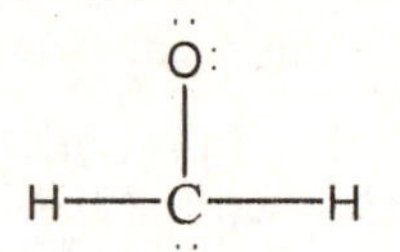

4. We can see that carbon and hydrogen have full shells, but oxygen is electron deficient. We will use the electron pair on carbon to form a double bond with oxygen:

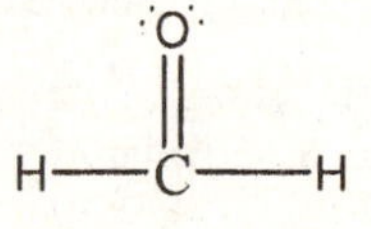

c. SO_2

1. Sulfur is the less EN atom, so we will put sulfur in the center and connect it to the oxygen atoms:

O—S—O

2. The total number of valence electrons is: 2×6(from oxygen)+6(from sulfur)=18.
3. Four electrons have been used to make the two bonds, so we have fourteen more to distribute:

:Ö—S̈—Ö:

4. We can see that one of the oxygen atoms is electron-deficient, so we will use a pair of electrons from silicon to form a double bond:

Ö═S̈—Ö:

d. NO_3^-

1. Nitrogen is the central atom since it is less EN than oxygen is:

2. The total number of valence electrons is: 3×6(from oxygen)+5(from nitrogen)+1(negatively charged ion) = 24.
3. We have used six of the valence electrons in bonding, so there are 18 left which we distribute as lone pairs.

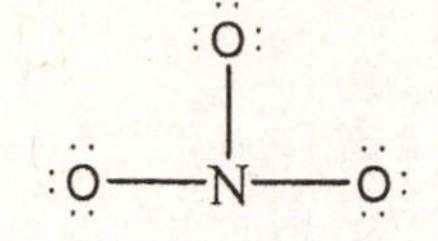

4. We see that the nitrogen atom does not have an octet, so we will form a double bond with one of the oxygen atoms. We will also use brackets around the ion and indicate the total charge:

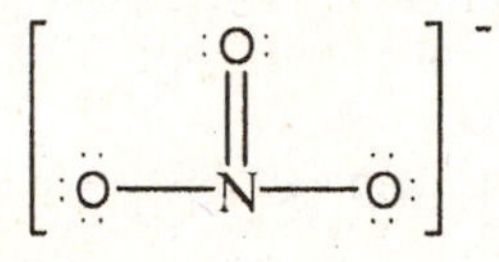

VIII. Resonance and Formal Charge

a. Some molecules have multiple equally valid Lewis structures.
   i. In nature, these molecules exist as an average between the possible structures; this average is a resonance hybrid.
b. Some molecules have more than one valid Lewis structure but the structures are not equivalent. In this case, the resonance hybrid is a weighted average favoring the more stable structures.
   i. A less stable structure will have a small contribution to the overall structure.
   ii. Formal charge can be used to determine which structure is most reasonable.
      1. Formal charge is the charge an atom would have if all of the bonds are perfectly covalent.

      Formal charge = (valence electrons)-(lone pair electrons + ½ bonding electrons)

      2. Small formal charges are preferable to large values (0 is best).
      3. Negative formal charges should be on the more EN atom.

EXAMPLE:

Determine the formal charge of each atom in the Lewis structures drawn in the previous example.

a. CH_4: Carbon has four valence electrons and is sharing 4 bonding pairs so the formal charge of carbon in CH_4 is $4-½(8) = 0$. Hydrogen has one valence electron and each is sharing a single pair of bonding electrons, so the hydrogen atoms all have a formal charge of $1-½(2)=0$.
b. H_2CO: Carbon has four valence electrons and is sharing four bonding pairs so the formal charge of carbon is $4-½(8)=0$. Hydrogen has one valence electron and each hydrogen atom is sharing a single pair of bonding electrons, so the hydrogen atoms both have a formal charge of $1-½(2)=0$. Oxygen has six valence electrons, has two lone pairs, and is sharing two bonding pairs so the formal charge of oxygen is $6-\{4+½(4)\}=0$.

c. SO_2: Sulfur has six valence electrons, is sharing six bonding electrons, and has one lone pair so the formal charge of sulfur is 6-{2+½(6)}=+1. The two oxygen atoms in this molecule are different because one is bonded to silicon with a single bond and one is bonded with a double bond. The single bonded oxygen has six valence electrons, three lone pairs, and one bonding pair so the formal charge is 6-{6+ ½(2)}=-1. The double bonded oxygen has six valence electrons, two lone pairs, and two bonding pairs so its formal charge is 6-{4+ ½(4)}=0. Note that the sum of the formal charges is equal to zero, which is consistent with the neutral charge on this molecule.

d. NO_3^-: Nitrogen has five valence electrons and is sharing four electron pairs so the formal charge of nitrogen is 5-½(8)=+1. There are two different types of oxygen atoms: two with a single bond and one with a double bond. The oxygen atom that is double bonded to nitrogen has six valence electrons, two lone pairs, and two bonding pairs so the formal charge is 6-{4+½(4)}=0. The two oxygen atoms that are bonded to nitrogen with a single bond have six valence electrons, three lone pairs, and one bonding pair so they have a formal charge of 6-{6+½(2)}=-1. Note that the sum of the formal charges is equal to -1, which is consistent with the charge of the ion.

EXAMPLE:

Draw resonance structures for SiO_2 and NO_3^-:

For SiO_2, we have two equivalent resonance structures since there are two identical oxygen atoms:

O═Si—O: ⟷ :O—Si═O

Notice that the oxygen atoms now have a –1/2 formal charge on average.

For NO_3^-, there will be three equivalent resonance structures since there are three identical oxygen atoms:

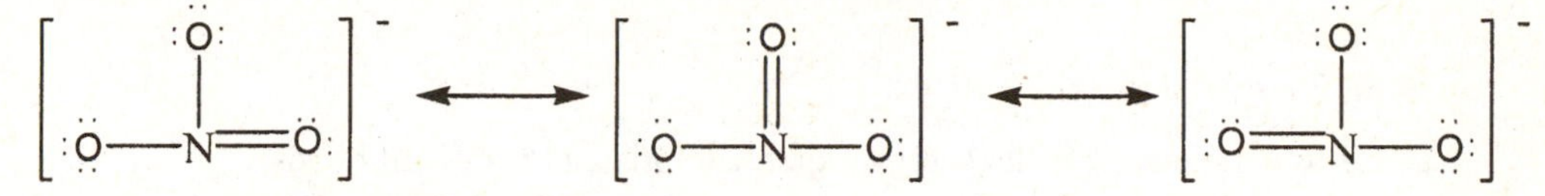

Notice that the oxygen atoms now have a -2/3 formal charge on average.

IX. Exceptions to the Octet Rule: Odd Electron Species, Incomplete Octets, and Expanded Octets

a. Free radicals have an odd number of electrons.

   i. Few radical compounds exist in nature and most are very reactive.

b. Some elements exist in compounds with fewer than eight electrons.

   i. Boron tends to have three electron pairs around it.

c. Some elements can have more than eight electrons around them in compounds.

   i. Only elements in the third row and higher (n>2) can have expanded octets since there are d orbitals available.

EXAMPLE:

Draw Lewis structures for the following molecules:

a. XeF_4

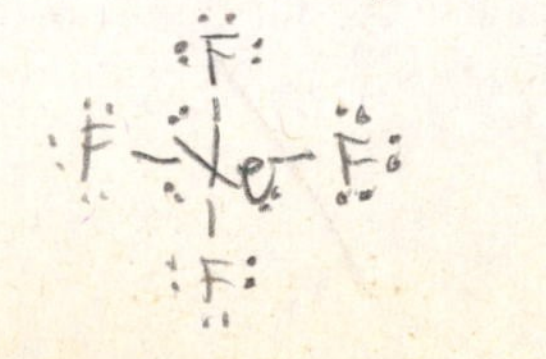

We follow the same four-step procedure as above with additional electron pairs added to xenon instead of fluorine because xenon is the fifth row of the periodic table:

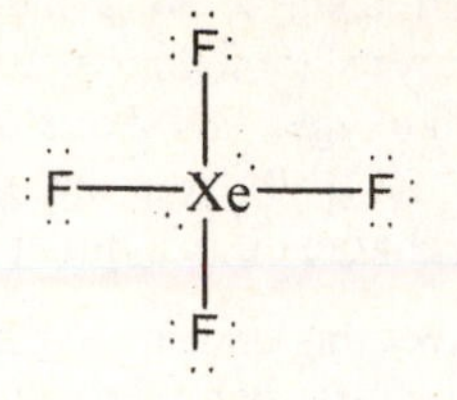

b. BH_3

Recognizing that boron can have fewer than eight electrons around it, we draw the structure:

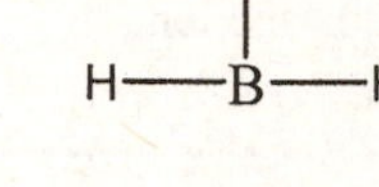

c. SCl_4

Again, we have more valence electrons than can be accommodated using the octet rule. Sulfur is the less EN atom and is the central atom of this compound; the extra pair of electrons will also be situated on the sulfur atom as halogens do not have more than eight electrons (unless they are the central atom):

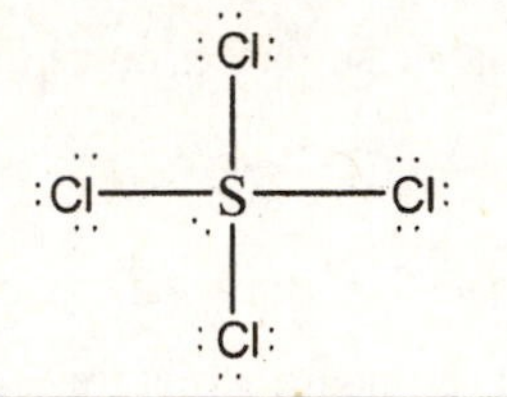

X. Bond Energies and Bond Lengths

a. The bond energy is the energy required to break one mole of bonds.

   i. Bond energy is always positive – energy is always required to break a bond.

   ii. The average bond energy is the average energy of a particular bond in many compounds.

   iii. Double and triple bonds have energies that are independent from and considerable greater than single bond energies.

b. The standard enthalpy change of a reaction can be estimated using individual bond energies.

ΔH = (energy required to break reactant bonds) - (energy released in making product bonds)

   i. A reaction is exothermic when weak bonds are broken and stronger bonds are formed.

   ii. A reaction is endothermic when strong bonds are broken and weaker bonds are formed.

EXAMPLE:

Estimate the enthalpy change for the combustion of methane using average bond energies listed in Table 9.3 of your textbook.

First we will write the balanced chemical reaction:

$$CH_4(g) + 2O_2(g) \rightarrow CO_2(g) + 2H_2O(g)$$

Now we will make a list of the bonds being broken along with the energy used for each:

4 C-H bonds in each methane molecule: $4 \times +414$ kJ/mol = +1656 kJ

1 O=O bond in each of two oxygen molecules: $2 \times +498$ kJ/mol = +996 kJ

The bonds being formed will have negative enthalpy change values:

2 C=O bonds in each carbon dioxide molecule: 2×-799 kJ/mol = -1598 kJ

2 O-H bonds in each of two water molecules: 4×-464 kJ/mol = -1856 kJ

The sum of these values gives the enthalpy change of the reaction:

$$\Delta H = (1656 + 996 + -1598 + -1856) \text{ kJ} = -802 \text{ kJ}$$

The negative sign is expected since combustion reactions release heat (are exothermic).

---

c. The average bond length is the average distance between two atoms in a large number of compounds.
    i. The bond length of a single bond is longer than a double bond, which is longer than a triple bond.

XI. Bonding in Metals: The Electron Sea Model

a. Metal atoms bond together to form a solid; in these metallic solids each metal atom donates its valence electrons to form a sea of electrons.
    i. Metals conduct electricity because electrons in the sea can move.
    ii. Metals conduct heat because thermal energy can easily be dispersed by motion of the sea of electrons.
    iii. Metals are malleable and ductile because there are no localized bonds.

Fill in the Blank Problems:

1. The potential energy of like charges is ________________, and it ________________ as the particles get farther apart.
2. The ability of an atom to attract electrons to itself in a chemical bond is called ________________.
3. Atoms that have a full outer shell have a(n) ________________ of electrons.
4. Bonds that occur between elements with large electronegativity differences are ________________.
5. When electrons are transferred from one species to another, the ions form ________________ bonds.
6. A(n) ________________ is the sharing or transfer of electrons to attain stable electron configurations.
7. ________________ represent molecules in which valence electrons are represented as dots.
8. A pair of electrons that is shared between two atoms is called a ________________.
9. The energy associated with forming a crystalline lattice of alternating positively and negatively charged ions is called the ________________.
10. The ________________ of an atom is the charge it would have if all bonding electrons were shared equally between the atoms.

11. The _________________ is a hypothetical series of steps that represent the formation of an ionic compound from its constituent elements.

12. The __ _______________ is the energy required to break 1 mole of the bond in the gas phase.

13. The average length of a bond between two particular atoms in a large number of compounds is the _________________.

14. A (n) _________________ bond is one in which electrons are shared unequally.

15. A s tream of water is bent by an electric field because water is a(n) _________________ molecule.

16. ____ _______________ is the most electronegative element.

17. ____ _______________ are species that have an odd number of electrons.

18. Expanded octets occur for elem ents in the _________________ row or below.

19. The lattice energ y _________________ when the size of the cation increases.

20. When t hree electron pairs are shared between two atoms, a _________________ forms.

Problems:

1. Each of the following substances contains both ionic and covalent bonds; indicate which bond is which, draw the Lewis structures for each, and assign formal charges to all atoms.
   a. $BaCO_3$
   b. KNO_3
   c. Na_2SO_4
   d. LiIO
   e. $Mg(ClO_3)_2$
2. Calculate the electronegativity difference, dipole moment, and percent ionic character for each of the bonds in the polyatomic ions from Question 1. Use average bond lengths from Table 9.4 in your book.
3. Draw resonance structures where appropriate for the polyatomic ions in Question 1.
4. Calculate the enthalpy change for the following reactions using the bond enthalpies given in Table 9.3 in your textbook. State whether the reactions are endothermic or exothermic.
   a. $H_2CO_3(g) \rightarrow H_2O(g) + CO_2(g)$
   b. $CO_2(g) + 4H_2(g) \rightarrow 2H_2O(g) + CH_4(g)$
   c. $2HNO_3(g) \rightarrow H_2O(g) + N_2O_5(g)$
5. Calculate the standard enthalpy change for each reaction in problem 4 using the standard enthalpies of formation from Appendix II. Compare the values that you calculated in problem 4 to the newly calculated standard enthalpy changes and suggest reasons for the differences.
6. Benzene, C_6H_6, is a six-membered ring with alternating double bonds.

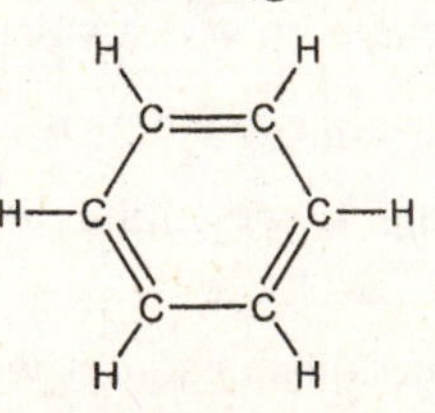

   a. What are the formal charges of all atoms in benzene?
   b. Draw all resonance structures for benzene.

   c. A carbon-carbon single bond has a bond length of 154 pm and a carbon-carbon double bond has a bond length of 134. Estimate the bond length(s) for all carbon-carbon bonds in benzene.

7. Calculate the lattice energy for the formation of the following ionic compounds (pertinent enthalpy change values can be found in your book).

   a. MgO

   b. RbCl

   c. CsI

Concept Questions:

1. In the beginning of this chapter, you were told that there is competition between attractions and repulsions when forming a bond. We can combine these statements into a single graph of potential energy:

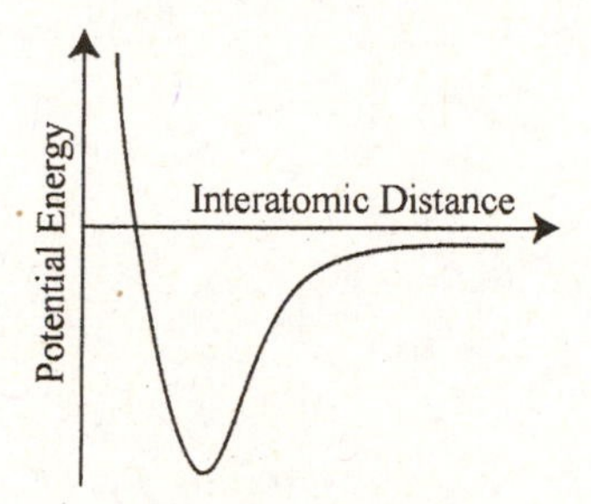

   a. Identify the region that represents the repulsions of electrons.

   b. Identify the region that represents the attraction of electrons to protons.

   c. Identify the region that represents the average bond energy.

   d. Notice that the potential energy approaches zero as the distance between atoms increases. Explain this behavior.

2. On page 371 of your textbook, you learned that ionic size and ion charge both have a large impact on the magnitude of the lattice energy. Consider the magnitude of the change for each factor and explain why the charge has a larger influence than size alone. You need to consider the equation for the Coulombic potential and any other factors that might be important.

3. In Chapter 2, we defined ionic substances as bonding between metals and nonmetals. In this chapter, we define an ionic bond as having an electronegativity value difference of greater than 2.0. Explain why both definitions are correct.

4. Carbon is the central atom for life. In fact, the chemistry of carbon-based compounds is the focus of an entire branch of chemistry called organic chemistry. Based on what you have learned about bonding in this chapter, why do you think that carbon is such a good candidate for the formation of so many different compounds?

5. As discussed in this chapter, the size of an atom and its electronegativity value are correlated. Explain this based on the principles that you learned in this chapter and the last.

6. We learned in this chapter that the least electronegative elements (except hydrogen) are most likely the central atom in a Lewis structure. Explain the physical basis of this rule of thumb.

7. Using the principle that electrons are shared in bonds, explain why boron commonly has only three bonds in stable compounds.

8. Certain elements can form molecules with an expanded octet.

   a. Which elements can have an expanded octet?

b. Why do you expect that only certain elements can have an expanded octet? Base your answer on the principles of quantum mechanics that you learned.

9. Water can be formed by reacting hydrogen gas and oxygen gas together. Using the fact that this reaction is exothermic and the principles of bond energies discussed in this chapter, discuss the relative bond strengths of the reactants and products in this reaction.

Chapter 10: Chemical Bonding II: Molecular Shapes, Valence Bond Theory, and Molecular Orbital Theory

Learning Objectives:

- Predict electron and molecular geometries using VSEPR theory.
- Determine molecular polarity.
- Use valence bond theory to explain molecular shape.
- Identify hybridized orbitals and determine the hybridization of the central atom in a molecule.
- Determine bond order and magnetic properties using molecular orbital theory.
- Construct molecular orbitals for diatomic molecules containing atoms in the first two rows of the periodic table.
- Understand electron delocalization and its implications.

Chapter Summary:

In this chapter, you will learn to predict three-dimensional shapes of molecules using three increasingly complex (and increasingly thorough) bonding theories. The discussion begins with the most simple of theories to explain molecular geometry, valence shell electron pair repulsion (VSEPR) theory. VSEPR theory uses the repulsions of electrons to explain electronic and molecular geometries. Using these geometries, you will learn how to determine if a given molecule is polar. A slightly more complex approach to bond formation, valence bond theory, will next be introduced in order to link the quantum mechanical orbitals that you learned about in Chapter 8 to the molecular shapes of VSEPR theory. In valence bond theory, atomic orbitals are hybridized in bonding and these hybrid orbitals lead to predictable shapes and relative energies. Through the discussion of hybrid orbitals, you will learn about different bonding orientations, sigma and pi, and relate these bonding types to physical properties of molecules. The most complex bonding theory treated here will be molecular orbital theory which is discussed in the final third of this chapter. Molecular orbital theory, you will see, is a more rigorous quantum mechanical description of bonding where electrons are described using molecular orbitals instead of atomic orbitals. You will see how this theory can be used to determine bond orders and magnetic properties (something that the other theories cannot do). Finally, electron delocalization in polyatomic atoms will be introduced.

Chapter Outline:

I. Artificial Sweeteners: Fooled by Molecular Shape
   a. Many biological processes are dependent on molecular shape in order to work.

II. VSEPR Theory: The Five Basic Shapes
   a. Valence shell electron pair repulsion (VSEPR) theory is a theory for molecular shapes based on the idea that electron groups repel one another.
      i. The shape of a molecule depends on the total number of groups around a central atom and whether those groups are lone pairs or bonding pairs.
   b. A linear geometry occurs when there are two electron groups around the central atom.
      i. The angle between groups is 180°.
   c. A trigonal planar geometry occurs when there are three electron groups around the central atom.
      i. The angle between groups is approximately 120°.

            1. Double bonds repel single bonds more than single bonds repel each other. For example, the angle between the double and single bond in a molecule that is trigonal planar is 121.9° and the angle between the single bonds is 116.2°.

    d. A tetrahedral geometry occurs when there are four electron groups around the central atom.

        i. The angle between the groups is approximately 109.5°.

    e. A trigonal bipyramidal geometry occurs when there are five electron groups around the central atom.

        i. There are two different "types" of electron groups in this geometry: the three that define a plane (equatorial) and the two that are perpendicular to that plane and are 180° from one another (axial).

            1. The angle between the equatorial groups is approximately 120°.

            2. The angle between the equatorial and axial groups is approximately 90°.

    f. An octahedral geometry occurs when there are six electron groups around the central atom.

        i. The angle between groups is approximately 90°.

III. VSEPR Theory: The Effect of Lone Pairs

    a. The electron geometry is the geometrical arrangement of electron groups around an atom.

    b. The molecular geometry is the geometrical arrangement of atoms around a central atom – the molecular geometry is related to the electron geometry but they are not always the same because of lone pairs.

    c. Lone pairs are less confined in space than bonding pairs are.

        i. Lone pairs exert a larger repulsion than bonding pairs do.

        ii. The angles in a molecule with lone pairs are not always the same as in a molecule with only bonding pairs – the geometry is distorted.

            1. For example, in the bent molecular geometry with trigonal planar electron geometry, the angle between bonds is <120°.

    d. The table below lists all of the possibilities for geometric arrangements.

        i. Note that the electron geometry is solely determined by the number of electron pairs while the molecular geometry depends on the number of lone pairs.

        ii. The wedge and dash method has been used to show three-dimensional structure.

        iii. A solid wedge indicates that the atom/electron pair is coming out of the page and a dash indicates that the atom/electron pair is going back into the page.

Electron Pairs	Electron Geometry	Bonding Pairs	Lone Pairs	Formula	Molecular Geometry	Example
2	Linear	2	0	AX_2	Linear	Cl——Be—

3	Trigonal Planar	3	0	AX_3	Trigonal Planar	BF_3 (F, B, F, F)
		2	1	AX_2	Bent	$[O\text{–}N\text{–}O]^-$
4	Tetrahedral	4	0	AX_4	Tetrahedral	CH_4 (H, C, H, H, H)
		3	1	AX_3	Trigonal Pyramidal	NH_3 (H, N, H, H)
		2	2	AX_2	Bent	H_2O (H, O, H)
5	Trigonal Bipyramidal	5	0	AX_5	Trigonal Bipyramidal	PF_5 (F, P, F, F, F, F)
		4	1	AX_4	See-Saw	SF_4 (F, S, F, F, F)
		3	2	AX_3	T-Shaped	ClF_3 (F, Cl, F, F)
		2	3	AX_2	Linear	XeF_2 (F, Xe, F)
6	Octahedral	6	0	AX_6	Octahedral	SF_6 (F, S, F, F, F, F, F)
		5	1	AX_5	Square Pyramidal	IF_5 (F, I, F, F, F, F)
		4	2	AX_4	Square Planar	XeF_4 (F, Xe, F, F, F)

EXAMPLE:

Predict the electron and molecular shapes for the following molecules:

a. CF_4

First we will draw the Lewis structure using the rules from Chapter 9:

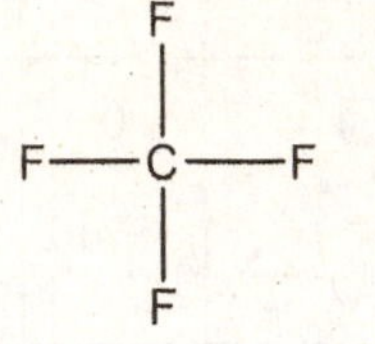

We see that the central atom (carbon) has four electron pairs around it, so it has a tetrahedral electron geometry. Since there are no lone pairs, this molecule also has a tetrahedral molecular geometry.

b. IBr_3

The Lewis dot structure is:

Br
:I—Br
Br

We see that the central atom (iodine) has five electron pairs around it, so it has a trigonal bipyramidal electron geometry. Since two of these electron pairs are lone pairs, the molecular geometry is t-shaped.

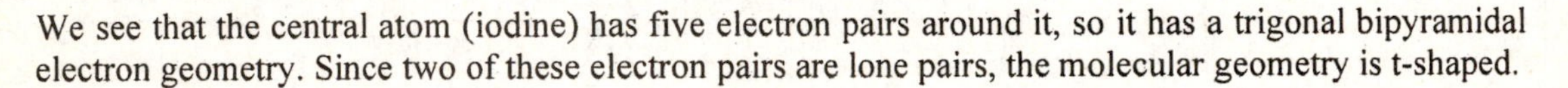

c. CO_2

The Lewis dot structure is:

$O=C=O$

The central atom (carbon) has two electron groups around it so the electron geometry is linear. Since both electron groups are bonding groups, the molecular geometry is also linear.

d. PH_3

The Lewis structure is:

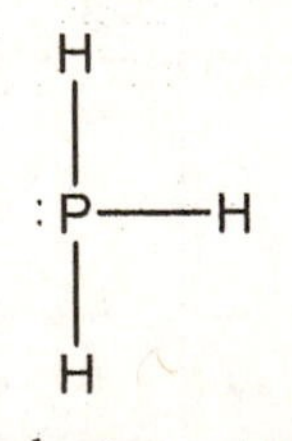

The central atom (phosphourus) has four electron groups around it so the electron geometry is tetrahedral. Since there is one lone pair, the geometry is trigonal pyramidal.

e. $AlCl_3$

The Lewis structure is:

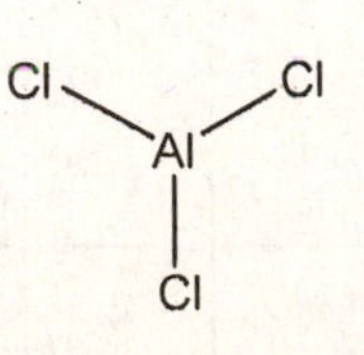

The central atom (aluminum) has three electron groups around it so the electron geometry is trigonal planar. Since there are no lone pairs on aluminum, the molecular geometry is also trigonal planar.

IV. VSEPR Theory: Predicting Molecular Geometries

a. Molecules that have more than one interior atom (no single central atom) will have shapes that are defined by the geometries of each interior atom.

EXAMPLE:

Predict the shape of each interior atom in the molecule shown below:

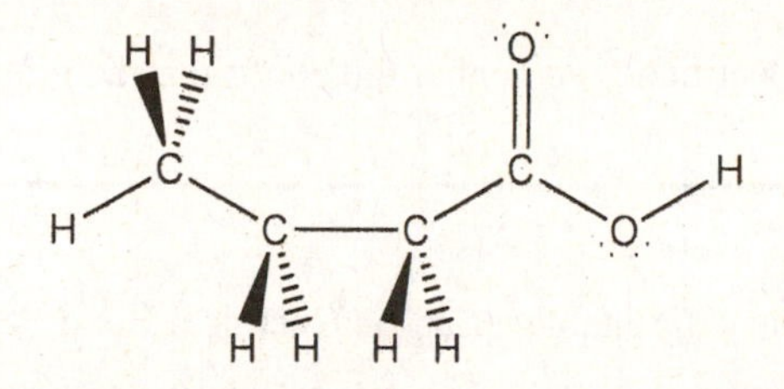

In order from left to right we can see that the first three carbon atoms are tetrahedral. The fourth carbon atom is trigonal planar. The last atom (oxygen) is bent.

V. Molecular Shape and Polarity

a. The polarity of a molecule depends on the polarity of its bonds and the molecular shape.

i. Dipole moments are directional, so they can add together or cancel each other out in a molecule.

ii. Look for symmetry in a molecule – a perfectly symmetric molecule will not have a dipole even if the bonds in it are polar.

EXAMPLE:

Which of the following molecules is polar?

a. SO_2

The Lewis structure is:

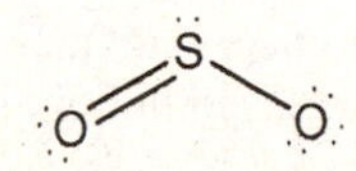

Oxygen is more electronegative than sulfur so the sulfur-oxygen bonds are polar. There will be a net dipole pointing down.

b. ClO_3^-

The Lewis structure is:

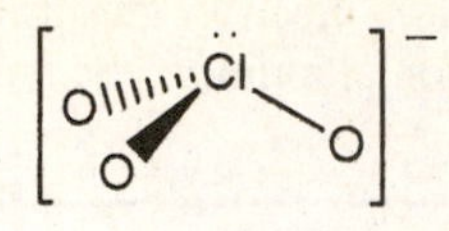

Oxygen is more electronegative than chlorine so each of the three bonds is polar. Since the molecule is asymmetrical, there will be a net dipole pointing down.

c. O_3

The Lewis structure for ozone is:

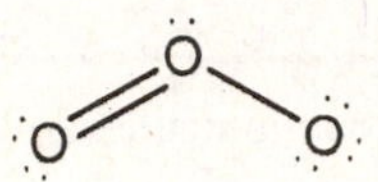

These are all perfectly covalent bonds so neither the bonds nor the molecule is polar.

VI. Valence Bond Theory: Orbital Overlap As Chemical Bond

a. Valence bond theory as discussed here is a qualitative discussion of quantum mechanical orbitals applied to bonding in molecules.

b. Electrons are treated as if they exist in atomic orbitals that are localized on individual atoms.

i. The atomic orbitals can be the simple s, p, d, or f orbitals discussed in Chapter 7 or they can be hybrid orbitals, discussed shortly.

c. When atoms approach each other, the electrons and nuclei interact.

i. According to valence bond theory, bonds occur when half-filled orbitals overlap with each other or when filled orbitals overlap with empty orbitals.

ii. The molecular geometry is a result of the orbital geometry.

VII. Valence Bond Theory: Hybridization of Atomic Orbitals

a. Standard atomic orbitals (s, p, d, and f) do not account for the bonding and shapes of all molecules.

b. Hybridization is a mathematical procedure in which standard atomic orbitals are combined to form hybrid orbitals.

i. The energy of the bond is minimized by constructing hybrid orbitals that overlap maximally.

1. It costs energy to make hybrid orbitals; they will only form if the energy released in bonding is increased.

2. The type of hybrid orbital that forms is the one that gives the lowest energy upon bond formation.

ii. We will assume that only the central atom in a molecule is hybridized.

iii. The total number of hybrid orbitals is equal to the total number of atomic orbitals combined to form them.

iv. The shapes and energies of hybrid orbitals are related to the shapes and energies of the atomic orbitals that comprise them.

c. sp^3 orbitals are made by combining one s orbital and three p orbitals:

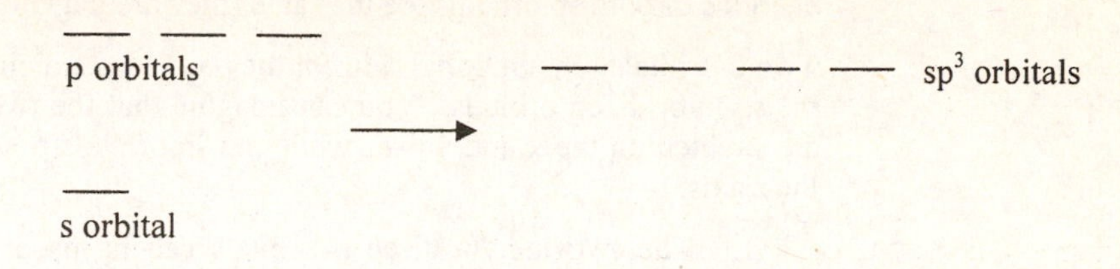

i. The energy of an sp^3 orbital is between the energy of the s and p orbitals and the energies of all four sp^3 orbitals is the same (they are degenerate).

ii. The geometry of the four sp^3 orbitals is tetrahedral; i.e., each orbital protrudes out into one of the four corners of a tetrahedron.

d. sp^2 orbitals are made by combining one s orbital and two p orbitals:

p orbitals — unhybridized p orbital

sp^2 orbitals

s orbital

i. The energy of an sp^2 orbital is between the energy of the s and p orbitals and the energies of all three sp^2 orbitals is the same (they are degenerate).

ii. One p orbital remains unhybridized and is perpendicular to the plane defined by the three sp^2 hybridized orbitals, which are oriented so that a lobe points in each corner of an equatorial triangle (trigonal planar arrangement)

iii. An unhybridized p orbital can be used to form a double bond.

1. Double bonds are the result of two bonding interactions: a sigma bond and a pi bond.

   a. A pi (π) bond occurs via the side to side overlap of atomic orbitals. Pi bonds form from overlapping p orbitals.

   b. A sigma (σ) bond occurs via the end to end overlap of atomic orbitals. Sigma bonds form from overlapping atomic or hybrid orbitals.

2. Double bonds do not rotate because in order to do so, the side to side overlap of the p orbitals must be disrupted – i.e., a pi bond would need to be broken.

   a. The rigidity results in cis/trans isomers with substituents on different sides of a double bond.

e. sp hybrid orbitals are made by combining one s and one p orbital:

p orbitals — unhybridized p orbitals

sp orbitals

s orbital

i. The energy of an sp orbital is between the energy of the s and p orbitals and the energies of both sp orbitals are the same (they are degenerate).

ii. Two p orbitals remain unhybridized; they are perpendicular to each other and to the sp hybridized orbitals. You can imagine that the two unhybridized orbitals are situated on the x and y axes while the sp orbitals point in either direction of the z axis.

1. The two unhybridized p orbitals can be used to form a triple bond, which is one σ bond and two π bonds.

f. sp^3d hybrid orbitals are made by combining one s, three p, and one d orbital:

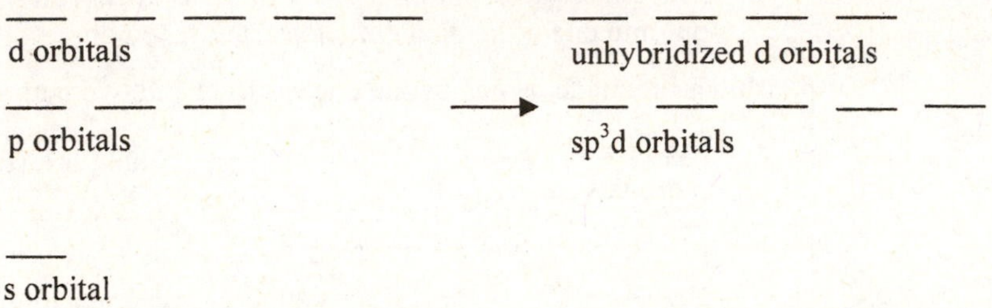

___ ___ ___ ___ ___ d orbitals → ___ ___ ___ ___ unhybridized d orbitals

___ ___ ___ p orbitals → ___ ___ ___ ___ ___ sp^3d orbitals

___ s orbital

i. The energies of the sp^3d orbitals are the same (they are degenerate) and four unhybridized d orbitals remain.

ii. The geometry of the sp^3d orbitals are in accordance with the trigonal bipyramidal arrangement.

g. sp^3d^2 hybrid orbitals are made by combining one s, three p, and two d orbitals:

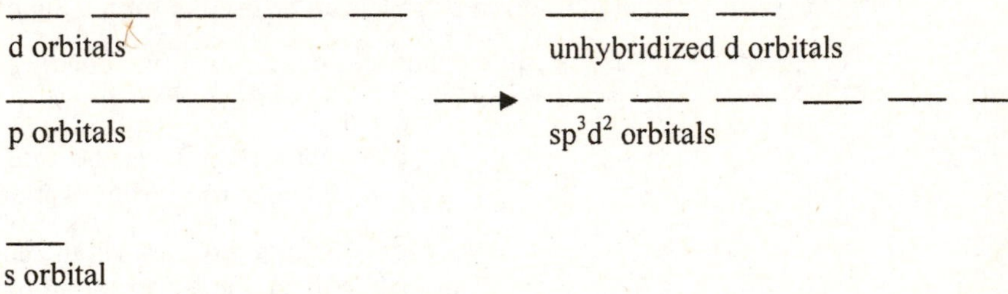

___ ___ ___ ___ ___ d orbitals → ___ ___ ___ unhybridized d orbitals

___ ___ ___ p orbitals → ___ ___ ___ ___ ___ ___ sp^3d^2 orbitals

___ s orbital

i. The energies of the sp^3d^2 orbitals are the same (they are degenerate) and three unhybridized d orbitals remain.

ii. The geometry of the sp^3d^2 orbitals are in accordance with the octahedral arrangement.

h. The particular hybridization can easily be determined by finding the electron geometry (using VSEPR) and then assigning an appropriate hybridization.

EXAMPLE:

What is the hydridization of each of the central atoms in the first example given in this chapter.

a. CF_4

This molecule has a tetrahedral electron geometry which corresponds to sp^3 hybridization.

b. IBr_3

This molecule has a trigonal bipyramidal electron geometry which corresponds to sp^3d hybridization.

c. CO_2

This molecule is linear with two double bonds which corresponds to sp hybridization.

d. PH_3

This molecule has a tetrahedral electron geometry which corresponds to sp^3 hybridization.

e. $AlCl_3$

This molecule has a trigonal planar electron geometry which corresponds to sp^2 hybridization.

VIII. Molecular Orbital Theory: Electron Delocalization

a. Valence bond theory treats electrons as if they exist on particular atoms and the best orbital will be the one with the lowest energy.

b. Molecular orbital theory uses orbitals that allow electrons to be distributed over the entire molecule.

c. Molecular orbitals are constructed using linear combinations of atomic orbitals.

i. All of the atomic orbitals in a molecule are used to construct molecular orbitals.

ii. Atomic orbitals are combined in an additive way (constructive interference of waves) to give bonding orbitals.

iii. Atomic orbitals are combined in a subtractive way (destructive interference of waves) to give anti-bonding orbitals.

iv. The total number of molecular orbitals is equal to the total number of bonding orbitals used to construct them.

d. The 1s orbitals on two atoms combine constructively to give σ_{1s} (bonding) and σ^*_{1s} (anti-bonding) orbitals.

e. The bond order of a molecule is:

$$\text{bond order} = \frac{(\text{number of electrons in bonding orbitals}) - (\text{number of electrons in antibonding orbitals})}{2}$$

A bond order greater than zero indicates that it is energetically favorable for a bond to form.

EXAMPLE:

What is the bond order of H_2^+?

We can construct the molecular orbitals from the two 1s orbitals:

σ^*_{1s} [ ]

σ_{1s} [↿]

We have two valence electrons total, one from each hydrogen atom and one electron is removed to form the cation. The bond order is thus ½ because only one electron is in the bonding orbital and there are zero electrons in the anti-bonding orbital.

---

f. 2s orbitals combine to form σ_{1s} and σ^*_{1s} orbitals.

g. 2p orbitals combine to form σ_{2p}, σ^*_{2p}, $2\,\pi_{1p}$, and $2\,\pi^*_{1p}$ orbitals.

i. The energetic ordering of the bonding and anti-bonding orbitals depends on which atoms are combined to form the molecular orbitals.

1. For B_2, C_2, and N_2:

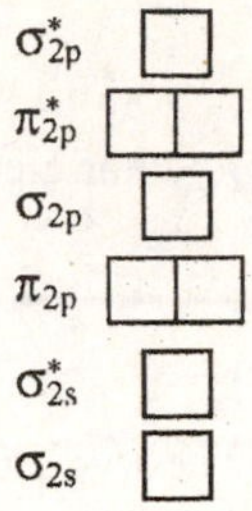

σ^*_{2p} □
π^*_{2p} □□
σ_{2p} □
π_{2p} □□
σ^*_{2s} □
σ_{2s} □

2. For O_2, F_2, and Ne_2:

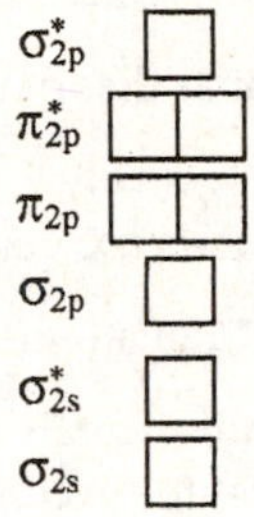

σ^*_{2p} □
π^*_{2p} □□
π_{2p} □□
σ_{2p} □
σ^*_{2s} □
σ_{2s} □

h. Electrons fill molecular orbitals from low to high energy with unpaired, parallel spins in degenerate orbitals (just as they fill atomic orbitals).

i. Paramagnetic species have unpaired electrons while diamagnetic species have all electrons paired.

i. When constructing heteronuclear diatomic molecules, the energy of the orbitals must be taken into account.

i. The more electronegative atom will have a lower energy and therefore it will make a greater contribution to the molecular orbital(s).

1. Electrons will be more localized on the more electronegative orbital.

ii. Nonbonding orbitals are atomic orbitals that are not used in the construction of molecular orbitals.

1. Nonbonding orbitals occur because there is poor overlap of the atomic orbitals from the two atoms.

EXAMPLE:

What is the bond order of N_2?

We use the first of the MO ordering schemes from above and fill in the ten valence electrons (five from each nitrogen atom) to give:

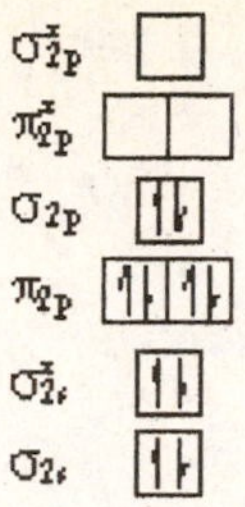

σ^*_{2p} [ ]
π^*_{2p} [ ][ ]
σ_{2p} [↑↓]
π_{2p} [↑↓][↑↓]
σ^*_{2s} [↑↓]
σ_{2s} [↑↓]

There are eight bonding electrons and two anti-bonding electrons so the bond order is:

$$\text{bond order} = \frac{(8) - (2)}{2} = 3$$

This is consistent with the Lewis structure that we would draw for N_2 (triple bond).

---

j. Polyatomic molecules can have molecular orbitals that allow electrons to be delocalized over an entire molecule.

Fill in the Blank Problems:

1. The hybridization of the central atom in formaldehyde COH_2 is ________________; there are ________________ sigma and ________________ pi bonds.
2. A molecule with unpaired spins is ________________.
3. There are ________________ degenerate sp^3d hybridized orbitals.
4. Central atoms that are surrounded by three electron groups have a(n) ________________ electron geometry.
5. Electron groups repel each other through ________________ forces.
6. A net dipole occurs from the ________________ distribution of electrical charge.
7. Sigma bonds result from ________________ overlap and pi bonds result from ________________ overlap.
8. Anti-bonding orbitals result from the ________________ interference of atomic orbitals.
9. The number of standard atomic orbitals that hybridize ________________ the number of hybrid orbitals formed.
10. In molecular orbital theory, electrons are ________________ over the entire molecule instead of being confined to atomic orbitals.
11. The angles bet ween atoms in a trigonal pyramidal geometry are ________________.

Problems:

1. Determine the electron geometry, molecular geometry, orbital hybridization, bond polarity, molecule polarity, and bond angles in the following molecules; then label all bonds as π or σ in each:

   a. N_2H_4

   b. HCN

   c. CF_3^-

   d. SiO_4^{4-}

e. PO_4^{3-}

f. CS_2

g. XeF_4

h. OCN^-

i. SO_2Cl_2

j. SCN^-

k. CH_3OH

l. CH_3OCH_3

m. BrI_3

2. Consider two molecules with carbon-carbon double bonds: $CH_2{=}C{=}CH_2$ and $CH_2{=}C{=}C{=}CH_2$.
   a. Draw these molecules and determine the hybridization at each carbon.
   b. Indicate the number of pi bonds and sigma bonds in each molecule.
   c. Sketch the orbitals for each molecule showing how they overlap with one another.
   d. Will the CH_2 groups at the ends of each molecule be parallel or perpendicular to each other?

3. Draw the molecular orbital diagrams for N_2, N_2^+, N_2^-.
   a. Which of these molecules is most stable?
   b. Which of these molecules is least stable?
   c. What is the bond order for each molecule?
   d. Which of these will be paramagnetic?

4. Draw the molecular orbital diagram for CO.
   a. CO is isoelectronic with N_2. Explain what this means in terms of a molecule.
   b. Do you expect the CO bond to be stronger, weaker, or the same strength as the N_2? Justify your answer.
   c. Look up the bond energies for CO and N_2 and compare them to your expectation. If the values do not agree with your prediction, explain the difference.

5. Draw the cis/trans isomers for difluoroethene CHFCHF.
   a. Determine the geometry at each carbon.
   b. Determine the hybridization of each carbon atom.
   c. Will either (or both) molecule have a dipole moment? Explain.

Concept Questions:

1. Electron energy varies with distance from the nucleus as we saw in our discussion of quantum mechanics. The energy also depends on the confinement of the electron: the more constrained an electron is, the higher its energy will be. Explain how this relates to delocalization and why pi bonding leads to a more energetically stable complex.

2. In benzene, there are six p orbitals that can be used to make the p bonding and anti-bonding orbitals; the lowest energy bonding orbital is shown on page 451 of your textbook. We know, from our discussion of hybrid orbitals, that when six atomic orbitals combine, they form six molecular orbitals. We also know that anti-bonding orbitals occur when there is a node (zero electron density) between atoms. Use this information to predict how the p orbitals combine in benzene to give six molecular orbitals. Hint: draw an outline of benzene and start introducing nodes and place p orbitals accordingly.

Chapter 11: Liquids, Solids, and Intermolecular Forces

Learning Objectives:

- Compare and contrast the properties of liquids, solids, and gases.
- Understand and identify the various types of intermolecular forces.
- Explain surface tension, viscosity, and capillary action using intermolecular forces.
- Understand the connections between intermolecular forces, vapor pressure, and phase change temperatures.
- Utilize the heats of vaporization and fusion in heating curve calculations.
- Sketch and identify important regions in phase diagrams.
- Identify and characterize unit cells for crystalline substances.
- Understand and identify the various types of crystalline solids as well as the physical properties associated with each.
- Use band theory to explain the conductivity of metals.

Chapter Summary:

This chapter is focused on the interactions between ions, atoms, and molecules. You will learn how these interactions lead to the phases of matter and the energy associated with changing from one phase to another. First, you will learn about the various types of intermolecular forces such as dispersion forces, dipole-dipole forces, hydrogen bonding, and ion-dipole forces. Your understanding of intermolecular forces will be used to explore properties such as surface tension, viscosity, and capillary action. Vapor pressure will then be explored as an indicator of the strength of intermolecular forces and the connection between temperature and vapor pressure will be quantified using the Clausius-Clapeyron equation. After a discussion of boiling and condensation you will learn about sublimation and fusion. With your knowledge of phase transitions, you will learn how to interpret and sketch the heating curve and phase diagram for a substance. Finally, you will explore the details of solid substances including definitions of the crystalline lattice and fundamental types of solids.

Chapter Outline:

I. Climbing Geckos and Intermolecular Forces
   a. Intermolecular forces are attractive forces that exist between all molecules.
   b. The state of matter depends on the magnitude of intermolecular forces relative to the thermal energy of the system.

II. Solids, Liquids, and Gases: A Molecular Comparison
   a. Gases, liquids, and solids differ in the freedom of movement of constituent atoms, ions, or molecules.
      i. Gases assume the shape and volume of their containers, are low in density, and are compressible.
      ii. Liquids have high densities compared to gases, assume the shape of their containers, and are not easily compressed.
      iii. Solids have high densities compared to gases, have definite volume and shape, are not compressible, and may be crystalline (ordered) or amorphous (disordered).

b. The phase of matter changes when the temperature and/or pressure changes.

   i. The stronger the intermolecular forces are, the higher the temperature needs to be in order to overcome them.

III. Intermolecular Forces: The Forces That Hold Condensed Phases Together

a. Intermolecular forces originate from interactions between charges, partial charges, and temporary charges between atoms, ions, and molecules.

   i. Just as bonds form in order to lower the potential energy of atoms, molecules interact in order to lower their collective potential energy.

   ii. The energy associated with intermolecular forces is smaller than the energy between atoms in a bond because the magnitudes of the charges are smaller and the distance between the particles is larger.

b. Dispersion forces (London forces) result from fluctuations in the electron distribution within molecules or atoms.

   i. All species have electrons and therefore all species experience dispersion forces.

   ii. Instantaneous or temporary dipoles that result from fluctuations will induce similar dipoles in surrounding species.

   iii. The magnitude of dispersion forces depends on the polarizability of an atom or molecule.

      1. The larger the atom, the more polarizable the particle is.

      2. Dispersion forces increase with increasing molar mass.

   iv. The stronger the dispersion forces are between species, the higher the melting and boiling points of that substance will be.

c. Dipole-dipole forces exist for polar molecules with permanent dipoles.

   i. The positive end of one molecule is attracted to the negative end of another molecule.

   ii. Dipole-dipole forces are stronger than dispersion forces when comparing species with the same molar mass; dispersion forces can become more important than dipole-dipole interactions when very large molecules are considered.

   iii. The larger the dipole moment of a molecule, the stronger the intermolecular forces will be and the higher the melting and boiling point will be.

   iv. Polarity influences the ability of two substances to form a homogeneous mixture (miscibility) – polar substances are miscible with other polar substances and non-polar substances are miscible with other non-polar substances.

d. Hydrogen bonding occurs for small electronegative atoms bonded to hydrogen.

   i. Hydrogen bonding is an extreme example of dipole-dipole interactions.

      1. Hydrogen bonding occurs in molecules that have HF, HO, and HN bonds.

   ii. Hydrogen bonds are stronger than dispersion forces and dipole-dipole forces.

e. Ion-dipole forces occur when an ionic compound is mixed with a polar compound.

   i. These are the strongest of the intermolecular forces discussed so far.

   ii. Ion-dipole forces are important for the solubility of ionic compounds in water.

1. Ionic species dissolve in water because the anions are surrounded by the positive end of water and the cations are surrounded by the negative end of water.

EXAMPLE:

Identify the type of intermolecular forces that dominate in the following examples.

a. Dichloromethane, CH_2Cl_2

In dichloromethane, there are two polar C-Cl bonds. Since the structure of the molecule is asymmetric, the molecule will have a net dipole and the individual molecules will orient themselves so that the dipoles align. In a sample of dichloromethane, the dipole-dipole interactions dominate.

b. Octane, $CH_3CH_2CH_2CH_2CH_2CH_2CH_2CH_3$

Octane has C-H bonds that are slightly polar, but because the molecule is symmetrical, these tiny dipoles cancel and the molecule is nonpolar. Octane will have strong dispersion forces because it is so large.

c. Octanol, $CH_3CH_2CH_2CH_2CH_2CH_2CH_2CH_2OH$

Octanol has a polar O-H group that can potentially form hydrogen bonds with other octanol molecules. Since it is so large, however, the dispersion forces will be the dominant intermolecular interactions.

d. Sodium chloride in water, NaCl(aq)

Sodium chloride dissociates into ions when dissolved in water. The sodium and chloride ions will interact strongly with the water through ion-dipole interactions.

e. Hydrogen fluoride, HF

Hydrogen fluoride forms a hydrogen bonding network since the H-F bond is very polar.

IV. Intermolecular Forces in Action: Surface Tension, Viscosity, and Capillary Action

a. Surface tension is the energy required to increase surface area by a unit amount.

i. Surface tension is the tendency of a liquid to minimize its surface area.

ii. Surface tension decreases as the intermolecular forces decrease.

b. Viscosity is the resistance of a liquid to flow.

i. The units of viscosity are poise (1 P = 1g/cm•s). Units of centiPoise (cP) are often used because the viscosity of water at room temperature is about 1 cP.

ii. The stronger the intermolecular forces are, the higher the viscosity will be.

iii. Viscosity also depends on shape since long or branched molecules can become tangled in each other.

iv. Viscosity is temperature dependent and generally decreases with increasing temperature.

c. Capillary action is the ability of a liquid to flow against gravity up a narrow tube.

i. Capillary action depends on the balance between adhesive forces (forces of attraction between the tube and the substance) and cohesive forces (intermolecular forces).

1. If the adhesion is greater than the cohesion, the liquid will spontaneously move against gravity up a tube.

2. If the cohesion is greater than the adhesion, the liquid will spontaneously move out of the tube when it is placed in the liquid.

ii. The meniscus of a liquid in a tube is an indicator of its capillary action.

V. Vaporization and Vapor Pressure

a. Vaporization is the escape of molecules from the liquid phase into the gas phase.

b. Higher temperatures lead to a higher average energy of a collection of molecules.

i. In vaporization, faster moving molecules escape from the liquid surface, where there are fewer intermolecular forces.

ii. In condensation, slower moving molecules in the gas phase are captured by the intermolecular forces of the liquid.

c. Increasing the temperature increases the number of molecules with enough energy to overcome the intermolecular forces.

d. Weaker intermolecular forces allow more molecules to escape from the liquid phase at a given temperature.

i. A volatile substance has weak intermolecular forces and therefore is vaporized at lower temperatures and pressures than a nonvolatile substance.

e. The heat of vaporization (ΔH_{vap}) is the amount of heat required to vaporize one mole of a liquid to a gas.

i. ΔH_{vap} is always positive since energy is always required for vaporization.

ii. Vaporization is always endothermic and condensation is always exothermic.

EXAMPLE:

Calculate the quantity of methane (in grams) that you would need to burn in order to boil 100.0 g of water. The heat of combustion for methane is -802.5 kJ/mol and the heat of vaporization for water is 40.7 kJ/mol.

We will first calculate the heat needed by the water in order for it to boil:

$$100.0\,\text{g}\,H_2O \times \frac{1\,\text{mol}\,H_2O}{18.02\,\text{g}\,H_2O} \times \frac{40.7\,\text{kJ}}{1\,\text{mol}\,H_2O} = 225.8\,\text{kJ}$$

Now we will calculate the quantity of methane that is needed in order to obtain this heat:

$$225.8\,\text{kJ} \times \frac{1\,\text{mol}\,CH_4}{802.5\,\text{kJ}} \times \frac{16.04\,\text{g}\,CH_4}{1\,\text{mol}\,CH_4} = 4.51\,\text{g}\,CH_4$$

So 4.51 g of methane needs to be burned in order to boil 100.0 g of water.

---

f. Vapor pressure is the pressure of a gas in dynamic equilibrium with its liquid.

i. Dynamic equilibrium occurs when the rate of condensation is equal to the rate of evaporation.

ii. Increasing the pressure above a liquid disturbs the equilibrium and results in condensation in order to reduce the pressure and reestablish equilibrium.

g. Boiling occurs when the vapor pressure of a substance is equal to the external pressure.

i. The normal boiling point is the temperature at which the vapor pressure is equal to 1 atm.

ii. The temperature of a liquid cannot be higher than the boiling point since all of the added energy goes into the phase transition.

h. The vapor pressure of a substance is temperature dependent:

$$P_{vap} = \beta \exp\left(-\frac{\Delta H_{vap}}{RT}\right)$$

where β is a constant that depends on the identity of the gas, R is the gas constant, and T is the absolute temperature.

i. This expression can be rearranged to give the Clausius-Clapeyron equation, which demonstrates a linear relationship between $\ln(P_{vap})$ and $1/T$:

$$\ln P_{vap} = -\frac{\Delta H_{vap}}{R}\frac{1}{T} + \ln \beta$$

ii. A two-point form of the Clausius-Clapeyron equation allows one to determine the vapor pressure of a substance at any temperature if the normal boiling point and the heat of vaporization are known:

$$\ln \frac{P_2}{P_1} = -\frac{\Delta H_{vap}}{R}\left(\frac{1}{T_2} - \frac{1}{T_1}\right)$$

EXAMPLE:

What is the normal boiling point of acetone if the vapor pressure at 20.0°C is 181.7 mmHg and the heat of vaporization is 29.1 kJ/mol?

We use the two-point Clausius-Clapeyron equation, rearranged to solve for the temperature:

$$\frac{1}{T_1} - \frac{R}{\Delta H_{vap}} \ln \frac{P_2}{P_1} = \frac{1}{T_2}$$

$$T_2 = \frac{1}{\frac{1}{T_1} - \frac{R}{\Delta H_{vap}} \ln \frac{P_2}{P_1}}$$

Now we can include the values given in the problem with the temperature converted to kelvin units, the pressure at the normal boiling point of 760 mmHg, and the heat of vaporization in J/mol:

$$T_2 = \frac{1}{\frac{1}{293\text{ K}} - \frac{8.314 \frac{\text{J}}{\text{mol}\cdot\text{K}}}{29100\text{ J/mol}} \ln \frac{760\text{ mmHg}}{181.7\text{ mmHg}}} = 332.9\text{ K}$$

So the normal boiling point is 59.7°C.

---

iii. In a closed container, the heating of a liquid will cause a gas to form; as the gas forms, the pressure increases, which tends to cause the gas to condense. Eventually, a new phase, a supercritical fluid, will form.

1. Supercritical fluids have properties between those of liquids and gases.
2. The temperature above which a liquid cannot exist at any pressure is called the critical temperature.

3. The critical pressure is the pressure required to bring about a transition to the liquid phase at the critical temperature.

VI. Sublimation and Fusion

a. Sublimation is the phase transition from solid to gas and deposition is the phase transition from gas to solid.

b. Melting is the phase transition from solid to liquid and freezing (or fusion) is the phase transition from liquid to solid.

   i. At the melting point, the molecules or atoms have sufficient thermal energy to overcome the intermolecular forces that hold them at stationary points.

   ii. Solids cannot be at temperatures higher than the melting point just as liquids cannot exist above the boiling temperature.

c. The heat of fusion (ΔH_{fus}) is the amount of heat required to melt one mole of a solid.

   i. Freezing is always exothermic and melting is always endothermic.

d. In general, $\Delta H_{vap} > \Delta H_{fus}$ because intermolecular forces must be totally overcome in order to transition to the gas phase.

VII. Heating Curve for Water

a. The heating curve for water is a graphical representation of how the temperature of water changes as heat is added:

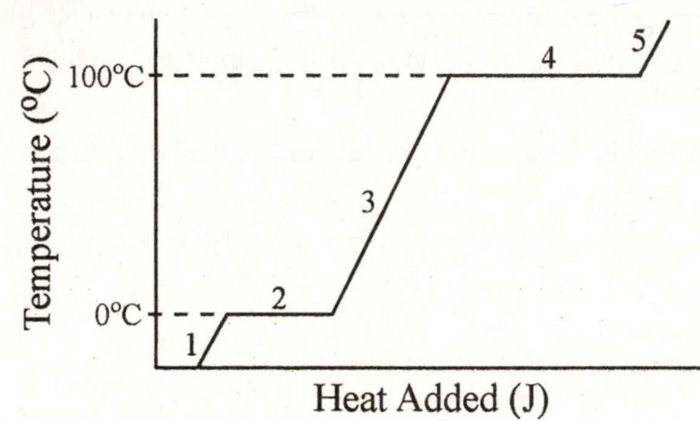

b. The heat added in each region of the curve can be calculated systematically.

   i. In region 1, ice is being heated; the added heat is dependent on the mass of ice, the specific heat of ice, and the temperature change:

$$q = m \cdot C_s \cdot \Delta T$$

   ii. In region 2, the ice is melting to form liquid water; the added heat is related to the quantity (in mol) and the heat of fusion:

$$q = n \cdot \Delta H_{fus}$$

   iii. In region 3, liquid water is being heated; the added heat is dependent on the mass of the liquid water, the specific heat of liquid water, and the temperature change:

$$q = m \cdot C_s \cdot \Delta T$$

   iv. In region 4, the liquid water is boiling to form water vapor; the added heat is related to the quantity (in mol) and the heat of vaporization:

$$q = n \cdot \Delta H_{vap}$$

v. In region 5, the water vapor is being heated; the added heat is dependent on the mass of water vapor, the specific heat of water vapor, and the temperature change:

$$q = m \cdot C_s \cdot \Delta T$$

EXAMPLE:

Calculate the amount of heat required to convert 10.0 g of ice at -10.°C to steam at 110.°C. The heat capacity of ice is 2.09 J/g·°C, the heat capacity of water is 4.184 J/g·°C, and the heat capacity of steam is 2.09 J/g·°C. The heat of fusion is 6.02 kJ/mol and the heat of vaporization is 40.7 kJ/mol.

We will calculate the heat required for each of the five regions of the heating curve.

First we calculate the heat needed to warm the ice from -10°C to the melting point at 0°C:

$$q = m \cdot C_{s,ice} \cdot \Delta T = 10.0\text{ g} \cdot 2.09\text{ J/ g}\cdot°\text{C} \cdot [0\text{-}(\text{-}10)]°\text{C} = 209\text{ J}$$

Next we calculate the heat required to convert the ice into liquid water:

$$q = n \cdot \Delta H_{fus} = 10.0\text{ g} \cdot \frac{1\text{ mol H}_2\text{O}}{18.02\text{ g H}_2\text{O}} \cdot 6.02\text{ kJ/mol} = 3.34\text{ kJ} \cdot \frac{1000\text{ J}}{1\text{ kJ}} = 3340\text{ J}$$

Now we calculate the heat needed to warm the liquid water at 0°C to the boiling point at 100°C:

$$q = m \cdot C_{s,liquid} \cdot \Delta T = 10.0\text{ g} \cdot 4.184\text{ J/ g}\cdot°\text{C} \cdot (100\text{-}0)°\text{C} = 4184\text{ J}$$

Now we calculate the heat required to vaporize the liquid water:

$$q = n \cdot \Delta H_{vap} = 10.0\text{ g} \cdot \frac{1\text{ mol H}_2\text{O}}{18.02\text{ g H}_2\text{O}} \cdot 40.7\text{ kJ/mol} = 22.59\text{ kJ} \cdot \frac{1000\text{ J}}{1\text{ kJ}} = 22590\text{ J}$$

Finally, we calculate the heat needed to warm the steam from 100°C to 110°C:

$$q = m \cdot C_{s,steam} \cdot \Delta T = 10.0\text{ g} \cdot 2.09\text{ J/ g}\cdot°\text{C} \cdot (110\text{-}100)°\text{C} = 209\text{ J}$$

The total heat required is the sum of these:

$$q_{total} = 209\text{ J} + 3340\text{ J} + 4184\text{ J} + 22590\text{ J} + 209\text{ J} = 30532\text{ J}$$

So 3.05×10^4 J are required to convert ice at -10°C to steam at 110°C.

---

c. Notice that the temperature does not change in regions 2 and 4, meaning that the kinetic energy of the molecules is constant in these regions.

d. The kinetic energy of a given phase is related to the heat capacity of that phase.

VIII. Phase Diagrams

a. A phase diagram is a map of the phase of a substance as a function of temperature and pressure:

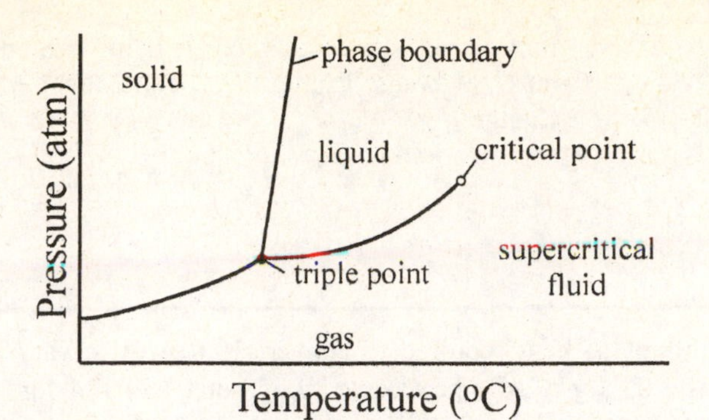

i. There are three main regions in the phase diagram: solid, liquid, and gas.
    1. The phase diagram shows the conditions where a particular phase is stable.

ii. Each line in a phase diagram is a set of temperatures and pressure at which the substance is in equilibrium between the two phases that are on either side of the line.

iii. The triple point is the set of conditions at which three phases are in equilibrium.

iv. The critical point shows the temperature and pressure above which the substance is a supercritical fluid.

b. Movement in the phase diagram represents a change in the temperature and/or pressure.

IX. Water: An Extraordinary Substance

a. Water has a high boiling point for its molar mass.

b. Water has a permanent dipole due to its bent geometry. The large dipole moment results in strong hydrogen bonds.

c. Water has a high heat capacity.

d. Water is unique for having a negative slope between the solid and liquid phase.

i. The negative slope means that liquid water is denser than solid water, which is due to the extended hydrogen bonding network in liquid water.

ii. Most organisms cannot survive freezing due to the expansion of water upon freezing.

X. Crystalline Solids: Determining Their Structure by X-Ray Crystallography

a. The long range order (arrangement) of solids can be studied using x-ray crystallography.

b. When x-rays are directed at crystals, they bounce back and result in a pattern of constructive and destructive interference on a screen.

i. The pattern of constructive and destructive interference is called a diffraction pattern.

ii. This pattern can be used to determine the location of atoms in the crystal using Bragg's law:

$$n\lambda = 2d \cdot \sin\theta$$

where d is the distance between atomic layers, λ is the wavelength of the incident light, n is the integer number of wavelengths, and θ is the angle of reflection.

EXAMPLE:

Find the distance between atomic layers of an unknown crystal given that the angle of maximum reflection is 42.1° when 184 pm light is shone upon it. Assume that n=1.

We rearrange the Bragg's law expression to solve for d:

$$d = \frac{n\lambda}{2 \cdot \sin\theta}$$

$$d = \frac{1 \cdot 184\,\text{pm}}{2 \cdot \sin(42.1)} = 137\,\text{pm}$$

XI. Crystalline Solids: Unit Cells and Basic Structures

a. A crystalline lattice is the regular arrangements of atoms in a crystalline solid.

b. A unit cell is a small collection of atoms or ions.

c. A lattice point is a point in space that is occupied by an atom or ion.

d. The coordination number is the number of atoms with which each atom is in direct contact.

e. The packing efficiency is the percentage of the volume of a cubic cell that is occupied by spheres (representing atoms or ions).

f. There are three main unit cells:

  i. The simple cubic unit cell is a cube that has an atom at each corner.

    1. One complete atom is found in each simple cubic unit cell.

    2. The coordination number is 6 and the packing efficiency is 52%.

  ii. The body-centered cubic unit cell is a cube with an atom at each corner and one in the center.

    1. Two complete atoms are found in each body-centered cubic unit cell.

    2. The coordination number is 8 and the packing efficiency is 68%.

  iii. The face-centered cubic unit cell is a cube with an atom at each corner, one in the center and one in the center of each face of the cube.

    1. Four complete atoms are found in each face-centered cubic unit cell.

    2. The coordination number is 12 and the packing efficiency is 74%.

EXAMPLE:

The density of iron solid is 7.87 g/mL. Calculate the radius of an iron atom given that it has a body-centered cubic closest-packing structure.

First we can calculate the mass of a single unit cell which is simply the number of atoms in a unit cell (2 in a body-centered cubic arrangement) and the mass of a single atom:

$$2\ \text{Fe atoms} \times \frac{1\ \text{mol Fe atoms}}{6.022 \times 10^{23}\ \text{Fe atoms}} \times \frac{55.85\ \text{g Fe}}{1\ \text{mol Fe atoms}} = 1.855 \times 10^{-22}\ \text{g Fe}$$

We can next determine the volume occupied by a single Fe atom using the mass and density:

$$1.855\times10^{-22}\ \text{g Fe}\times\frac{1\ \text{ml}}{7.87\ \text{g Fe}}\times\frac{1\ \text{cm}^3}{1\ \text{mL}}\times\left(\frac{1\ \text{m}}{100\ \text{cm}}\right)^3=2.357\times10^{-29}\ \text{m}^3$$

The edge length for a body-centered cubic unit cell is given in terms of the atomic radius:

$$\ell=\frac{4r}{\sqrt{3}}$$

Since the length is the cube root of the volume of a unit cell, we can solve for the radius:

$$\sqrt[3]{V}=\frac{4r}{\sqrt{3}}$$

$$r=\frac{\sqrt{3}}{4}\sqrt[3]{V}=\frac{\sqrt{3}}{4}\sqrt[3]{2.357\times10^{-29}}=1.24\times10^{-10}\ \text{m}$$

Converting to a more standard atomic radius unit of pm gives our solution:

$$1.24\times10^{-10}\ \text{m}\times\frac{10^{12}\ \text{pm}}{1\ \text{m}}=124\ \text{pm}$$

---

g. Closest-packed structures are the ways in which atoms or ions form layers stacked on one another. We can represent the layer types as A, B, and C:
   i. Simple cubic packing involves one layer directly on top of another: AAAAA...
   ii. Hexagonal closest-packing involves shifting one layer by ½ an atom from the layer below it. The third layer is the same as the first: ABABAB...
   iii. Cubic closest-packing involves shifting one layer by ½ of an atom from the layer below it and the layer below that. The fourth layer is the same as the first: ABCABC...

XII. Crystalline Solids: The Fundamental Types

a. Molecular solids are held together by intermolecular forces and have low melting points.

b. Ionic solids are held together by coulombic interactions.
   i. The melting points of ionic solids are higher than molecular solids because of the stronger forces.
   ii. The coordination number of an ionic solid depends on the relative size of the cations and anions.
      1. A larger size difference results in a cubic cell with a smaller coordination number.

c. Atomic solids are species in which atoms are the composite unit; there are three types of atomic solids:
   i. Non-bonding ionic solids are held together by dispersion forces and have low melting points.
   ii. Metallic solids are held together by metallic bonds (as discussed in Chapter 9) and have high melting points.
   iii. Network covalent solids are held together by covalent bonds and have very high melting points.

1. Carbon forms two common covalent solids: diamond and graphite.
    a. Diamond has carbon atoms that are bonded to four other carbon atoms in a tetrahedral arrangement.
    b. Graphite has carbon atoms that are bonded to three other carbon atoms in a trigonal planar arrangement.
2. Silicates like quartz form network covalent bonds.

XIII. Crystalline Solids: Band Theory

a. When metals bond together, their atomic orbitals interact in such a way as to form delocalized molecular orbitals over the entire crystal.
    i. The bonding orbitals are collectively referred to as the valence band.
    ii. The antibonding orbitals are collectively referred to as the conduction band.
b. In general, metals can conduct electricity when electrons from the valence band can be promoted into the conduction band.
c. Species in which there is a large jump in energy between the valence band and the conduction band will be insulators (they will not conduct electricity).
d. Species in which there is a small jump in energy between the valence band and the conduction band will be semiconductors.
    i. Semiconductors act like conductors when energy is supplied to them and act like insulators when there is not sufficient energy for an electron to be promoted to the conduction band.
e. Doping involves the replacement of atoms in a metal with other atoms in order to alter the conductivity.
    i. N-type doping increases conductivity by introducing extra electrons into the semiconductor.
    ii. P-type doping introduces holes (missing electrons) into the semiconductor, which also increases conductivity.
f. A p-n junction is a spot in a semiconductor that is p-type on one side and n-type on the other side.
    i. Junctions like these are used in diodes: when the extra electron combines with the hole, energy is released as light.

Fill in the Blank Problems:

1. When the pressure of a gas in dynamic equilibrium with its liquid phase is ________________ the gas will condense to form more liquid.
2. The point in the phase diagram called the ________________ is the temperature and pressure above which a supercritical fluid exists.
3. The amount of heat required to melt one mole of a solid is called the ________________.
4. According to Coulomb's law, the potential energy between like charges ________________ with increasing distance between them.
5. Hydrogen bonds occur when hydrogen is bonded to one of three atoms: ________________.
6. The energy difference between the conduction band and the valence band in a semiconductor is called the ________________.
7. The asymmetric distribution of electrons in a molecule leads to a(n) ________________.

8. The energy required to increase a liquid's surface area by a unit amount is its ________________.
9. The ability of a liquid to flow against gravity up a narrow tube is ________________.
10. A tomic solids that are held together by covalent bonds are called ________________.
11. Polar molecule s have ________________ that interact with neighboring molecules.
12. The percentage of the volume of a unit cell occupied by spheres is called the ________________.
13. The heat change associated with vaporization is always ________________ and the heat change associated with condensation is always ________________.
14. The ________________ is the temperature at which its vapor pressure is equal to the external pressure.
15. The phase transitio n from a solid directly to a gas is called ________________.
16. The densit y of liquid water is ________________ than solid water (ice).
17. Solids in which the atoms or molecules that compose them are arranged in a well-ordered array are called ________________.
18. Dispersion forces ________________ with increasing molar mass.
19. Solids that are composed of ions are called ________________; they have ________________ melting points.

Problems:

1. Identify the dominant intermolecular force in each of the following molecules, atoms, or ions:
   a. Br_2
   b. Butane, $CH_3CH_2CH_2CH_3$
   c. Sulfuric acid solution, $H_2SO_4(aq)$
   d. Hydrogen sulfide, H_2S
   e. Methyl amine, CH_3NH_2
   f. Xenon, Xe
   g. Dimethyl ether, CH_3OCH_3
   h. Magnesium chloride, $MgCl_2(aq)$
2. Rank the following in order of increasing vapor pressure:
   a. He, NH_3, NF_3, NaCl
   b. NF_3, NCl_3, NBr_3, NH_3
   c. HF, F_2, ClF
   d. $CH_3CH_2CH_3$, $CH_3CH_2CH_2CH_2CH_2CH_2CH_2CH_3$, $CH_3CH_2CH_2CH_2CH_3$
3. Rank the following in order of increasing melting point:
   a. NaCl, $AlCl_3$, MgS, NaBr
   b. CO, CO_2, CCl_4
   c. SO_2, CO_2, O_2
   d. Al_2O_3, SiO_2, P_2O_3, SO_2
4. Consider taking a sample of dry ice, CO_2, at -35.0°C and heating it until it is a gas at 123°C. The pressure is held at 1 atm for all questions.

a. At what temperature and phase during this process will the sample have the largest amount of kinetic energy? Explain your answer.

b. At what temperature and phase during this process will the sample have the largest internal energy? Explain your answer.

c. The normal (1 atm) sublimation point for CO_2 is -78.5°C, the triple point is -56.6°C and 5.11 atm, and the critical point is 31.0°C and 73.0 atm. The solid phase is denser than the liquid phase. Sketch the phase diagram of CO_2.

d. Is it possible for the liquid phase of CO_2 to be stable at -60°C? Explain why or why not.

e. What is the vapor pressure of solid CO_2 at -78.5°C? Explain how you arrived at your answer.

f. A sample of solid CO_2 is originally at a pressure of 1 atm and a temperature of -80.0°C. The temperature is increased to 0.0°C while the pressure is held constant. Is the sample still a solid?

g. A sample of solid CO_2 is originally at a pressure of 1 atm and a temperature of -80.0°C. The temperature is increased to 0.0°C while the pressure is increased to 50. atm. Will the phase changes (if they occur) be the same as in part f? Why or why not?

5. Three 50.0-mL containers each containing an equimolar sample of one of the following gases are all held at constant temperature. Use the data below and the ideal gas law's assumption that molecules have no intermolecular forces to answer the following questions.

	Pressure of CO_2 (mmHg)	Pressure of NH_3 (mmHg)	Pressure of H_2 (mmHg)
Ideal gas prediction	489	489	489
Real gas measurement	1881	202	489

a. Explain how the intermolecular forces cause NH_3 to deviate from ideal behavior.

b. Which gas do you expect to have the highest boiling point?

c. A supercritical fluid is often modeled using the ideal gas law. The idea is that the constituents have enough kinetic energy to overcome all intermolecular forces. Use this idea to determine which gas has the lowest critical temperature.

6. Consider the phase diagram for water shown below:

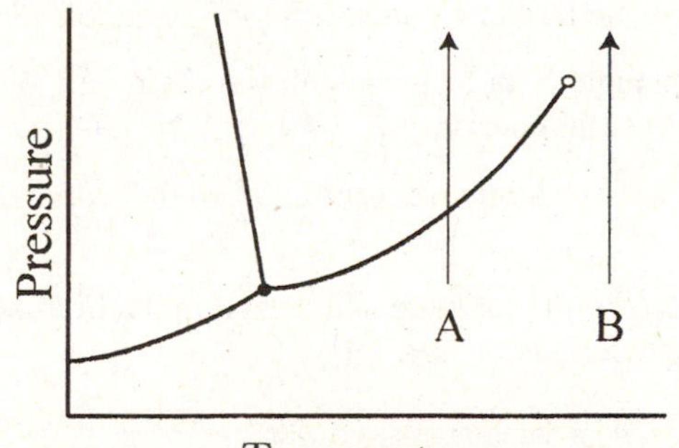

a. Indicate the liquid, solid, gas, and supercritical fluid regions on the diagram.

b. An isotherm is a curve showing a change that occurs while maintaining a constant temperature. For example, A and B are isotherms for water. Sketch a graph of the pressure vs. volume for each of these isothermal changes. Hint: you need to consider how the pressure and volume are related at a phase change.

c. Using your plots of pressure vs. volume, give another definition for an ideal gas.

7. A cooling curve indicates how the temperature of a substance changes as heat is released.

a. The vapor pressure of ammonia is 0.409 bar at -50°C. Given that ammonia's normal boiling point is -33.3°C, calculate the heat of vaporization for ammonia.

b. The melting point of ammonia is -77.3°C at 1 atm. The specific heat of liquid ammonia is 4.52 J/g·K and the specific heat of solid and gaseous ammonia is 0.028 J/g·K. Sketch the heating curve for ammonia starting with a temperature of -100.0°C and ending at a temperature of 0.0°C.

c. Calculate the amount of heat that must be supplied to a 10.0 g sample of solid ammonia at -100.0°C to convert it to gaseous ammonia at 0.0°C. The heat of fusion is 5.63 kJ/mol.

8. Calculate the amount of empty space in each of the closest-packing structures discussed in this chapter and relate this to the density of solids with each configuration. Using the internet, search for metals that utilize each of the packing arrangements and confirm or reassess your conclusion.

9. Consider a substance that has the band diagram shown below. The shading represents the occupation of the band with electrons. The lower energy band is completely filled with electrons and the higher energy band is partially filled.

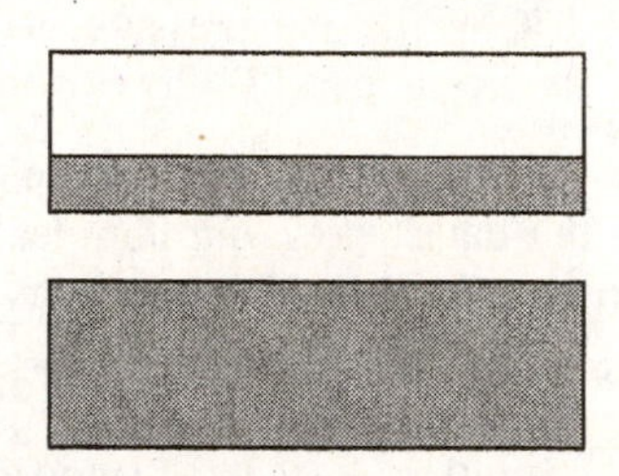

a. Based on the diagram shown, do you expect this substance to be a conductor, semi-conductor, or insulator? Explain your answer.

b. What do you expect to happen to the electrical properties of this substance if all of the electrons are removed from the conduction band? Explain your answer.

c. How would you expect the conductivity from part b to change when the temperature is increased? Explain your answer.

Concept Questions:

1. Two beakers are placed side by side in a 25°C room. One beaker contains 2.5 L of pentane, $CH_3CH_2CH_2CH_2CH_3$, and one beaker contains 2.5 L of pentanol, $CH_3CH_2CH_2CH_2CH_2OH$.

   a. If left out for long enough, both beakers will eventually empty due to evaporation. Which beaker will be empty first? Explain your answer.

   b. Which beaker will have a larger concentration of its vapor above the solution? Explain your answer.

   c. Both beakers are placed on a hot plate and heated up. Which beaker will begin boiling at a lower temperature? Explain your answer.

2. In DNA, the two strands are held together by hydrogen bonding. Explain why you think that nature utilizes hydrogen bonds instead of covalent bonds. Why do you think that dispersion forces aren't as well suited to this job?

3. Table 11.7 in your textbook shows the heats of vaporization for some common substances. Explain the trend in the heats of vaporization as they relate to the intermolecular forces that dominate in each substance.

4. Based on your understanding of intermolecular forces, do you think that a sodium chloride solution will have a higher or lower boiling point than pure water? Explain your answer.

5. The heating curve of a substance can be related to a slice of the phase diagram at a given temperature. Each region of the heating curve, therefore, corresponds to a region of the phase diagram. Identify the connections between the five regions of the heating curve and the phase diagram for water.

Chapter 12: Solutions

Learning Objectives:

- Identify the solute and solvent in a solution.
- Understand the physical process of solution formation and the energetic considerations behind it.
- Understand how temperature affects the solubility of solids and gases in a liquid solvent.
- Understand the relationship between pressure and gas solubility.
- Identify a solution as unsaturated, saturated, or supersaturated.
- Qualitatively and quantitatively express solution concentrations.
- Understand why the vapor pressure of a solution is different from the vapor pressure of a pure solvent and quantify the relationship using Raoult's law.
- Understand and calculate the freezing point depression, boiling point elevation, and osmotic pressure for a given solution.
- Identify a colloid based on particle size and understand the properties of a colloidal dispersion.

Chapter Summary:

This chapter is focused on the formation and properties of solutions. You will begin by defining a solution and identifying the different components that comprise it. You will then explore the energy associated with solution formation using the ideas of intermolecular forces discussed in Chapter 11. Unsaturated, saturated, and supersaturated solutions will be defined in order to explain solubility; changes in the solubility as a function of temperature and pressure will then be discussed. Qualitative and quantitative descriptions of solution concentration will be introduced next. With a firm grasp of basic solution properties, you will be presented with the more complex ideas of colligative properties that result from solution formation including vapor pressure depression, boiling point elevation, freezing point depression, and osmotic pressure. Finally, you will learn about colloidal dispersions which differ from traditional solutions because of the size of the particles.

Chapter Outline:

I. Thirsty Solutions: Why You Should Not Drink Seawater
   a. A solution is a homogeneous mixture of two or more substances.
      i. The majority component is the solvent and the minority component is the solute.

II. Types of Solutions and Solubility
   a. A solution is any homogeneous mixture and can be more than one gas, liquid(s) and gas(es), liquid(s) and solid(s), more than one liquid, or more than one solid.
   b. An aqueous solution is one in which water is the solvent and a solid, liquid, or gas is the solute.
   c. Solubility is the amount of a substance that will dissolve in a given amount of solvent.
   d. Solutions form because of the increased entropy that results, not solely because of a decrease in potential energy.
      i. Entropy is a measure of the energy dispersion or randomization.
   e. Whether or not a solution forms will depend on the balance between the interactions broken and interactions formed as well as the increased entropy.

i. A solution always forms when the solvent-solute interactions are stronger than solute-solute and solvent-solvent interactions; both entropy and potential energy drive the formation of this type of solution.

ii. A solution always forms when the solvent-solute interactions are about the same as the solute-solute and solvent-solvent interactions; entropy is the driving factor in this type of solution formation.

iii. A solution will sometimes form when the solvent-solute interactions are weaker than the solute-solute and solvent-solvent interactions; whether or not the solution forms will depend on how much weaker the solvent-solute interactions are.

1. A good rule of thumb is that "like dissolves like" meaning that solvents and solutions with similar intermolecular interactions will form solutions.

EXAMPLE:

Predict whether or not the following molecules will form a solution based on their intermolecular interactions.

a. $CH_3CH_2CH_2CH_3$ and CH_3OH

Butane is nonpolar and methanol is polar; these compounds will not, therefore, form a solution.

b. C_6H_6 and LiOH

Benzene is nonpolar and lithium hydroxide is ionic; these compounds will not, therefore, form a solution.

c. $CH_3CH_2CH_2CH_2OH$ and $CH_3CH_2CH_2CH_2CH_2CH_3$

Butanol has a polar end and a nonpolar carbon-hydrogen chain while hexane is nonpolar; these compounds will mix because of their similar intermolecular forces.

III. Energetics of Solution Formation

a. We can think of the formation of a solution as three energetic steps:

i. Separation of solute into particles; this requires the intermolecular forces of the solute particles to be overcome: $\Delta H_{solute}>0$.

ii. Separation of solvent particles to make room for the solute that will be introduced; again, intermolecular forces must be overcome: $\Delta H_{solvent}>0$.

iii. Enthalpy change associated with mixing; here intermolecular forces will be introduced so the enthalpy change is negative: $\Delta H_{mix}<0$.

$$\Delta H_{solution} = \Delta H_{solute} + \Delta H_{solvent} + \Delta H_{mix}$$

b. The enthalpy change associated with forming a solution will be endothermic or exothermic depending on the magnitude of the enthalpies associated with each step.

c. For an aqueous solution, the enthalpy of hydration is $\Delta H_{hydration} = \Delta H_{solvent} + \Delta H_{mix}$ and the enthalpy of the solute is $\Delta H_{solute} = -\Delta H_{lattice}$. So for aqueous solutions:

$$\Delta H_{solution} = -\Delta H_{lattice} + \Delta H_{hydration}$$

IV. Solution Equilibrium and Factors Affecting Solubility

a. Dissolution of a solid in water is an equilibrium process in which the solid and aqueous phases are in dynamic equilibrium.

   i. The rates of dissolution and deposition are equal at equilibrium.

$$AB(s) \leftrightarrows A^+(aq) + B^-(aq)$$

   ii. A saturated solution is one in which dynamic equilibrium is established; when additional solid is added to the solution it will fall to the bottom as solid.

   iii. An unsaturated solution contains less than the equilibrium amount of dissolved solute; when additional solid is added to the solution it will dissolve.

   iv. A supersaturated solution contains more than the equilibrium amount of dissolved solute; a supersaturated solution is unstable and solid will precipitate out of it spontaneously to form a saturated solution.

b. The solubility of most solids increases with increased temperature.

   i. Recrystallization is a purification technique that relies on the increased solubility of solids at higher temperatures. When the saturated solution is slowly cooled, pure crystals of the solute result.

c. The solubility of gases decreases with temperature and increases with pressure.

   i. The pressure dependence is quantified using Henry's law:

$$S_{gas} = k_H P_{gas}$$

   where S_{gas} is the solubility of the gas, k_H is Henry's law constant, and P_{gas} is the pressure of the gas above the solution.

EXAMPLE:

How many moles of gaseous nitrogen are in a 2.000 L container holding 0.523 L of a 0.00152 M solution of water and nitrogen gas at 25°C?

We need to determine the pressure of the gas that must be over the solution using Henry's law. The Henry's law constant for nitrogen is 6.1×10^{-4} M/atm:

$$0.00152\ \text{M} = (6.1\times10^{-4}\ \text{M/atm})P_{gas}$$

$$P_{gas} = 2.49\ \text{atm}$$

The pressure of the gas is related to the number of moles of gas through the ideal gas law. The volume available to the gas is the volume of the container minus the volume of the solution: 1.477 L.

$$n = \frac{PV}{RT} = \frac{(2.49\ \text{atm})(1.477\ \text{L})}{\left(0.0821\frac{\text{L}\cdot\text{atm}}{\text{K}\cdot\text{mol}}\right)(298\ \text{K})} = 0.150\ \text{mol}$$

V. Expressing Solution Concentration

a. A qualitative description of solution concentration uses terms such as dilute (low concentration) and concentrated (high concentration).

b. A quantitative description of solution concentration can be formulated in multiple ways.

i. Molarity is the amount of solute (in moles) in 1 L of solution:

$$M = \frac{\text{moles of solute}}{1\,\text{L of solution}}$$

ii. Molality is the amount of solute (in moles) in 1 kg of solvent:

$$m = \frac{\text{moles of solute}}{1\,\text{kg of solvent}}$$

iii. Percent by mass is the mass of solute over the mass of solution:

$$\text{percent by mass} = \frac{\text{g of solute}}{1\,\text{g of solution}} \times 100\%$$

iv. Percent by volume of the solute over the volume of solution and is generally used when two liquids are mixed to form a solution:

$$\text{percent by volume} = \frac{\text{L of solute}}{1\,\text{L of solution}} \times 100\%$$

v. Parts per million is the mass of the solute over the mass of the solution multiplied by 1,000,000:

$$\text{parts per million} = \frac{\text{g of solute}}{1\,\text{g of solution}} \times 1{,}000{,}000$$

vi. Parts per billion is the mass of the solute over the mass of the solution multiplied by 1,000,000,000:

$$\text{parts per billion} = \frac{\text{g of solute}}{\text{g of solution}} \times 1{,}000{,}000{,}000$$

vii. Mole fraction (χ_{solute}) is the moles of solute over total moles:

$$\chi_{\text{solute}} = \frac{\text{mol of solute}}{1\,\text{mol of solution}}$$

viii. Mole percent is the mole fraction expressed as a percentage:

$$\text{mole percent} = \chi_{\text{solute}} \times 100\%$$

EXAMPLE:

A sodium chloride solution is prepared by dissolving 2.48 g of NaCl in 425 mL of water. The solution has a density of 1.0067 g/mL. Calculate the following concentration units assuming that no volume change occurs when the salt is added to the water.

a. molarity

We need to convert the grams of NaCl given to moles of NaCl:

$$2.48\,\text{g NaCl} \times \frac{1\,\text{mol NaCl}}{58.44\,\text{g NaCl}} = 0.0424\,\text{mol NaCl}$$

This is the number of moles of solute in 0.425 L of solution, so the molarity is:

$$M = \frac{\text{moles of solute}}{1\,\text{L of solution}} = \frac{0.0424\,\text{mol NaCl}}{0.425\,\text{L}} = 0.0999\,\text{M}$$

b. molality

Instead of dividing the moles of solute by the volume of solution, we will convert the solution volume to the mass of solution (in kg):

$$425\text{ mL solution} \times \frac{1.0067\text{ g solution}}{1\text{ mL solution}} \times \frac{1\text{ kg}}{1000\text{ g}} = 0.428\text{ kg solution}$$

We divide the moles of solute by the mass of the solution in order to calculate molality:

$$m = \frac{\text{moles of solute}}{1\text{ kg of solution}} = \frac{0.0424\text{ mol NaCl}}{0.428\text{ kg of solution}} = 0.0991\,m$$

c. percent by mass

The mass of the NaCl is given as 2.48 g and the mass of the solution is 428 g. The percent by mass is:

$$\text{percent by mass} = \frac{\text{g of solute}}{\text{g of solution}} \times 100\% = \frac{2.48\text{ g NaCl}}{428\text{ g solution}} \times 100\% = 0.579\%$$

d. parts per million

Again we divide the solute mass by the solution mass and multiply by 1,000,000:

$$\text{parts per million} = \frac{\text{g of solute}}{\text{g of solution}} \times 1{,}000{,}000 = \frac{2.48\text{ g NaCl}}{428\text{ g solution}} \times 1{,}000{,}000 = 5790\text{ ppm}$$

VI. Colligative Properties: Vapor Pressure Lowering, Freezing Point Depression, Boiling Point Elevation, and Osmotic Pressure

a. Properties that depend on the amount of solute and not on the solute type are called colligative properties.

i. Vapor Pressure of Solutions

1. When a non-volatile solute is dissolved in a solution, the vapor pressure of the solution is lower than the vapor pressure of the pure solvent.

a. We can explain this by considering that fewer solvent molecules are at the surface of the liquid. Since only particles at the surface are able to escape into the gas phase, a reduced number of solvent particles at the surface results in fewer molecules that can escape the solution.

b. We can also explain this using the tendency of particles to mix. The increase in entropy that results when a solution forms causes the solution to be more stable than the pure solvent so more energy is required to remove the solvent from the liquid phase.

2. Vapor pressure lowering (ΔP) is the difference between the vapor pressure of the solvent and the vapor pressure of the solution:

$$\Delta P = P^{\circ}_{solvent} - P_{solution}$$

3. Raoult's law quantifies the vapor pressure of a solution using the mole fraction:

$$P_{\text{solution}} = \chi_{\text{solvent}} P^{\circ}_{\text{solvent}}$$

where P° is the vapor pressure of the pure substance.

4. Ionic solutes dissociate into ions so that the number of particles present in solution is higher than the number of solute particles dissolved.

EXAMPLE:

What is the vapor pressure of a solution prepared by dissolving 42.5 g of K_2O in enough water to prepare 1.25 L of solution at 25°C? Assume that the density of the solutions is 1.00 g/mL and that the vapor pressure of pure water is 23.8 torr at 25°C.

We were given the solute mass, so we will calculate the solution mass using the density:

$$1.25\,\text{L solution} \times \frac{1000\,\text{mL}}{1\,\text{L}} \times \frac{1\,\text{g solution}}{1\,\text{mL solution}} = 1250\,\text{g solution}$$

The mass of the solvent is the difference between the mass of the solution and the mass of the solute. The mass of the solvent is then: 1250g-42.5g = 1207.5 g solvent.

Converting both the solute and solvent masses to moles gives:

$$42.5\,\text{g}\,K_2O \times \frac{1\,\text{mol}\,K_2O}{94.20\,\text{g}\,K_2O} \times \frac{3\,\text{mol ions}}{1\,\text{mol}\,K_2O} = 1.3535\,\text{mol ions}$$

Notice that in order to calculate the moles of solute in the solution, we had to convert from the ionic formula to the number of ions using a conversion factor.

$$1207.5\,\text{g}\,H_2O \times \frac{1\,\text{mol}\,H_2O}{18.016\,\text{g}\,H_2O} = 67.0238\,\text{mol}\,H_2O$$

Now we will calculate the mole fraction of the solute:

$$\chi_{\text{solute}} = \frac{\text{moles ions}}{\text{moles ions} + \text{moles of } H_2O} = \frac{1.3535\,\text{mol ions}}{1.3535\,\text{mol ions} + 67.0238\,\text{mol}\,H_2O} = 0.0198$$

The mole fraction of water (the solvent) is simply one minus the mole fraction of ions (solute):

$$\chi_{H_2O} = 1 - \chi_{\text{solute}} = 1 - 0.0198 = 0.9802$$

We can now calculate the vapor pressure of the solution using Raoult's law:

$$P_{\text{solution}} = \chi_{\text{solvent}} P^{o}_{\text{solvent}} = (0.9802)(23.8\,\text{torr}) = 23.3\,\text{torr}$$

5. Ideal solutions are those that obey Raoult's law exactly. In these solutions, the energy associated with the solvent-solute interactions are the same as solvent-solvent and solute-solute interactions. For a two-component system:

$$P_A = \chi_A P^{o}_A$$

$$P_B = \chi_B P^{o}_B$$

$$P_{\text{total}} = P_A + P_B$$

6. Deviations from ideal behavior occur because the interactions between the solvent and solute are not equal to solute-solute or solvent-solvent interactions.

a. A positive deviation from ideal behavior occurs when the solute-solvent interactions are weaker than solute-solute and solvent-solvent interactions.

b. A negative deviation from ideal behavior occurs when the solute-solvent interactions are stronger than solute-solute and solvent-solvent interactions.

EXAMPLE:

Calculate the total pressure above 1.0 L of a 1.42 M CH_3OH solution at 25°C. The vapor pressure of pure methanol at 25°C is 127.1 torr and the vapor pressure of pure water is 23.8 torr at 25°C. You may assume that the density of the solution is 1.00 g/mL.

We first need to calculate the mass of the solute in the solution so that we can determine the mass of the solvent. We will assume that we have 1.00 L of the solution.

$$1.00\text{ L solution} \times \frac{1.42\text{ mol } CH_3OH}{1\text{ L solution}} \times \frac{32.04\text{ g } CH_3OH}{1\text{ mol } CH_3OH} = 45.50\text{ g } CH_3OH$$

We find the mass of the solution using the density:

$$1.00\text{ L solution} \times \frac{1000\text{ mL}}{1\text{ L}} \times \frac{1\text{ g solution}}{1\text{ mL solution}} = 1000\text{ g solution}$$

The mass of the solvent is the difference between these values, so there are 954.50 g solvent.

We will now calculate the moles of the solute and moles of solvent:

$$1.00\text{ L solution} \times \frac{1.42\text{ mol } CH_3OH}{1\text{ L solution}} = 1.42\text{ mol } CH_3OH$$

$$954.50\text{ g } H_2O \times \frac{1\text{ mol } H_2O}{18.02\text{ g } H_2O} = 52.97\text{ mol } H_2O$$

Now we will calculate the mole fraction of each:

$$\chi_{CH_3OH} = \frac{\text{moles of } CH_3OH}{\text{moles of } CH_3OH + \text{moles of } H_2O} = \frac{1.42\text{ mol } CH_3OH}{1.42\text{ mol } CH_3OH + 52.97\text{ mol } H_2O} = 0.0261$$

The mole fraction of water is simply one minus the mole fraction of methanol:

$$\chi_{H_2O} = 1 - \chi_{CH_3OH} = 1 - 0.0261 = 0.9739$$

We can now calculate the vapor pressure of the solute and solvent using Raoult's law:

$$P_{H_2O} = \chi_{H_2O} P^{o}_{H_2O} = (0.9739)(23.8\text{ torr}) = 23.2\text{ torr}$$

$$P_{CH_3OH} = \chi_{CH_3OH} P^{o}_{CH_3OH} = (0.0261)(127.1\text{ torr}) = 3.32\text{ torr}$$

The total pressure is the sum of the partial pressures, so:

$$P_{total} = P_{CH_3OH} + P_{H_2O} = (3.32 + 23.2)\text{ torr} = 26.4\text{ torr}$$

---

ii. Freezing Point Depression, Boiling Point Elevation, and Osmosis

1. Because the vapor pressure of a solution is lower than that of the solvent, the freezing point will be lower and the boiling point will be higher for a solution.

   a. The freezing point depression depends on the solvent properties:

$$\Delta T_f = m \times K_f$$

   where m is the molality of the solution and K_f is the freezing point depression constant for the solvent.

EXAMPLE:

Calculate the freezing point of a solution that results when 2.5 g of sugar ($C_6H_{12}O_6$) is dissolved in 500 mL of water. Note that this is the same solution as in the previous example.

We already calculated the molality, so we can calculate the change in the freezing point directly using the freezing point depression constant for water which is 1.86°C/*m*:

$$\Delta T_f = m \times K_f = (0.02775\,m) \times (1.86^{\circ}\,C/m) = 0.0516\ ^{\circ}C$$

The normal freezing point of water is 0.00°C, so the freezing point of the solution is 100.05°C.

---

2. The boiling point elevation also depends on a constant:

$$\Delta T_b = m \times K_b$$

   where m is the molality of the solution and K_b is the boiling point elevation constant for the solvent.

EXAMPLE:

Calculate the boiling point of a solution that results when 2.5 g of sugar ($C_6H_{12}O_6$) is dissolved in 500 mL of water.

We will first calculate the molality of the solution:

$$\frac{2.5\,g\,C_6H_{12}O_6}{500\,mL\,H_2O} \times \frac{1\,mol\,C_6H_{12}O_6}{180.16\,g\,C_6H_{12}O_6} \times \frac{1\,mL\,H_2O}{1\,g\,H_2O} \times \frac{1000\,g}{1\,kg} = 0.02775\,m\,C_6H_{12}O_6 \text{ solution}$$

The boiling point elevation constant for water is 0.512°C/*m*, so the change in the boiling point is:

$$\Delta T_b = m \times K_b = (0.02775\,m) \times (0.512\ ^{\circ}C/m) = 0.0142\ ^{\circ}C$$

The normal boiling point is 100.00°C, so the boiling point of the solution is 100.01°C.

---

3. Osmosis is the flow of solvent from a solution of high solvent concentration to low solvent concentration.

   ► When considering the direction of solvent flow, remember that the area of high solvent concentration is the area of low solute concentration. In osmosis, the solutions are trying to equalize the concentrations.

a. Osmosis occurs because of nature's tendency to mix.

b. A semipermeable membrane allows some substances to pass through but not others (usually based on size differences).

c. Osmosis is quantified as the external pressure that needs to be applied in order to stop osmotic flow through a semipermeable membrane. This pressure is the osmotic pressure (Π):

$$\Pi = MRT$$

where M is the molarity, R is the gas constant, and T is the absolute temperature.

EXAMPLE:

Calculate the osmotic pressure required to keep the sugar solution from the previous two examples from flowing through a semipermeable membrane at 25°C. Assume that the volume of the solution is equal to the volume of the water used.

We first need to calculate the molarity of the solution.

$$\frac{2.5\,g\,C_6H_{12}O_6}{500\,mL\,solution} \times \frac{1\,mol\,C_6H_{12}O_6}{180.16\,g\,C_6H_{12}O_6} \times \frac{1000\,mL}{1\,L} = 0.02775\,M\,C_6H_{12}O_6\text{ solution}$$

We can now calculate the osmotic pressure of this solution:

$$\Pi = MRT = (0.02775\,M) \times \left(0.0821\frac{L \cdot atm}{mol \cdot K}\right) \times (298.15\,K) = 0.68\,atm$$

---

i. Hyperosmotic solutions are those with an osmotic pressure greater than body fluids.

ii. Hyposmotic solutions are those with an osmotic pressure less than body fluids.

iii. Isosmotic (or isotonic) solutions have an osmotic pressure equal to that of body fluid.

VII. For ionic solutions, the number of particles depends on the number of ions produced when the ionic solute is dissolved.

a. The number of particles is represented by the van't Hoff factor (i):

$$i = \frac{\text{moles of particles in solution}}{\text{moles of formula units dissolved}}$$

b. Boiling point elevation, melting point depression, and osmotic pressure expressions must be adjusted for ionic solutions to include the van't Hoff factor.

EXAMPLE:

Calculate the boiling point of a solution that results when 2.5 g of NaCl is dissolved in 500 mL of water.

We will first calculate the molality of the solution:

$$\frac{2.5\,g\,NaCl}{500\,mL\,H_2O} \times \frac{1\,mol\,NaCl}{58.44\,g\,NaCl} \times \frac{1\,mL\,H_2O}{1\,g\,H_2O} \times \frac{1000\,g}{1\,kg} = 0.08556\,m\text{ NaCl solution}$$

The boiling point elevation constant for water is 0.512, so the change in the boiling point is:

$$\Delta T_b = i \times m \times K_b = 2 \times (0.08556\, m) \times (0.512\ ^{\circ}C/m) = 0.0876\ ^{\circ}C$$

The normal boiling point is 100.00°C, so the boiling point of the solution is 100.09°C.

VIII. Colloids

a. A colloidal dispersion (colloid) is a mixture in which a dispersed substance is finely divided in a dispersing medium.

   i. Colloids are defined according to the solute-like particle size. Colloids have particles that are between 1 nm – 1000 nm dispersed in them.

b. Soap solutions are colloid particles.

   i. Soap molecules have a polar end and a non-polar end.

      1. Micelles form when soap is put into water. A micelle is a spherical structure in which the polar ends of the molecules are sticking out toward the water and the non-polar ends of the molecules congregate with each other in the center.

c. The scattering of light by a colloidal dispersion is called the Tyndall effect.

Fill in the Blank Problems:

1. When two substances are soluble in each other in all proportions they are ________________.
2. ________________ relates the solubility of a gas to the pressure of the gas above a solution of it.
3. A(n) ________________ solution is one in which water is the solvent.
4. ________________ is the pressure that must be applied in order to stop the flow of solvent through a semipermeable membrane.
5. A colloidal dispersion scatters light according to the ________________.
6. The boiling point of a solution is ________________ than the boiling point of the pure solvent.
7. ________________ is a measure of energy randomization or energy dispersal.
8. The vapor pressure of a solution with a nonvolatile solute is ________________ than the vapor pressure of the pure solvent.
9. A(n) ________________ solution is one where the solid phase and the aqueous phase are in equilibrium with one another.
10. A sol ution is a(n) ________________ mixture of two or more substances.
11. A(n) ________________ follows Raoult's law exactly.
12. A real solution that has solvent-solute interactions that are stronger than the solute-solute and solvent-solvent interactions will deviate in a ________________ from Raoult's law.
13. The __ ________________ is the sum of the heat of hydration and the negative lattice energy.
14. The ________________ must be included in colligative property calculations for solutions of ionic substances.
15. The solubility of gases ________________ with an increase in temperature and ________________ with an increase in the pressure of the gas over the solution.

16. The ________________ of a solution is the amount of solute (in moles) over the mass of the solvent (in kilograms).

Problems:

1. Which of the following substances do you expect to dissolve to a significant extent in water? Explain your choices and suggest a different solvent for those that you expect not to dissolve in water.
   a. NaCl(s)
   b. CO_2(g)
   c. $CH_3CH_2CH_2CH_2CH_2CH_3$
   d. LiOH
   e. NO_3^-
2. Ammonium chloride is a common cold pack ingredient.
   a. Solid ammonium chloride is added to an ammonium chloride solution. Some of the solid dissolves while the rest falls to the container floor. Was the original solution saturated, unstaturated, or supersaturated? What about the solution after adding ammonium chloride? Explain your answer.
   b. If ammonium chloride is used for a cold pack, what are the relative magnitudes of the heat of hydration and lattice energy?
   c. The lattice energy of ammonium chloride is smaller than that of sodium chloride. Why do you think that this is?
3. What is the molar solubility of O_2 in water when 5.0 mol of O_2 is pumped into a 1.0 L container holding 525 mL of water at 25°C? Henry's law constant for O_2 is 0.0013 M/atm. Hint: you will need to use the ideal gas law and express the moles of gas in terms of the solubility.
4. The density of a 20.% by volume solution of ethanol, CH_3CH_2OH, in water is 0.9687 g/mL.
   a. Suggest why the density of the ethanol solution is less than the density of water.
   b. Given that the density of a pure ethanol solution is 0.7893 g/mL do you expect this solution to behave ideally? If not, will it deviate in a positive or negative fashion?
   c. Calculate the molarity of the solution.
   d. Calculate the molality of the solution.
   e. Calculate the percent by mass of ethanol in the solution.
   f. Calculate the mole fraction of ethanol in the solution.
   g. Calculate the mole fraction of water in the solution.
5. The vapor pressure of liquid ammonia is 10.2 atm at 25°C.
   a. Calculate the partial pressure of ammonia over of a 1.2 M solution of ammonia in water at 25°C. Assume that the density of the ammonia solution is the same as the density of water.
   b. Calculate the partial pressure of water over this same solution.
   c. Do you expect that the actual vapor pressure of ammonia over the solution will be higher or lower than the value that you calculated in part a? Explain your answer.
6. Calculate the vapor pressure, freezing point, boiling point, and osmotic pressure of a solution prepared by dissolving the following substances in 1.0 L of ethanol at 25°C. The vapor pressure of pure ethanol is 59.0 torr. Assume no volume change occurs upon dissolution and that liquid volumes are additive.
   a. 5.45 g $CaCl_2$

b. 32.7 mL H_2O

c. 89.25 mg of LiCl

d. 63.8 g of SiO_4

e. 398 mL of a 0.10 M methanol (CH_3OH) solution

7. Calculate the amount of salt that you would need to dissolve in water in order to change its temperature by 1°C. Based on your answer, do you think that salt is added to cooking water in order to change its temperature or for flavor? Justify your answer.

Concept Questions:

1. When defining solution types, every phase could be combined with every other phase except gases and solids. Why do you think that these phases are not included?
2. There is a rule of thumb for the solubility of alcohols in water. It states that when less than six carbon atoms are present in the alcohol it will be soluble in water and if there are more than six carbon atoms in the alcohol it will be soluble in hexane. Explain the reasoning behind this rule of thumb.
3. Explain why the solubility of gases decreases with increasing temperature using the Boltzmann distribution of molecular speeds.
4. The K_f values listed in your book vary a great deal. For example, carbon tetrachloride has a freezing point depression constant of 29.9°C/*m* while diethyl ether has a freezing point depression constant of 1.79°C/*m*.

    a. Which of these solutions will have a larger freezing point change when 1.0 mol of NaCl is dissolved in equal volumes of each of them?

    b. Suggest one reason for the large difference in K_f values.

    c. The boiling point elevation constants do not vary by such a large amount. Suggest one reason for this observation.

5. We generally assume that the volume of a solution is equal to the volume of the solvent.

    a. This is based on the idea that the solute-solvent interactions are approximately equal to the solvent-solvent interactions. Why does this assumption lead to a negligible volume change?

    b. Is this always a valid assumption? If so why? If not, suggest a system that would deviate from this assumption and state the direction of the deviation.

    c. What do you expect to be true about the vapor pressures of solutions that have a volume that is smaller than the solvent volume?

6. NaCl is dissolved in two beakers. One beaker contains pentane, $CH_3CH_2CH_2CH_2CH_3$, and one beaker contains pentanol, $CH_3CH_2CH_2CH_2CH_2OH$.

    a. In which beaker do you expect more NaCl to dissolve? Explain.

    b. Both substances will have a new boiling point upon the addition of NaCl. Explain why the boiling point changes.

    c. The boiling point changes by a different amount for the two beakers. Explain why this is and which solution will have a higher boiling point elevation.

7. There are pumps in your cells that operate to remove sodium ions from the cell. These pumps operate against the concentration gradient meaning that they pump sodium ions from a region of low concentration to a region of high concentration. Is this process energetically favorable? Explain your answer using the concepts of entropy discussed in this chapter.
8. Consider two non-ideal solutions: one made by mixing water with ethanol (CH_3CH_2OH) and one made by mixing water with nitric acid (HNO_3).

a. Does the ethanol/water solution deviate from Raoult's law in a positive or negative fashion?

b. Based on your answer in part a, do you think that the actual change in the boiling point for the solution compared to the solvent will be greater or less than in an ideal solution?

c. Does the nitric acid/water solution deviate from Raoult's law in a positive or negative fashion?

d. Based on your answer in part b, do you think that the actual change in the boiling point for the solution compared to the solvent will be greater or less than for an ideal solution?

e. Do you expect these solutions to have a similar or different change in their boiling point elevations? Explain your answer.

9. Soap solutions are great because they help to dissolve dirt and oil in water; this is how your laundry detergent gets your clothes clean. Explain how soap works to dissolve nonpolar substances such as dirt and oil in the polar water that is used in washing machines.

Chapter 13: Chemical Kinetics

Learning Objectives:

- Calculate the average rate of a reaction and predict how it changes over time.
- Determine the rate law and reaction order for a reaction using the method of initial rates.
- Identify the rate of a reaction using the integrated rate law on the basis of graphed data.
- Understand the dependence of the reaction rate on temperature and compare reactions to one another based on this dependence.
- Sketch a potential energy diagram for a reaction.
- Identify reasonable reaction mechanisms for chemical reactions.
- Understand the role of catalysts in chemical reactions.

Chapter Summary:

Chemical kinetics is the study of the rates of chemical reactions and the underlying physical processes that define them. In this chapter, you will learn to appreciate the broad spectrum of reaction speeds, types, and molecular events. First, you will carry out the most basic reaction rate calculation, the determination of an average rate; then you will learn a more sophisticated calculation: the determination of the instantaneous rate. Experimentally, the method of initial rates is used to determine the rate law of a chemical reaction, which you will learn to solve for and interpret. You will then be able to use the rate law (rate as a function of reactant concentration) and the integrated rate law (concentration as a function of time) in conjunction with experimental data to determine the order of a reaction. The pathway of a chemical reaction is described using a series of elementary steps which, taken together, comprise the reaction mechanism; you will learn to identify possible reaction mechanisms using the experimental rate law and your understanding of chemical phenomena. The temperature dependence of the reaction rate will be examined in the context of collision theory. You will learn how to use an Arrhenius plot to determine the activation energy and the collision requirements of a particular chemical reaction. Finally, you will learn how a catalyst alters a reaction pathway to one with a lower barrier and therefore causes the reaction to occur faster.

Chapter Outline:

I. The Rate of a Chemical Reaction

a. The rate of a chemical reaction is a measure of how fast the reaction occurs. The rate is equal to the ratio of the change in reactant concentration to the change in time:

$$\text{rate} = -\frac{\Delta[\text{reactant}]}{\Delta\text{time}}$$

Since Δ[reactant] is always negative and Δtime is positive, the negative sign in the definition results in a rate that is always a positive number

b. The rate reflects the reaction stoichiometry. For the reaction:

$$aA + bB \rightarrow cC + dD$$

the rate is:

$$\text{rate} = -\frac{1}{a}\frac{\Delta[A]}{\Delta t} = -\frac{1}{b}\frac{\Delta[B]}{\Delta t} = \frac{1}{c}\frac{\Delta[C]}{\Delta t} = \frac{1}{d}\frac{\Delta[D]}{\Delta t}$$

- ▶ The inclusion of the stoichiometric coefficients ensures that each reaction has a single defined rate associated with it.

c. Reactant concentrations decrease over time while product concentrations increase over time.

d. The average rate of a reaction can be calculated using any two data sets.

e. The average rate of a reaction decreases over time as the reactants turn into products.

EXAMPLE:

Consider the generic chemical reaction: $A + B \rightarrow 2C$

a. Using the data in the table below, calculate the average reaction rate over three different time intervals:

[A] (M)	Time (s)
0.052	0.00
0.024	0.10
0.014	0.20

The rate is given by the expression:

$$\text{rate} = -\frac{1}{a}\frac{\Delta[A]}{\Delta t} = -\frac{1}{b}\frac{\Delta[B]}{\Delta t} = \frac{1}{c}\frac{\Delta[C]}{\Delta t} = \frac{1}{d}\frac{\Delta[D]}{\Delta t}$$

So the rate of the reaction for the first time interval is:

$$\text{rate} = -\frac{[A]_2 - [A]_1}{t_2 - t_1} = -\frac{0.024 - 0.052}{0.10 - 0.00} = 0.28\text{M/s}$$

The rate of the reaction for the second time interval is:

$$\text{rate} = -\frac{[A]_3 - [A]_2}{t_3 - t_2} = -\frac{0.014 - 0.024}{0.20 - 0.10} = 0.10\text{M/s}$$

The rate of the reaction for the overall time interval is:

$$\text{rate} = -\frac{[A]_3 - [A]_1}{t_3 - t_1} = -\frac{0.014 - 0.052}{0.20 - 0.00} = 0.19\text{M/s}$$

b. Comment on the differences between the three reaction rates calculated.

The reaction rate for the first time interval is fastest and for the second time interval is slowest. This is consistent with the idea that the reaction rate decreases over time as a consequence of a decrease in the amount of reactant(s) available for reaction. The reaction rate calculated for the third time interval is the least meaningful since it is an average over a longer period of time – the reaction rate is not actually 0.19 M/s over this whole range.

c. Predict the concentration of B after 0.10 s if the initial concentration is 0.052 M (the same as the initial concentration of A).

Since the rate of the reaction is the same for all reactants and products and the stoichiometry indicates a 1:1 ratio of A to B, the concentration of B will change at the same rate as the concentration of A changes. Since we start off with the same amount of A and B, at some time later the amounts of A and B will remain the same. Therefore, the concentration of B after 0.10 s will be 0.024 M.

This can also be solved mathematically:

$$-\frac{[A]_2-[A]_2}{t_2-t_1}=-\frac{[B]_2-[B]_1}{t_2-t_1} \rightarrow \qquad -\frac{0.024-0.052}{0.10-0.00}=-\frac{[B]_2-0.052}{0.10-0.00}$$

---

f. The instantaneous rate of a reaction is the rate at a given point in time and can be calculated using the slope of the line tangent to the plot of the rate at a given time.

EXAMPLE:

Again, consider the generic chemical reaction: A + B → 2C

Data is collected over a period of time and plotted on the graph as shown below:

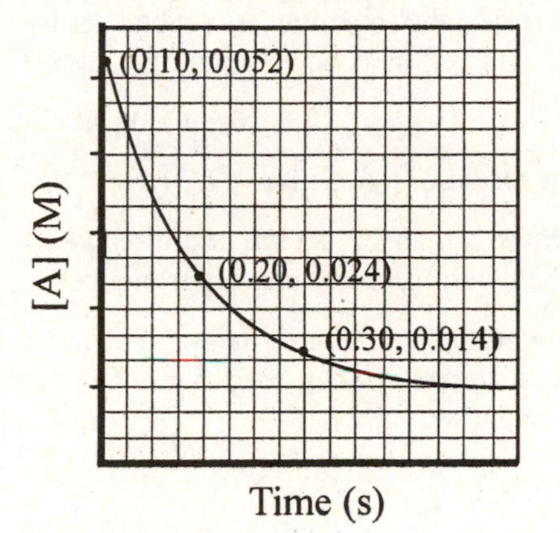

Calculate the instantaneous rate at 0.20 s.

We can draw the line tangent to the plotted data at 0.20 s and select two points on that line:

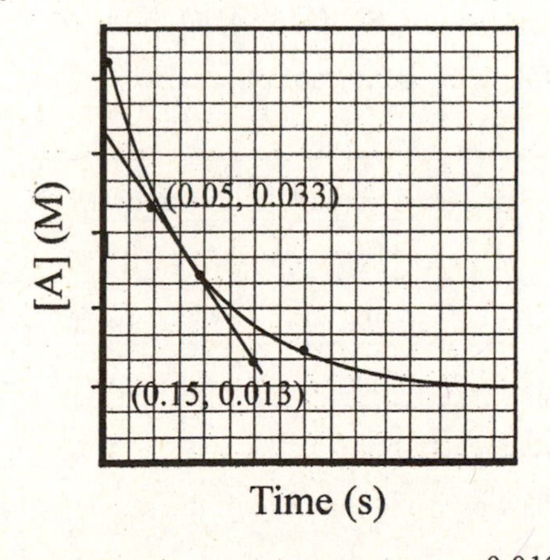

Using the two points selected, we can calculate the slope of the line: $\frac{0.013-0.033}{0.15-0.05}=-0.20$ M/s which is the instantaneous rate at 0.20 M/s.

---

g. Reaction rates are measured experimentally using a number of techniques such as polarimetry, spectroscopy, and pressure changes.

II. The Rate Law: The Effect of Concentration on Reaction Rate

a. For a reaction A → B, the rate law relates the rate of the reaction to the concentration of the reactant raised to some power.

$$\text{rate} = k[A]^n$$

The proportionality constant k is called the rate constant, and the exponent n is the order of the reaction.

i. When n=0, the reaction is zero order and the rate of the reaction is independent of the concentration of A:

$$\text{rate} = k$$

ii. When n=1, the reaction is first order and the rate of the reaction is directly proportional to the concentration of A:

$$\text{rate} = k[A]$$

iii. When n=2, the reaction is second order and the rate of the reaction is proportional to the concentration of A squared:

$$\text{rate} = k[A]^2$$

b. Determining the Order of a Reaction

i. Rate laws cannot be determined directly from the reaction stoichiometry; they must be determined through experiments.

ii. The method of initial rates measures the rate at the beginning of the reaction for several different starting conditions. The data is used to determine the exponent in the rate law.

EXAMPLE:

The method of initial rates is used to determine the rate law for the reaction:

$$2N_2O_5 \rightarrow 4NO_2 + O_2$$

At a constant temperature, the following data was collected:

$[N_2O_5]$ (M)	rate (M/s)
0.100	5.1×10^{-5}
0.200	1.0×10^{-4}
0.300	1.5×10^{-4}

a. Determine the rate law for the reaction.

We can determine the rate law by using the ratio of the rate laws for two data points:

$$\frac{\text{rate}_1}{\text{rate}_2} = \frac{k[N_2O_5]_1^n}{k[N_2O_5]_2^n} \qquad \rightarrow \qquad \frac{0.100}{0.200} = \frac{k(5.1\times10^{-5})^n}{k(1.0\times10^{-4})^n}$$

Since the rate constant does not change (constant temperature), we can cancel it out. We can also collect the concentration terms and simplify:

$$\frac{0.100}{0.200} = \left(\frac{5.1\times10^{-5}}{1.0\times10^{-4}}\right)^n \qquad \rightarrow \qquad 0.500 = (0.51)^n \qquad \rightarrow \qquad n=1$$

So the rate law is: rate = k [N_2O_5].

b. What is the order of the reaction with respect to N_2O_5?

The reaction is first order with respect to N_2O_5.

c. Calculate the rate constant.

The rate constant can be calculated by using one of the data sets in the rate law:

$$5.1*10^{-5} = k \times 0.100 \rightarrow \qquad k = \frac{5.1\times10^{-5}}{0.100}$$

So k = 5.1×10^{-4} s^{-1} (be sure to include correct units).

---

c. Reaction Order for Multiple Reactants

i. For more complicated reactions, such as:

$$aA + bB \rightarrow cC + dD$$

the rate is given by the general rate law:

$$rate = k[A]^m[B]^n$$

where m is the order of the reaction with respect to A, n is the order of the reaction with respect to B, and (m+n) is the overall reaction order.

► The determination of n and m is the same as in the above example with an additional step.

EXAMPLE:

The following data was collected for the reaction of NO with Cl_2:

$$2NO + Cl_2 \rightarrow 2NOCl$$

[NO] (M)	[Cl_2] (M)	Rate (M/min)
0.100	0.100	0.18
0.100	0.200	0.35
0.200	0.200	1.45

a. Determine the rate law for the reaction.

The rate law is of the form: rate = $k[NO]^n[Cl_2]^m$. We can determine m using the first and second experiments (notice that since [NO] remains constant it cancels out of the expression):

$$\frac{rate_1}{rate_2} = \frac{k[NO]_1^n[Cl_2]_1^m}{k[NO]_2^n[Cl_2]_2^m} \rightarrow \qquad \frac{0.18}{0.35} = \frac{k(0.100)^n(0.100)^m}{k(0.100)^n(0.200)^m}$$

We can now collect like terms and simplify:

$$0.51 = (0.50)^m \qquad \rightarrow \qquad m = 1$$

And now we can use the second and third experiments to determine n (this time [Cl_2] remains unchanged and therefore cancels out of the expression):

$$\frac{rate_2}{rate_3} = \frac{k[NO]_2^n[Cl_2]_2^m}{k[NO]_3^n[Cl_2]_3^m} \rightarrow \qquad \frac{0.35}{1.45} = \frac{k(0.100)^n(0.200)}{k(0.200)^n(0.200)}$$

We now collect like terms and simplify:

$$0.24 = (0.500)^n \quad \rightarrow \quad n = 2$$

So the rate law is: rate = $k[NO]^2[Cl_2]$.

b. Calculate the rate constant for the reaction.

We can calculate the rate constant using one set of data points and the rate law that we have just identified:

$$1.45 = k\,(0.200)^2(0.200) \quad \rightarrow \quad k = 181\ (M^2 \cdot min)^{-1}$$

III. The Integrated Rate Law: The Dependence of Concentration on Time

a. The integrated rate law for a chemical reaction is the relationship between concentrations of reactants and time. In the following expressions, $[A]_0$ is the initial concentration of A and $[A]_t$ is the concentration of A after some time, t.

i. Zero-order integrated rate law:

$$[A]_t = -kt + [A]_0$$

ii. First-order integrated rate law:

$$\ln[A]_t = -kt + \ln[A]_0$$

iii. Second-order integrated rate law:

$$\frac{1}{[A]_t} = kt + \frac{1}{[A]_0}$$

EXAMPLE:

Use the plotted data shown below to determine the reaction order (n), the rate constant (k), the initial concentration ($[A]_0$), and the concentration of A after 60 s ($[A]_{60}$):

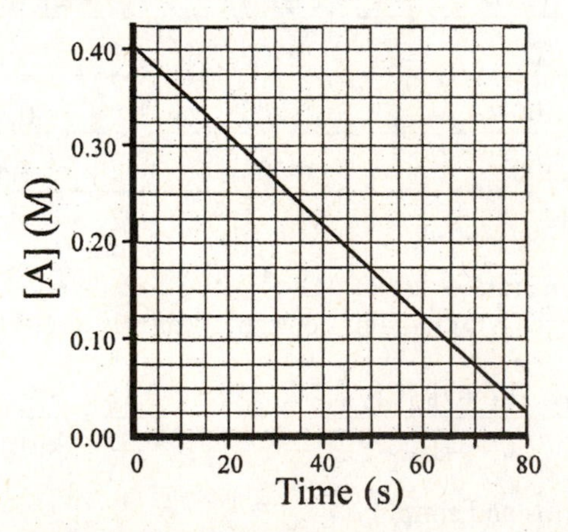

Since the data is linear when the concentration is plotted versus time, it is a zero-order reaction. We can compare the integrated rate law to the equation of a line (y=mx+b) to determine the rate constant and the initial concentration:

$$[A]_t = -kt + [A]_0$$

$$y = [A]_t,\ m = -k,\ x = t,\ \text{and}\ b = [A]_0$$

Inspection of the graph shows that the slope is $\frac{\Delta y}{\Delta x}=\frac{(0.40-0.125)}{(0.00-60)}=-0.00458$ M/s and the y-intercept is 0.40 M. So the rate constant is 0.0046 M/s and the initial concentration is 0.40 M. We can find the concentration of A after 60 seconds by solving the equation:

$$[A]_{60}=-(0.00458)(60)+0.40 \quad \rightarrow \quad [A]_{60}=0.125 \text{ M}$$

We could also obtain the concentration after 60 s from the graph and would arrive at the same answer.

b. The half-life of a reaction is the time required for the concentration of a reactant to be reduced to one half of its original amount ($[A]_t = \frac{1}{2}[A]_0$).

i. Zero-order reaction half-life:

$$t_{1/2}=\frac{[A]_0}{2k}$$

The half-life decreases as the concentration decreases.

ii. First-order reaction half-life:

$$t_{1/2}=\frac{0.693}{k}$$

The half-life is independent of the concentration.

iii. Second-order reaction half-life:

$$t_{1/2}=\frac{1}{k[A]_0}$$

The half-life increases as the concentration decreases.

EXAMPLE:

Carbon-14 decays via first-order kinetics and has a half life of 5,730 years. If 1.0 g of ^{14}C is produced in a campfire, how many years will it take for only 7.8×10^{-3} g to remain?

We can solve this problem by calculating the number of half-lives required. To do this, we take the starting amount, 1.0 g, and divide by two until we have 0.0078 g remaining:

1.0/2 → 0.50/2 → 0.25/2 → 0.125/2 → 0.0625/2 → 0.03125/2 → 0.015625/2 → 0.0078125

Seven half-lives are required to reduce 1.0 g of ^{14}C to 0.0078 g of ^{14}C. Seven half-lives is a total time of $7\times5{,}730$ years = 40,110 years.

EXAMPLE:

Dinitrogen pentoxide, N_2O_5, decomposes according to first-order kinetics with a half-life of 4.03×10^3 s.

a. What is the rate constant for the decomposition of N_2O_5?

We can solve for the rate constant by using the half-life equation and the given value:

$$t_{1/2}=\frac{0.693}{k} \quad \rightarrow \quad 4.03\times10^3=\frac{0.693}{k} \quad \rightarrow \quad k=1.72\times10^{-4}$$

b. How much N_2O_5 will remain after two days, if you start with 10.0 g?

In order to solve this problem, we need to determine how many half-lives two days are:

$$2\text{days} \times \frac{24\text{hr}}{1\text{day}} \times \frac{60\text{min}}{1\text{hr}} \times \frac{60\text{s}}{1\text{min}} \times \frac{1\text{ half-life}}{4030\text{s}} = 43\text{ half-lives}$$

So if we begin with 10 g, we will end up with $10/2^{43}$ g or 1.14×10^{-12} g.

IV. The Effect of Temperature on Reaction Rate

a. A good rule of thumb is that the rate of a reaction approximately doubles for each 10°C increase in temperature.

b. Rates depend on temperature because the rate constant, k, is temperature-dependent.

c. The Arrhenius equation relates k to the temperature through an exponential factor:

$$k = Ae^{-E_a/RT}$$

where k is the rate constant at a temperature T (in Kelvin), E_a is the activation energy, A is the frequency factor, and R is the gas constant.

d. The activation energy, E_a, is the amount of energy required to go from reactants to the activated complex or transition state; it is the minimum amount of energy required in order for a reaction to occur. The higher the activation energy is, the slower the reaction rate will be.

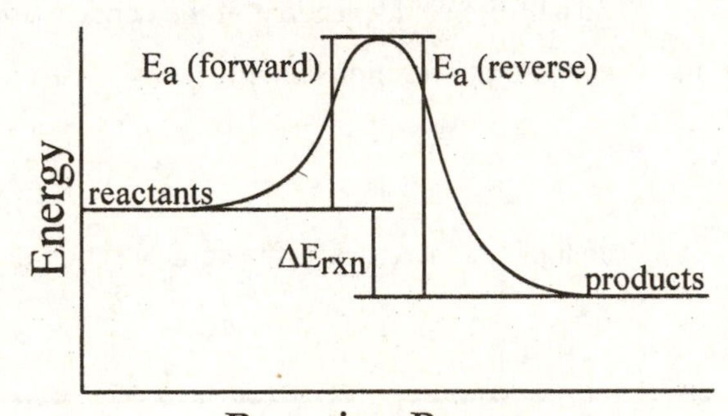

e. Increasing the temperature increases the energy of reacting species, which increases the fraction of reactant species with enough energy to react. This is easily seen by inspecting the Boltzmann distribution for the fraction of molecules with energy greater than a certain value (for $T_2 > T_1$).

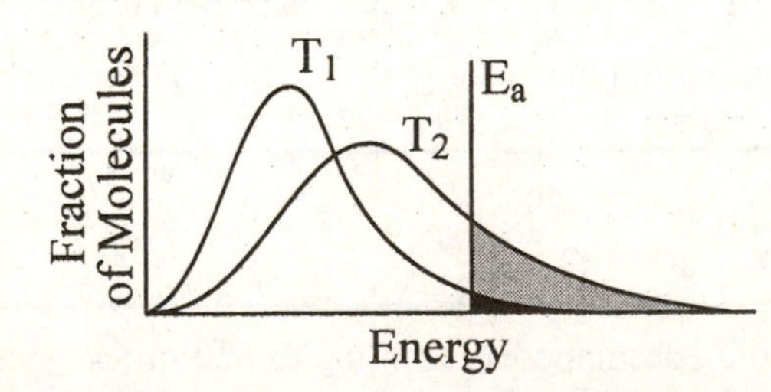

f. When the rate constant is measured at a variety of temperatures, an Arrhenius plot can be used to find the activation energy and frequency factor for a given reaction.

$$\ln k = -\frac{E_a}{R}\left(\frac{1}{T}\right) + \ln A$$

EXAMPLE:

The conversion of cyclopropane to propene occurs at 300°C with a rate constant of $2.41\times10^{-10}\ s^{-1}$ and occurs at 400°C with a rate constant of $1.16\times10^{-6}\ s^{-1}$.

a. Calculate the activation energy of the reaction.

We need to convert the temperature to the Kelvin scale and then we can use the equation above and substitute in the given information twice:

For the first set of conditions: $\ln(2.41\times10^{-10}) = -\frac{E_a}{R}\left(\frac{1}{573.15}\right) + \ln A$

For the second set of conditions: $\ln(1.16\times10^{-6}) = -\frac{E_a}{R}\left(\frac{1}{673.15}\right) + \ln A$

Now we can subtract the second equation from the first equation and simplify:

$$\ln(2.41\times10^{-10}) - \ln(1.16\times10^{-6}) = -\frac{E_a}{R}\left(\frac{1}{573.15}\right) - \left(-\frac{E_a}{R}\left(\frac{1}{673.15}\right)\right)$$

$$\ln\frac{2.41\times10^{-10}}{1.16\times10^{-6}} = \frac{E_a}{8.3145}\left(\frac{1}{673.15} - \frac{1}{573.15}\right)$$

$$-8.4791 = \frac{E_a}{8.3145}(-0.00025919)$$

E_a = 272000 J/mol or 272 kJ/mol.

b. Calculate the rate constant at 500°C.

We have solved for the activation energy, but we have not solved for the frequency factor. We will, therefore, need to use the two-point form of the Arrhenius equation as we did above:

$$\ln\frac{k}{2.41\times10^{-10}} = \frac{272000}{8.3145}\left(\frac{1}{573.15} - \frac{1}{773.15}\right) \rightarrow \ln\frac{k}{2.41\times10^{-10}} = 14.7649$$

$$\ln k = 14.7649 + \ln(2.41\times10^{-10}) \rightarrow \ln k = -7.3813$$

$$k = e^{-7.3813}$$

So $k = 6.23\times10^{-4}\ s^{-1}$ which makes sense – it is larger than the value of k at 600°C (a lower temperature).

c. Sketch the potential energy diagram for this reaction given that the overall reaction is exothermic with a ΔH° = -33 kJ/mol.

We can sketch the reaction profile using the activation energy (reaction barrier) calculated and the enthalpy of reaction given:

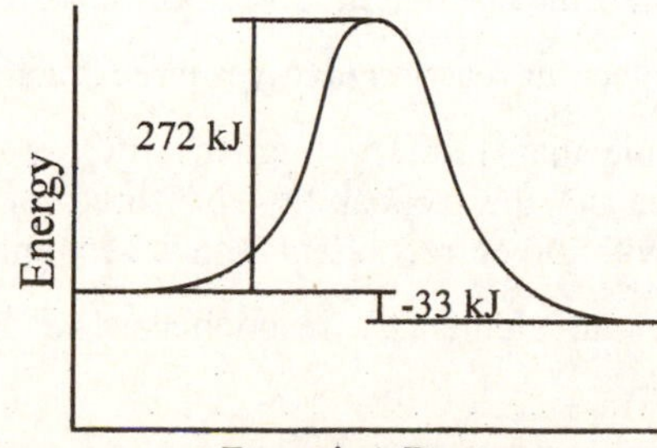

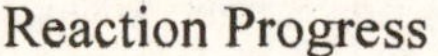

g. The frequency factor, A, can be interpreted using collision theory.

   i. Collision theory assumes that a reaction occurs when reactant molecules collide in the proper orientation with enough energy to overcome the barrier. The orientation criteria is embedded in the frequency factor:

   $$A = pz$$

   where p is the orientation factor and z is the collision frequency.

   ii. The orientation factor takes into account the fact that some reactions need a specific collision geometry in order to occur.

EXAMPLE:

If all of the reactions are carried out at the same temperature, which reaction do you expect to have the largest value of A?

a. $H_2 + Cl_2 \rightarrow 2HCl$

b. $Na^+ + Cl^- \rightarrow NaCl$

c. $CH_2{=}CH_2 + H_2 \rightarrow CH_3\text{-}CH_3$

Since all of the reactions are carried out at the same temperature, we can assume that the total number of collisions are the same (temperature is proportional to kinetic energy). This means that the main factor influencing the value of A is the orientation factor, p. We can conclude that reaction b. $Na^+ + Cl^- \rightarrow NaCl$ will have the largest A value since it is a reaction of two spherically symmetric species that can collide with any orientation and have a successful reaction outcome.

V. Reaction Mechanisms

a. Reactions are usually written as an overall reaction and do not provide information about how the reaction progresses.

b. A reaction mechanism is a sequence of simple, elementary reactions that add up to give the overall reaction.

   i. Elementary reactions show the microscopic (molecular) events as they are thought to happen.

c. Reaction intermediates are species that are formed in an elementary step and consumed in a subsequent step. Intermediates do not appear in the overall chemical reaction.

d. The molecularity of an elementary step is the number of species reacting in that step – it is the overall order of the elementary reaction.

   i. A unimolecular reaction involves a single reactant species.

   ii. A bimolecular reaction involves two reactant species.

   iii. A termolecular reaction involves three reactant species.

e. Very few termolecular (and no higher order molecularity) reactions exist; this is presumably because the probability of three or more reactant species colliding simultaneously with the correct orientation is very small.

f. The rate law for an elementary reaction can be deduced directly from the balanced reaction.

g. The rate-determining step is the slowest step in a reaction mechanism and determines the rate of the overall reaction.

EXAMPLE:

The reaction between nitrogen dioxide and fluorine gas is believed to occur via the following mechanism:

$NO_2 + F_2 \rightarrow NO_2F + F$ Slow

$NO_2 + F \rightarrow NO_2F$ Fast

a. What is the overall reaction that is occurring?

The overall reaction is the sum of the steps with any species that appear on both sides eliminated:

$2NO_2 + F_2 \rightarrow 2NO_2F$

b. Identify any intermediates in the reaction mechanism.

A reaction intermediate is any species that is produced in one step of the reaction mechanism and is subsequently consumed. F is produced in the first step, consumed in the second step, and does not appear in the overall reaction – F is therefore a reaction intermediate.

c. What is the rate law predicted by the reaction mechanism?

The rate law can be written directly from the slow elementary step:

rate = $k[NO_2][F_2]$

d. What is the molecularity of the first step?

The first step is bimolecular since there are two species that must collide in that step.

EXAMPLE:

The reaction between hydrochloric acid and oxygen gas is believed to occur in three steps:

$HCl + O_2 \rightarrow HOOCl$

$HOOCl + HCl \rightarrow 2\ HOCl$

$HOCl + HCl \rightarrow H_2O + Cl_2$

If the experimentally observed rate law is: rate = $k[HCl][O_2]$, what is the rate-determining step of the reaction mechanism?

Since the rate law can be written directly from the first step of the reaction mechanism, we can assume that the first step is the slow, or rate-determining, step.

EXAMPLE:

Consider the reaction mechanism shown below:

$NO + NO \rightleftharpoons N_2O_2$ fast equilibrium

$N_2O_2 + O_2 \rightarrow 2NO_2$ slow

a. What is the overall reaction?

The overall reaction is the sum of the steps with any species that appear on both sides eliminated:

$2NO + O_2 \rightarrow 2NO_2$

b. What is the rate law for the reaction?

The rate law is determined from the slow step:

$$\text{rate} = k_2[N_2O_2][O_2]$$

but the rate law should not include intermediates like N_2O_2. So we need to consider the equilibrium expression:

$$\text{rate}_{forward} = k_1[NO][NO]$$

$$\text{rate}_{reverse} = k_{-1}[N_2O_2]$$

at equilibrium the reverse and forward rates are equal so $\text{rate}_{forward} = \text{rate}_{reverse}$ and we solve for $[N_2O_2]$:

$$k_1[NO][NO] = k_{-1}[N_2O_2]$$

$$\frac{k_1}{k_{-1}} = \frac{[N_2O_2]}{[NO][NO]} \quad \rightarrow \quad [N_2O_2] = \frac{k_1}{k_{-1}}[NO]^2$$

Substituting this into the rate expression gives:

$$\text{rate} = \frac{k_2 k_1}{k_{-1}}[O_2][NO]^2$$

Since k_2, k_1, and k_{-1} are all constants, we can replace them with a single constant, k:

$$\text{rate} = k[O_2][NO]^2$$

---

h. Reaction mechanisms are never proven to be true, but are only shown to agree with experimental evidence. In order for a proposed mechanism to be reasonable it must:

   i. Have elementary steps that add up to give the correct overall reaction.

   ii. Must give a rate law that is consistent with the experimentally determined rate law.

      1. The rate law should be written in terms of reactant species that appear in the overall reaction. Intermediate species do not appear in the rate law.

   iii. Should be physically plausible.

EXAMPLE:

The reaction between NO_2 and F_2 was studied. The experimentally determined rate law is:

$$2NO_2 + F_2 \rightarrow 2NO_2F \qquad \text{rate} = k[NO_2][F_2]$$

Which of the following mechanisms is most likely?

a. $2NO_2 + F_2 \rightarrow 2NO_2F$ — single step

b. $NO_2 + F_2 \rightarrow NO_2F + F$ — fast

   $NO_2 + F \rightarrow NO_2F$ — slow

c. $F_2 \rightarrow 2F$ — slow

   $F + NO_2 \rightarrow NO_2F$ — fast

d. $NO_2 + F_2 \rightarrow NO_2F + F$ — slow

   $NO_2 + F \rightarrow NO_2F$ — fast

We can first look at the rate laws for each of the reaction mechanisms:

a. $\text{rate} = k[NO_2]^2[F_2]$

b. rate = $k[NO_2][F]$

c. rate = $k[F_2]$

d. rate = $k[NO_2][F_2]$

We see that the rate law for mechanism d is consistent with the experimentally determined rate law. Since it also adds up to the overall reaction and appears to be physically reasonable, mechanism d is most likely.

VI. Catalysis

a. A catalyst increases the rate of a chemical reaction without being consumed by it.

   i. A homogeneous catalyst exists in the same phase as the reacting species.

   ii. A heterogeneous catalyst exists in a different phase as the reacting species.

b. Catalysts increase reaction rates by providing an alternate reaction mechanism (pathway) with a lower activation energy than the uncatalyzed reaction.

EXAMPLE:

The decomposition of hydrogen peroxide to produce water and oxygen is an exothermic reaction. The reaction can be catalyzed using iodide ions; the catalyzed reaction is believed to occur via the following two-step mechanism:

$$H_2O_2 + I^- \rightarrow OI^- + H_2O \qquad \text{slow}$$

$$H_2O_2 + OI^- \rightarrow O_2 + H_2O + I^- \qquad \text{fast}$$

Sketch the potential energy diagram of the catalyzed and uncatalyzed reaction using the information given.

The uncatalyzed reaction can be drawn as our starting point. We are given that the reaction is exothermic so the products are lower in energy than the reactants.

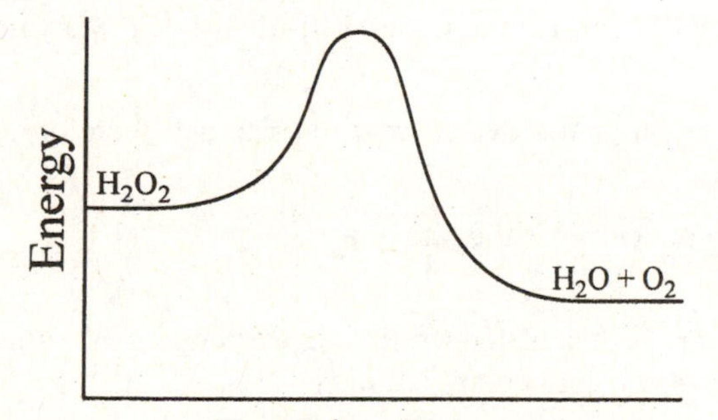

We can then sketch in the catalyzed reaction. It occurs in two steps and we know that the activation energy of the two steps is lower than that of the uncatalyzed reaction. Further, we know that the activation energy of the first step is larger than the activation energy of the second step because the first step is slower.

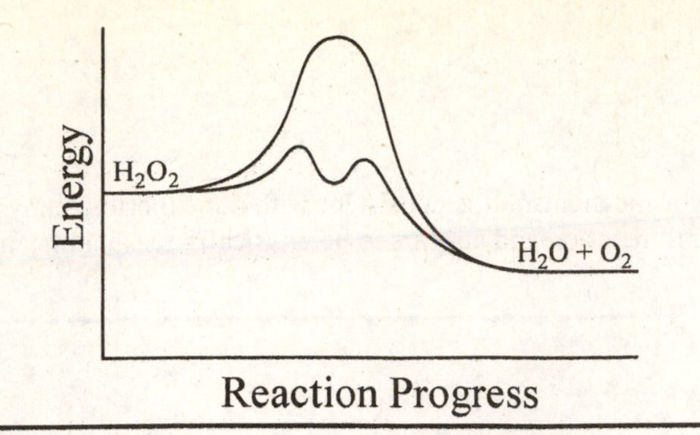

c. Enzymes are biological catalysts that catalyze the biochemical reactions that occur in organisms.

   i. Enzymes are very selective in the reactions that they catalyze.

   ii. The active site of an enzyme is the region of the enzyme that binds to the reactant molecule, called the substrate.

   iii. Enzyme kinetics are generally described by the following mechanism:

$E + S \leftrightarrows ES$	fast equilibrium
$ES \rightarrow E + P$	slow

Fill in the Blank Problems:

These problems are intended to ensure that you are familiar with the common terms and their definitions from this chapter.

1. The relationship between the reaction rate and the concentration of the reactant species is given by the ________________.
2. In a ________________-order reaction, the half-life of the reacting species is independent of concentration.
3. The ________________ limits the overall reaction rate and therefore determines the rate law for a given chemical reaction.
4. A good rule of thumb to remember is the rate of a reaction ________________ with each temperature increase of 10°C.
5. The instantaneous rate can be calculated by finding the slope of the line ________________ to the concentration vs. time data at a given point in time.
6. Reaction mechanisms can never be ________________ but can only be shown to be consistent with experimental evidence.
7. A(n) ________________ speeds up a chemical reaction by providing a lower energy pathway, but it is not consumed in the reaction.
8. The integrated rate law is an expression of the concentration as a function of ________________.
9. A(n) ________________ is a species that is produced in an elementary step, consumed in a subsequent step, and does not appear in the overall chemical reaction.
10. ________________ is the study of the rates of chemical reactions and the underlying microscopic steps that comprise them.
11. The reactant concentration ________________ over time in a chemical reaction.

12. In a second-order reaction, the half-life of the reacting species ________________ as the concentration of the reactant decreases.

13. The __ ________________ is given by the sum of the exponents of the reacting species in the rate law.

14. The rate law for a(n) ________________ can be written directly from its balanced reaction.

15. The ________________ is the minimum amount of energy required in a collision in order for a reaction to occur.

16. ____ ______________ are biological catalysts that catalyze specific biochemical reactions.

17. The frequency factor contains two terms: ________________, which determines the number of collisions with the correct geometry, and ________________, which indicates the total number of collisions.

18. When a plot of _____ ____________ vs. time is linear, the reaction can be identified as second order.

19. The rate of a che mical reaction has units of ________________ over time.

20. The ________________ indicates the number of species that must come together in a single elementary step.

21. A first-order reaction is one in which the rate is ________________ the concentration of the reacting species.

22. A (n) ________________ provides a molecular view of the individual steps that comprise a reaction.

23. The primary reason that temperature affects reaction rates is that the ________________ increases with an increase in temperature.

24. The slope of an Arrhe nius plot can be used to determine the ________________ of a reaction.

25. Reaction rates ________ _________ as the concentration of reactants decrease over time.

Problems:

1. Use the reaction shown below and the experimental data gathered in the table to determine the rate law for the reaction, the rate for the fourth experiment, and the rate constant.

$$2H_2(g) + 2NO(g) \rightarrow N_2(g) + 2H_2O(g)$$

Experiment	$[H_2]$ (M)	[NO] (M)	Rate (M/s)
1	0.010	0.020	1.64×10^{-6}
2	0.020	0.020	3.28×10^{-6}
3	0.020	0.040	6.56×10^{-6}
4	0.020	0.040	

2. Which of the following mechanisms do you think is most plausible for the reaction in question 1? Provide a 1–2 sentence justification for your answer.

Mechanism 1:

$H_2 + NO \rightarrow H_2O + N$ slow
$N + NO \rightarrow N_2 + O$ fast
$O + H_2 \rightarrow H_2O$ fast

Mechanism 2:

$H_2 + 2NO \rightarrow N_2O + H_2O$ slow
$N_2O + H_2 \rightarrow N_2 + H_2O$ fast

Mechanism 3:

$2NO \leftrightarrows N_2O_2$ fast equilibrium
$N_2O_2 + H_2 \rightarrow N_2O + H_2O$ slow
$N_2O + H_2 \rightarrow N_2 + H_2O$ fast

3. Consider the reaction of hydrogen and iodine gases to form hydrogen iodide:

   $H_2(g) + I_2(g) \rightarrow 2HI(g)$

   This reaction is thought to occur via the following reaction mechanism:

   $I_2(g) \leftrightarrows 2I(g)$ fast equilibrium

   $H_2(g) + 2I(g) \rightarrow 2HI(g)$ slow

   Determine the rate law and sketch the potential energy diagram for the reaction based on the mechanism given.

4. 2-Butene has two isomers: cis-2-butene and trans-2-butene. The cis conformation is 4 kJ/mol higher in energy than the trans form and, therefore, the trans form is dominant in a mixture of the compound. Cis-2-butene is converted to trans-2-butene using iodine gas as a catalyst according to the following five step mechanism:

   1st step: I_2 dissociates into 2I atoms.

   2nd step: I bonds to a carbon atom of cis-2-butene, breaking the carbon-carbon double bond.

   3rd step: The carbon-carbon bond, now a single bond, rotates freely.

   4th step: The I atom dissociates from its carbon, recreating the carbon-carbon double bond.

   5th step: I_2 is regenerated from 2I atoms.

   Given that the uncatalyzed reaction occurs in a single step with an activation energy of 262 kJ/mol and that the 3rd step of the catalyzed reaction has the highest activation energy (115 kJ) of the steps in the catalyzed mechanism, sketch the potential energy diagram for the catalyzed and uncatalyzed reactions.

5. Phenol acetate reacts with water to form acetic acid and phenol. This reaction is carried out and the concentration of phenol acetate is monitored over time:

[phenol acetate] (M)	Time (s)
0.55	0
0.42	15
0.31	30
0.23	45
0.17	60
0.12	75
0.082	90

   Graph the data in the following three ways:

   1. [phenol acetate] vs. time
   2. 1/[phenol acetate] vs. time
   3. ln[phenol acetate] vs. time

   Use these plots to determine the order of the reaction with respect to the concentration of phenol acetate.

6. Consider an exothermic reaction with an activation energy of 125 kJ/mol and a frequency factor of $4.0\times10^{13}s^{-1}$.

   a. What is the rate constant of the reaction at 25°C?

b. What will the rate constant be at 25°C if a catalyst is used which reduces the activation energy to 75 kJ/mol?

c. At what temperature would the uncatalyzed reaction have to be carried out in order for the rate constant to be equal to that of the catalyzed reaction at 25°C?

7. Radon is a radioactive noble gas that is sometimes found in basements of homes. Radon-222 decays according to first-order kinetics and has a half-life of 3.82 days. If 1.0 L of air contains 2.5×10^{13} atoms of radon, how many atoms of radon-222 will remain in 1.0 L of air in your basement after 30 days?

8. The decomposition of N_2O to N_2 and O_2 is a first-order reaction with a half-life of 3580 min at 730°C. If the initial pressure of N_2O is 2.1 atm at 730°C, calculate the total gas pressure inside the vessel after one half-life. Assume that the total volume and temperature remain constant.

Concept Questions:

1. A successful chemical reaction requires that reacting species collide, that they collide with the correct geometry, and that the reacting species have sufficient energy to overcome the reaction barrier. Discuss how temperature affects each of these criteria.

2. A series of three experiments were carried out on a single reaction at two different temperatures. The results of the experiments are plotted below.

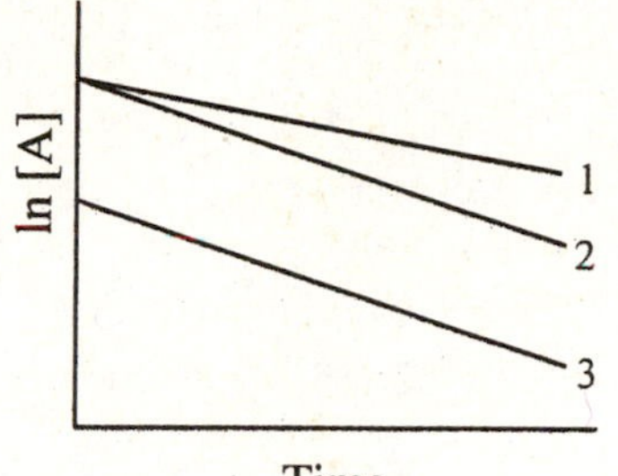

Explain how the experimental conditions of 1, 2, and 3 are different and how you can tell.

3. A glow stick can be "saved" by putting it in the freezer. How does this "save" it?

4. Consider the combustion of ethane (C_2H_6):

$$2C_2H_6(g) + 7O_2(g) \rightarrow 4CO_2(g) + 6H_2O(g)$$

Do you think that this reaction occurs in a single elementary step or multiple steps? Explain the reasoning behind your answer in 2–3 sentences.

5. In this chapter, we learned that increasing the temperature increases the rate of a chemical reaction. Some reactions are reversible; that is, they go in the forward and in the reverse direction. If heating up the reaction always increases the rate, then we should expect that both the forward and reverse reactions are sped up if the reaction is reversible. Using an Arrhenius plot, predict which is more sensitive to temperature, the forward or reverse reaction, for a reaction that is exothermic in the forward direction.

6. Termolecular reactions usually involve the association or combination of two particles with the help of a third, as shown in the reaction below:

$$2A + M \rightarrow B + M$$

In this reaction, M is a particle whose role is to remove the excess energy that is produced upon the bond formation that results in B. M is not a spectator since its presence is necessary in order for the reaction to occur. Using the information you have learned in this chapter, explain why it is reasonable that most termolecular reactions are of this type.

7. Three mechanisms are proposed for the reaction between carbon monoxide and nitrogen dioxide:

Mechanism 1:	$NO_2 + CO \rightarrow CO_2 + NO$	
Mechanism 2:	$NO_2 + NO_2 \rightarrow NO + NO_3$	slow
	$NO_3 + CO \rightarrow NO_2 + CO_2$	fast
Mechanism 3:	$NO_2 + NO_2 \leftrightarrows NO + NO_3$	fast equilibrium
	$NO_3 + CO \rightarrow NO_2 + CO_2$	slow

Suggest an experiment that would allow you to differentiate between the three mechanisms.

8. Rate laws can be written directly from the rate-determining elementary step. We can understand this by considering the concentration as being directly related to the number of collisions between reacting species. For example, the rate law: rate=k[A][B] suggests that the rate increases as the number of A and B molecules increase due to the increase in the number of collisions. Use the same type of logic to explain the exponent in the rate law: rate = $k[A]^2$.

9. An important aspect of catalysts is that they are not consumed in the reaction that they catalyze. For example, a single enzyme can be used to catalyze millions of reactions. Suggest why this might be a biologically important aspect to enzyme kinetics.

10. Su bstance X is found to be carcinogenic and have a half-life of 1.4 days. Considering the relationship between half-life and reaction order, after 1 week will substance X have a higher concentration if it decays via zero-, first-, or second-order kinetics?

Chapter 14: Chemical Equilibrium

Learning Goals:

- Understand the concept of dynamic equilibrium.
- Write equilibrium constant expressions for balanced chemical reactions in terms of concentrations and pressures.
- Understand the meaning of equilibrium constant values in terms of reactant- or product-favored reactions.
- Manipulate equilibrium constants alongside manipulations of chemical reactions.
- Calculate equilibrium concentrations given the equilibrium constant for a reaction.
- Predict the direction of chemical change using the reaction quotient.
- Utilize Le Châtelier's principle to predict the direction of a chemical reaction that has undergone a disturbance.

Chapter Summary:

Up until this point in the book, we have mostly written chemical reactions as proceeding in a single direction, from reactants to products. In this chapter, you will learn that chemistry is not generally that simple. Reactions often proceed in both the forward and reverse directions; when the rates of the forward and reverse reactions are equal, a dynamic equilibrium is established where the concentrations no longer appear to change. The position of the equilibrium is quantified by the equilibrium constant, which you will learn to write in terms of concentrations and pressures. You will learn how to understand the meaning of the equilibrium constant qualitatively and how to calculate its value quantitatively. By comparing non-equilibrium conditions to equilibrium conditions, you will learn to predict the direction in which a chemical reaction will proceed in order to establish equilibrium. Given the value of the equilibrium constant, you will be able to calculate equilibrium concentrations for all species using exact and approximate methods. Finally, you will learn to utilize Le Châtelier's principle to understand and predict the ways that equilibrium positions change when systems are pushed away from equilibrium.

Chapter Outline:

I. The Concept of Dynamic Equilibrium

   a. Reversible reactions are reactions that proceed in both the forward and reverse directions. Two oppositely facing arrows are used to denote reversible reactions:

   $$\text{Reactants} \leftrightarrows \text{Products}$$

   b. A dynamic equilibrium for a chemical reaction is the condition at which the rates of the forward and reverse reactions are equal.

      i. The concentrations of reactants and products do not change once equilibrium is established even though both forward and reverse reactions continue to take place.

II. The Equilibrium Constant (K)

   a. The equilibrium constant (K) quantifies equilibrium product and reactant concentrations.

   b. The equilibrium constant (K) is the ratio of product concentrations (in units of molarity) raised to their stoichiometric coefficients divided by the reactant concentrations (in units of molarity) raised to their stoichiometric coefficients.

$$aA + bB \leftrightarrows cC + dD$$

$$K = \frac{[C]^c[D]^d}{[A]^a[B]^b}$$

EXAMPLE:

Write equilibrium constants for the following reactions:

a. $N_2O_4(g) \leftrightarrows 2NO_2(g)$

$$K = \frac{[NO_2]^2}{[N_2O_4]}$$

b. $PCl_5(g) \leftrightarrows PCl_3(g) + Cl_2(g)$

$$K = \frac{[PCl_3][Cl_2]}{[PCl_5]}$$

c. $CH_4(g) + 2O_2(g) \leftrightarrows 2H_2O(g) + CO_2(g)$

$$K = \frac{[H_2O]^2[CO_2]}{[CH_4][O_2]^2}$$

d. $CH_3CO_2H(aq) + NH_3(aq) \leftrightarrows NH_4^+(aq) + CH_3CO_2^-(aq)$

$$K = \frac{[CH_3CO_2^-][NH_4^+]}{[CH_3CO_2H][NH_3]}$$

---

c. There are three regions in which K can be qualitatively interpreted:

   i. When $K \gg 1$, the reaction favors product formation at equilibrium.

   ii. When $K \ll 1$, the reaction favors reactant formation at equilibrium.

   iii. When $K \approx 1$, neither the product nor the reactant formation is favored at equilibrium.

   ▶ Note that an equilibrium constant of 1 does not mean that the concentrations of reactants and products are equal; it only means that the ratio of concentrations raised to their proper coefficients is equal to one.

EXAMPLE:

Determine if each of the following reactions favor the formation of products, reactants, or neither.

a. $2H_2O(l) \leftrightarrows H_3O^+(aq) + OH^-(aq)$ $\quad K = 1\times10^{-14}$ at 25°C

We see here that K is a number that is much smaller than 1, so this is a reactant-favored reaction.

b. $N_2(g) + O_2(g) \leftrightarrows 2NO(g)$ $\quad K = 5.0\times10^{-4}$ at 1627°C

We see again that K is a number that is much less than 1, so this is a reactant-favored reaction.

c. $ClNO_2(g) + NO(g) \leftrightarrows NO_2(g) + ClNO(g)$ $\quad K = 1.3\times10^{4}$ at 25°C

In this case, K is a number much larger than 1, so this is a product-favored reaction.

---

d. When manipulating a chemical equation, the value of the equilibrium constant changes in predictable ways:

   i. When the reaction is reversed, the new equilibrium constant is the inverse of the original equilibrium constant.

   ii. When the reaction is multiplied by a factor, the new equilibrium constant is the original equilibrium constant raised to the power of the factor.

   iii. When reactions are added together to give an overall reaction, the equilibrium constant of the overall reaction is the multiplicative product of the equilibrium constants for each step in the reaction.

EXAMPLE:

Determine the equilibrium constant for the reaction: $HCl(g) \leftrightarrows \frac{1}{2}Cl_2(g) + \frac{1}{2}H_2(g)$ given that the equilibrium constant for the reaction $Cl_2(g) + H_2(g) \leftrightarrows 2HCl(g)$ is 4×10^{31} at 300 K.

We first must reverse the reaction and take the inverse of the equilibrium constant:

$$2HCl(g) \leftrightarrows Cl_2(g) + H_2(g) \qquad K = \frac{1}{4\times10^{31}} = 2.5\times10^{-32}$$

Now we need to multiply the stoichiometric coefficients by ½ and raise K to the ½ power:

$$HCl(g) \leftrightarrows \tfrac{1}{2}Cl_2(g) + \tfrac{1}{2}H_2(g) \qquad K = (2.5\times10^{-32})^{1/2} = 1.6\times10^{-16}$$

III. Expressing K in Terms of Pressure

a. Up to now, we have expressed the equilibrium constant in terms of concentration. When the equilibrium constant is expressed in these terms, it is called K_c.

b. For a gas phase reaction, we can express the equilibrium constant in terms of partial pressures and will give the equilibrium constant the symbol K_p.

$$aA(g) + bB(g) \leftrightarrows cC(g) + dD(g)$$

$$K_P = \frac{P_C^c P_D^d}{P_A^a P_B^b}$$

EXAMPLE:

Write equilibrium constants in terms of pressure for the following reactions:

a. $N_2O_4(g) \leftrightarrows 2NO_2(g)$

$$K_P = \frac{{P_{NO_2}}^2}{P_{N_2O_4}}$$

b. $PCl_5(g) \leftrightarrows PCl_3(g) + Cl_2(g)$

$$K_P = \frac{P_{PCl_3} P_{PCl_2}}{P_{PCl_5}}$$

c. $CH_4(g) + 2O_2(g) \leftrightarrows 2H_2O(g) + CO_2(g)$

$$K_P = \frac{P_{H_2O}{}^2 P_{CO_2}}{P_{CH_4} P_{O_2}{}^2}$$

---

c. K_c and K_p are only equal when there is no change in the number of gas phase particles during the reaction. When the number of gas phase particles changes by Δn, K_p, and K_c are related through the expression:

$$K_p = K_c(RT)^{\Delta n}$$

EXAMPLE:

Determine K_p for the following reactions:

a. $PCl_3(g) + Cl_2(g) \leftrightarrows PCl_5(g)$ $\qquad K_c = 98.3$ at 400. K

There is one mole of gas on the product side of the reaction and two moles of gas on the reactant side, so the change in moles of gas is -1. We can use this in the expression for K_p:

$$K_p = 98.3(0.0821 \times 400)^{-1}$$

$$K_p = 2.99$$

b. $C_2H_6(g) + Cl_2(g) \leftrightarrows C_2H_5Cl(g) + HCl(g)$ $\qquad K_c = 0.10$ at 283 K

There are two moles of gas on the product side of the reaction and two moles of gas on the reactant side, so the change in moles of gas is 0. We can use this in the expression for K_p:

$$K_p = 0.10(0.0821 \times 283)^0$$

$$K_p = 0.10$$

We see that in this case K_p=K_c because the number of moles of gas didn't change in the course of the reaction.

c. $N_2O_4(g) \leftrightarrows 2NO_2(g)$ $\qquad K_c = 11$ at 373 ^{K}

There are two moles of gas on the product side of the reaction and one mole of gas on the reactant side, so the change in moles of gas is 1. We can use this in the expression for K_p:

$$K_p = 11(0.0821 \times 373)^1$$

$$K_p = 340$$

d. $SO_2(g) + \frac{1}{2}O_2(g) \leftrightarrows SO_3(g)$ $\qquad K_c = 2.2$ at 700. K

There is one mole of gas on the product side of the reaction and one and a half moles of gas on the reactant side, so the change in moles of gas is -½. We can use this in the expression for K_p:

$$K_p = 2.2(0.0821 \times 700)^{-1/2}$$

$$K_p = 0.29$$

---

d. The units of K are generally dropped because they are variable; units of molarity are assumed for the terms in K_c and units of atm are assumed for the terms in K_p.

IV. Heterogeneous Equilibrium: Reactions Involving Solids and Liquids

a. Concentrations of pure solids and liquids do not change in the course of a reaction as long as there is always some of the pure liquid and/or solid present.

b. Pure liquids and solids are not included in the equilibrium constant expression.

i. Since the concentrations of pure solids and liquids do not change in the course of a reaction, they are incorporated into the value of K_c or K_p.

EXAMPLE:

Write equilibrium constants for the following reactions:

a. $CO_2(g) + H_2(g) \leftrightarrows CO(g) + H_2O(l)$

$$K = \frac{[CO]}{[CO_2][H_2]}$$

b. $AgCl(s) \leftrightarrows Ag^+(aq) + Cl^-(aq)$

$$K = [Ag^+][Cl^-]$$

c. $CuO(s) + H_2(g) \leftrightarrows Cu(s) + H_2O(g)$

$$K = \frac{[H_2O]}{[H_2]}$$

V. Calculating the Equilibrium Constant from Measured Equilibrium Concentrations

a. The equilibrium constant can be calculated by inserting equilibrium concentrations into the equilibrium constant expression for a reaction.

b. The value of the equilibrium constant is always the same for a given temperature, but the concentration of reactant and product species can vary.

EXAMPLE:

Calculate the equilibrium constant, K_c, at 100°C for the reaction $N_2O_4(g) \leftrightarrows 2NO_2(g)$ given the following equilibrium concentrations:

a. $[N_2O_4]$=1.45 M, $[NO_2]$=3.99 M

First we will write down the equilibrium constant expression for the reaction:

$$K = \frac{[NO_2]^2}{[N_2O_4]}$$

We can now compute K by using the equilibrium values given in the problem:

$$\frac{3.99^2}{1.45} = 11.0$$

$K_p = K_c(RT)^{\Delta n}$

b. $[N_2O_4]=1.5\times10^{-4}$ M, $[NO_2]=4.1\times10^{-2}$ M

Since we already have the equilibrium constant expression, we can simply use the new equilibrium values to solve for K:

$$\frac{(4.1\times10^{-2})^2}{(1.5\times10^{-4})}=11$$

c. $[N_2O_4]=8.95\times10^{8}$ M, $[NO_2]=9.92\times10^{4}$ M

$$\frac{(9.92\times10^{4})^2}{8.95\times10^{8}}=11.0$$

d. $[N_2O_4]=386$ M, $[NO_2]=65.2$ M

$$\frac{65.2^2}{386}=11.0$$

We see from these four examples that the equilibrium constant is always the same number when the equilibrium concentrations are measured at 100°C for this reaction.

EXAMPLE:

The following reaction is carried out with initial concentrations of $[PCl_5]=4.580$ M, $[Cl_2]=5.870$ M, and $[PCl_3]=1.280$ M.

$$PCl_3(g) + Cl_2(g) \leftrightarrows PCl_5(g)$$

Given that the equilibrium concentration of PCl_5 is 5.847 M, calculate K_c.

We can use a table of the concentrations in order to solve this problem:

Reaction Condition	$[PCl_3]$ (M)	$[Cl_2]$ (M)	$[PCl_5]$ (M)
Initial	1.280	5.870	4.580
Change	-x	-x	+x
Equilibrium	1.280-x	5.870-x	4.580+x

The number of moles that react is x; x comes from the reaction stoichiometry: for every molecule of PCl_3 that reacts, one molecule of Cl_2 will react and one molecule of PCl_5 will be produced.

We are given that the equilibrium concentration of PCl_5 is 5.847 M, so we can solve for x:

$$4.580+x = 5.847 \quad \rightarrow \quad x = 1.267$$

From this we can calculate the equilibrium concentrations of PCl_5 and Cl_2:

$$[PCl_3] = 1.280\text{-}x = 0.013 \text{ M}$$

$$[Cl_2] = 5.870\text{-}x = 4.603 \text{ M}$$

Now that we have the equilibrium concentrations we can use them in the equilibrium constant expression:

$$K_c = \frac{[PCl_5]}{[PCl_3][Cl_2]}$$

$$K_c = \frac{5.847}{0.013\times4.603}$$

$K_c = 98$

VI. The Reaction Quotient: Predicting the Direction of Chemical Change

a. The reaction quotient is of the same form as the equilibrium constant expression, but uses non-equilibrium concentrations and can be used to predict the direction in which a chemical reaction proceeds:

   i. If $Q<K$, the reaction will proceed toward products until equilibrium is established.

   ii. If $Q>K$, the reaction will proceed toward reactants until equilibrium is established.

   iii. If $Q=K$, the reaction is at equilibrium.

EXAMPLE:

Consider the reaction between sulfur dioxide and nitrogen dioxide:

$$SO_2(g) + NO_2(g) \leftrightarrows SO_3(g) + NO(g) \qquad K_c = 8.8 \text{ at } 1000 \text{ K}$$

Given the following concentrations, predict the direction in which the reaction will proceed:

a. $[SO_2]$=64.0 M, $[NO_2]$=12.4, $[SO_3]$=18.5 M, $[NO]$=87.5 M

We will start by writing the expression for Q and then inserting the concentration values into it:

$$Q = \frac{[SO_3][NO]}{[SO_2][NO_2]}$$

$$Q = \frac{18.5 \times 87.5}{64.0 \times 12.4}$$

$$Q = 2.04$$

We see that $Q<K$, so the reaction will proceed toward products until equilibrium is reached.

b. $[SO_2]$=1.73 M, $[NO_2]$=0.581, $[SO_3]$=4.89 M, $[NO]$=158.2 M

We will use the same expression for Q and calculate using the revised concentrations:

$$Q = \frac{4.89 \times 158.2}{1.73 \times 0.581}$$

$$Q = 770.$$

In this case, $Q>K$ so the reaction will proceed in the reverse direction toward reactants until equilibrium is reached.

c. $P_{SO_2} = 4.87$ atm, $P_{NO_2} = 4.87$ atm, $P_{SO_3} = 10.4$ atm, $P_{NO} = 10.4$ atm

We need to rewrite Q in terms of partial pressures and calculate K_p:

$$Q = \frac{P_{SO_3} P_{NO}}{P_{SO_2} P_{NO_2}}$$

$$K_p = K_c(RT)^{\Delta n} \quad \rightarrow \quad K_p = K_c(RT)^0 \quad \rightarrow \quad K_p = K_c$$

Now we can use our expression for Q as before:

$$Q = \frac{10.4 \times 10.4}{4.87 \times 4.87}$$

$$Q = 4.56$$

Since Q<K, the reaction will proceed toward products until equilibrium is established.

VII. Finding Equilibrium Concentrations

a. Given the equilibrium constant and all but one of the equilibrium concentrations, the final equilibrium concentration can be easily identified.

EXAMPLE:

The equilibrium constant for the reaction $N_2O_4(g) \leftrightarrows 2NO_2(g)$ is 11.0 at 373 K. Calculate the equilibrium concentration of NO_2 given that the equilibrium concentration of N_2O_4 is 4.87 M.

We use the equilibrium constant expression and rearrange it to solve for the unknown variable:

$$K = \frac{[NO_2]^2}{[N_2O_4]} \quad \rightarrow \quad K[N_2O_4] = [NO_2]^2 \quad \rightarrow \quad \sqrt{K[N_2O_4]} = [NO_2]$$

And now we use the values given to solve for $[NO_2]$:

$$[NO_2] = \sqrt{11.0 \times 4.87} \quad \rightarrow \quad [NO_2] = 7.32 \text{ M}$$

---

b. Given the equilibrium constant and the initial concentrations or pressures, an ICE table can be used to determine equilibrium concentrations, using the variable x to represent the change in concentration required to reach equilibrium.

EXAMPLE:

The production of nitrogen monoxide gas from nitrogen and oxygen gases is reactant-favored at 2500 K:

$$N_2(g) + O_2(g) \leftrightarrows 2NO(g) \quad K = 5.0\times10^{-4} \text{ at } 2500 \text{ K}$$

If you set up this reaction with a 2.5 M sample of $N_2(g)$ and a 2.5 M sample of $O_2(g)$, what will the concentration of NO(g) be when the reaction reaches equilibrium?

We will set up an ICE table for this reaction using the information given and the reaction stoichiometry:

Reaction Condition	$[N_2]$ (M)	$[O_2]$ (M)	[NO] (M)
Initial	2.5	2.5	0
Change	-x	-x	+2x
Equilibrium	2.5-x	2.5-x	2x

We see in the table that for each mole of nitrogen gas that reacts, one mole of oxygen gas will react and two moles of nitrogen monoxide gas will be produced.

Now we will write the equilibrium constant expression for the reaction and then insert the equilibrium concentrations from the ICE table into it:

$$K = \frac{[NO]^2}{[N_2][O_2]}$$

$$5.0 \times 10^{-4} = \frac{(2x)^2}{(2.5-x)(2.5-x)}$$

$$5.0 \times 10^{-4} = \frac{(2x)^2}{(2.5-x)^2}$$

We can take the square root of both sides of this equation in order to simplify the expression:

$$\sqrt{5.0 \times 10^{-4}} = \frac{(2x)}{(2.5-x)}$$

We will now rearrange the expression and solve for x:

$$0.055902 - 0.022361x = 2x$$

$$0.055902 = 2.022361x$$

$$x = 0.02764$$

Since the concentration of NO at equilibrium is equal to 2x, [NO]=0.55 M.

EXAMPLE:

CO_2 and H_2 are placed in a closed container and react at a constant temperature according to the equation below. Calculate the equilibrium concentrations of all species when the initial pressures are: $P_{CO_2} = 2.00$ atm and $P_{H_2} = 1.00$ atm.

$$CO_2(g) + H_2(g) \leftrightarrows CO(g) + H_2O(g) \qquad K_p = 0.64 \text{ at } 900 \text{ K}$$

We will set up our ICE table using partial pressures and express changes in terms of pressure. This is reasonable since the pressure is directly proportional to the number of moles when the volume and temperature are held constant.

Reaction Condition	$[CO_2]$ (atm)	$[H_2]$ (atm)	[CO] (atm)	$[H_2O]$ (atm)
Initial	2.00	1.00	0	0
Change	-x	-x	+x	+x
Equilibrium	2.00-x	1.00-x	x	x

We can now write the equilibrium constant expression and then use the partial pressures of all species at equilibrium in order to solve for x:

$$K_p = \frac{P_{CO} P_{H_2O}}{P_{CO_2} P_{H_2}}$$

$$0.64 = \frac{(x)(x)}{(2.00-x)(1.00-x)}$$

We will now multiply out the terms in the denominator to get all values in the numerator of the equation:

$$0.64 = \frac{x^2}{(2.00 - 3.00x + x^2)}$$

$$1.28 - 1.92x + 0.64x^2 = x^2$$

Now we will put our expression in the form needed to use the quadratic formula:

$$1.28 - 1.92x - 0.36x^2 = 0$$

$$0.36x^2 + 1.92x - 1.28 = 0$$

We have a=0.36, b=1.92, and c=–1.28 for the quadratic formula:

$$x = \frac{-1.92 \pm \sqrt{(1.92)^2 - 4(0.36)(-1.28)}}{2(0.36)}$$

$$x = 0.60 \text{ or } x = -5.9$$

The solution x=–5.9 atm is physically unreasonable since it is negative, and the amount of product cannot increase from the starting conditions. The answer is, therefore, x=0.60 atm. Now we need to put this back into our expressions for the equilibrium partial pressures to answer the question asked:

$$P_{CO_2} = 2.00 - 0.60 = 1.40 \text{ atm}$$

$$P_{H_2} = 1.00 - 0.60 = 0.40 \text{ atm}$$

$$P_{CO} = 0.60 \text{ atm}$$

$$P_{H_2O} = 0.60 \text{ atm}$$

To check the answer, plug these pressures into the equilibrium ratio: $(0.60\times0.60)/(1.40\times0.40) = 0.64$

---

c. When the equilibrium constant is very small, the reaction will not proceed very far towards products and an approximation can be used to determine the equilibrium concentrations.

   i. The approximation is only valid when <5% of the original reactant concentration reacts.

      ▶ In cases where the equilibrium constant is 100 times smaller than the initial concentration of the reactant, this approximation is generally valid.

EXAMPLE:

Acetic acid dissociates in water according to the equation:

$$HC_2H_3O_2(aq) + H_2O(l) \leftrightarrows C_2H_3O_2^-(aq) + H_3O^+(aq) \quad K=1.8\times10^{-5} \text{ at } 25^\circ C$$

If you begin with a 0.476 M solution of acetic acid, what will the concentrations of all species be when the solution reaches equilibrium?

We will set up an ICE table for the reaction. In the ICE table we will ignore H_2O since pure liquids do not appear in the equilibrium constant expression:

Reaction Condition	$[HC_2H_3O_2]$ (M)	$[H_2O]$ (M)	$[C_2H_3O_2^-]$ (M)	$[H_3O^+]$ (M)
Initial	0.476	-	0	0

Change	-x	-	+x	+x
Equilibrium	0.476-x	-	x	x

We can now write the equilibrium constant expression and then use the concentrations of all species at equilibrium in order to solve for x:

$$K = \frac{[C_2H_3O_2^-][H_3O^+]}{[HC_2H_3O_2]}$$

$$1.8\times10^{-5} = \frac{(x)(x)}{(0.476-x)}$$

Here we could rearrange and solve the quadratic formula. Since K is so small in this case, however, we know that the reaction will not go very far to form products; we make the approximation that 0.476-x will be approximately 0.476 and solve the much simpler expression:

$$1.8\times10^{-5} = \frac{x^2}{0.476}$$

$$8.568\times10^{-6} = x^2$$

$$x = 0.00293$$

In order to determine if our approximation is valid, we will calculate the percent of the original concentration that has reacted:

$$\frac{0.00293}{0.476}\times100\% = 0.615\%$$

Since this is much less than 5%, our approximation is valid and we can calculate the equilibrium concentrations of all species:

$$[HC_2H_3O_2] = 0.473\text{ M} - 0.00293\text{ M} = 0.47\text{ M}$$

$$[C_2H_3O_2^-] = 0.0029\text{ M}$$

$$[H_3O^+] = 0.0029\text{ M}$$

---

ii. When >5% of the original reactant concentration reacts, the method of successive approximation can be used.

EXAMPLE:

Calculate the percentage of phosphorous pentachloride that dissociates when 0.05 mol of PCl_5 is placed in a closed container at 250°C and 2.00 atm pressure. Note that the volume of the container does not change during the course of the reaction.

$$PCl_5(g) \leftrightarrows PCl_3(g) + Cl_2(g) \qquad K_p = 1.78 \text{ at } 250°C$$

Since the volume of the container is not permitted to change, the pressure in the container will change over time. It will be simpler to use the K_c expression, so we need to calculate the initial concentration of PCl_5 and the value of K_c at 250°C:

$$\frac{n}{V} = \frac{P}{RT} \quad \rightarrow \quad \frac{n}{V} = \frac{2.00\text{ atm}}{0.0821\frac{\text{L}\cdot\text{atm}}{\text{mol}\cdot\text{K}}\cdot 523\text{ K}} \quad \rightarrow \quad \frac{n}{V} = 0.046578\text{ M}$$

$$K_p = K_c(\text{RT})^{\Delta n} \quad \rightarrow \quad \frac{K_p}{(\text{RT})^{\Delta n}} = K_c \quad \rightarrow \quad K_c = \frac{1.78}{(0.0821 \cdot 523)^1} \quad \rightarrow \quad K_c = 0.041455$$

Now we can set up an ICE table for the reaction:

Reaction Condition	$[PCl_5]$ (M)	$[PCl_3]$ (M)	$[Cl_2]$ (M)
Initial	0.04658	0	0
Change	-x	+x	+x
Equilibrium	0.04658 – x	x	x

Now we will insert these values into the equilibrium constant expression:

$$K = \frac{[PCl_3][Cl_2]}{[PCl_5]}$$

$$0.04155 = \frac{(x)(x)}{(0.04658 - x)}$$

We will approximate the value of 0.04658-x to be 0.04658 and solve for x:

$$0.04155 = \frac{x^2}{(0.04658)}$$

$$0.0019354 = x^2$$

$$x = 0.04399$$

This is much larger than 5% of the initial PCl_5 concentration – it is actually 94.4%. So we cannot stop our calculation here. Instead, we will use this value to get a new approximation. Instead of letting 0.04658-x=0.04658, we will insert our calculated value of x:

$$0.04155 = \frac{x^2}{(0.04658 - 0.04399)} \quad \rightarrow \quad x = 0.01047$$

We will again substitute this value into the denominator of our equilibrium constant expression because it is very different from our previous result:

$$0.04155 = \frac{x^2}{(0.04658 - 0.01047)} \quad \rightarrow \quad x = 0.03873$$

Substituting in with our new value of x gives:

$$0.04155 = \frac{x^2}{(0.04658 - 0.03873)} \quad \rightarrow \quad x = 0.01806$$

Substituting in with our new value of x gives:

$$0.04155 = \frac{x^2}{(0.04658 - 0.01806)} \quad \rightarrow \quad x = 0.03442$$

We will continue to substitute in the new value of x until our answers converge to a single value. The numerous substitutions are not shown here – only the beginning and end of the process are shown.

Substituting in with our new value of x gives:

$$0.04155 = \frac{x^2}{(0.04658 - 0.02789)} \quad \rightarrow \quad x = 0.02787$$

Substituting in with our new value of x gives:

$$0.04155 = \frac{x^2}{(0.04658 - 0.02787)} \quad \rightarrow \quad x = 0.02788$$

So we can now calculate the equilibrium values:

$[PCl_5] = 0.0187$ M

$[PCl_3] = 0.0279$ M

$[Cl_2] = 0.0279$ M

Note that in this problem, the quadratic formula would have been a much simpler (and faster) way to solve for x. A good rule of thumb is that the method of successive approximations is only more efficient when the percent dissociation calculated with the first approximation is around 10% or less.

VIII. Le Châtelier's Principle: How a System at Equilibrium Responds to Disturbances

a. Le Châtelier's principle states that when a chemical system at equilibrium is disturbed, the system shifts in a direction that minimizes the disturbance.

    i. The system will find a new equilibrium position after a disturbance.

b. Concentration changes affect the equilibrium position in a manner predicted by the reaction quotient, Q.

    i. An increase in the reactant concentrations results in a shift toward products.

    ii. An increase in the product concentrations results in a shift toward reactants.

    iii. A decrease in the reactant concentrations results in a shift toward reactants.

    iv. A decrease in the product concentrations results in a shift toward products.

EXAMPLE:

Aqueous iron ions are yellow and form a red, complex ion ($FeSCN^{2+}$) when combined with the colorless thiocyanate ion (SCN^-):

$$Fe^{3+}(aq) + SCN^-(aq) \leftrightarrows FeSCN^{2+}(aq)$$

How will the color of the solution change when the following changes are carried out:

a. A solution of $Fe(NO_3)_3$ is added to the reaction.

When $Fe(NO_3)_3$ is added to the reaction, the concentration of Fe^{3+} increases. Since Fe^{3+} is a reactant, the reaction will shift toward the formation of products until equilibrium is re-established. The formation of more products will result in a solution that is more red.

b. Solid KSCN is added to the solution.

When KSCN is added to the reaction the concentration of SCN^- increases. Since SCN^- is a reactant, the reaction will shift toward the formation of products until equilibrium is re-established. The formation of more products will result in a solution that is more red.

c. Fe^{3+} is removed from the flask.

Since Fe^{3+} is a reactant, its removal will result in the formation of more reactants in order to re-establish equilibrium. The reduction in the product concentration will result in a solution that is less red.

---

c. The effect of a change in volume or pressure change on equilibrium depends on the number of gas phase species on each side of the equation.

   i. A decrease in volume (or increase in pressure) will cause the reaction to shift toward the side of the reaction with fewer gas phase particles.

   ii. An increase in volume (or decrease in pressure) will cause the reaction to shift toward the side of the reaction with more gas phase particles.

   iii. If there are the same number of gas phase particles on either side of the reaction, the equilibrium position will not change when the volume (or pressure) is changed.

   iv. When an inert gas is added to a reaction vessel, the partial pressures of all species remain unchanged and there is, therefore, no change in the equilibrium position.

EXAMPLE:

When solid carbon is placed in a closed container with water vapor, hydrogen gas and carbon monoxide gas are produced according to the following equation:

$$C(s) + H_2O(g) \leftrightarrows CO(g) + H_2(g)$$

The reaction is allowed to reach equilibrium. Predict how the concentration of water vapor will change when each of the following adjustments are made:

a. Carbon monoxide gas is added to the reaction vessel.

   Since CO is a product in the reaction, its addition will cause the reaction to shift toward reactants, increasing the concentration of water vapor.

b. The volume of the container is doubled.

   When the volume of the container is doubled, the pressure is cut in half. In order for the equilibrium to be re-established the pressure will increase through the formation of more gas molecules. Since the product side of the reaction has more gas particles, the reaction will shift to form more products until equilibrium is established and thereby reduce the concentration of water vapor.

c. 5.0 mol of argon gas is added to the container.

   When an inert gas is added to the container, the partial pressures of the gases remain unchanged. Since there is no change in the partial pressures of the gas, the system is not moved from the equilibrium position and the concentration of water vapor will remain unchanged.

d. Solid carbon is added to the reaction vessel.

   When solid carbon is added to the container, the equilibrium remains unchanged since the concentration of a solid is never changed (so long as there is some available). Consequently, the concentration of water vapor will remain unchanged.

e. Liquid water is added to the reaction vessel.

   When liquid water is added to the reaction vessel, the pressure of water vapor increases, meaning the equilibrium is disturbed and the reaction will shift to the right in order to produce more products. The concentration of the water vapor will decrease as a result, but it will be higher than it was in the original equilibrium mixture.

---

d. When the temperature of a reaction is changed, the equilibrium constant changes.

   i. In order to predict the direction of the equilibrium constant change, we can think of heat as either a reactant or product in the reaction.

      1. In an exothermic reaction, heat can be thought of as a product. Increasing the temperature will cause the reaction to shift toward the reactants and decreasing the temperature will cause the reaction to shift toward the products.

      2. In an endothermic reaction, heat can be thought of as a reactant. Increasing the temperature will cause the reaction to shift toward the products and decreasing the temperature will cause the reaction to shift toward the reactants.

EXAMPLE:

The reaction of hydrogen gas with iodine gas to form hydrogen iodide is endothermic:

$$H_2(g) + I_2(g) \leftrightarrows 2HI(g) \qquad \Delta H = +52 \text{ kJ/mol}$$

How will placing the reaction vessel in an ice bath change the concentration of HI at equilibrium?

Since the reaction is endothermic, we can consider heat to be a reactant in the reaction. If we place the reaction vessel in ice water, heat will flow from the reaction vessel into the surroundings meaning that heat will be removed from the system. Removing heat (a reactant) will cause the reaction to shift toward the reactants until equilibrium is re-established; the concentration of HI will therefore decrease.

Fill in the Blank:

1. Pure solids and liquids do not appear in the equilibrium constant expression because their concentrations ________________.
2. K_p and K_c are equal to one another one when the ________________ does not change in the course of a chemical reaction.
3. In an exothermic reaction, raising the temperature will cause the reaction to shift toward the ________________.
4. The ________________ is an expression similar to the equilibrium constant expression, but may utilize non-equilibrium concentrations.
5. Dynamic equilibrium is the condition for a chemical reaction under which the rate of the forward reaction is ________________ to the rate of the reverse reaction.
6. When K is much larger than 1, the reaction favors the formation of ________________ at equilibrium.
7. When a chemical reaction is reversed, the value of the equilibrium constant is ________________.
8. The ________________ is a way to quantify the relationships between concentrations of reactants and products at equilibrium.
9. Le Châtelier's principle states that a system at equilibrium, when disturbed, will shift in the direction that ________________ the disturbance.
10. A reaction that can proceed in both the forward and reverse directions is said to be ________________.

Problems:

1. Use the reactions provided below to determine the value of the equilibrium constant, K_c, for the reaction of hydrosulfuric acid and water:

$$2H_2S(aq) + 3H_2O(l) \leftrightarrows 3H_3O^+(aq) + S^{2-}(aq) + HS^-(aq)$$

Reaction 1: $H_2S(aq) + H_2O(l) \leftrightarrows H_3O^+(aq) + HS^-(aq)$ $K_c = 1.1\times10^{-3}$

Reaction 2: $H_3O^+(aq) + S^{2-}(aq) \leftrightarrows HS^-(aq) + H_2O(l)$ $K_c = 1.0\times10^{19}$

2. At 500°C the equilibrium constant for the formation of ammonia is measured:

$$N_2(g) + 3H_2(g) \leftrightarrows 2NH_3(g) \qquad K_p = 0.0610$$

If analysis of the concentrations of all species finds that 3.0 mol of N_2, 2.0 mol of H_2, and 0.50 mol of NH_3 are present in a 1.0-L reaction flask, is the reaction at equilibrium? If not, in which direction will the reaction proceed in order to reach equilibrium?

3. When PCl_5 is put into a closed 2.0-L container at 556 K, it decomposes according to the reaction:

$$PCl_5(g) \leftrightarrows Cl_2(g) + PCl_3(g) \qquad K_p = 4.96$$

a. What is the value of K_c for this reaction?

b. If the total pressure in the reaction vessel is 0.50 atm, what fraction of PCl_5 has decomposed at equilibrium?

c. If the total pressure is 1.00 atm, what fraction of PCl_5 has decomposed at equilibrium?

d. How do your answers in part b and c relate to Le Châtelier's principle?

4. A 200.0-g sample of NH_4HS is placed into a 5.0-L reaction vessel at 25°C. If the only reaction that takes place is:

$$NH_4HS(s) \leftrightarrows NH_3(g) + H_2S(g) \qquad K_p = 0.110$$

a. What are the equilibrium partial pressures of NH_3 and H_2S?

b. What percentage of NH_4HS has decomposed at this temperature?

c. What would the equilibrium partial pressures of NH_3 and H_2S be if 0.500 mol of NH_3 is added to the flask after it has reached equilibrium?

5. The photosynthesis reaction is endothermic (the energy from the sun is used) and results in the creation of sugar:

$$6CO_2(g) + 6H_2O(l) \leftrightarrows C_6H_{12}O_6(aq) + 6O_2(g) \qquad \Delta H = +2802 \text{ kJ}$$

Suppose that this reaction is at equilibrium in a closed container. How will the equilibrium shift if the following changes are made?

a. The partial pressure of O_2 is increased.

b. The system is compressed.

c. The temperature is increased.

d. Helium gas is added to the container at constant pressure.

e. Helium gas is added to the container at constant volume.

f. Water is added.

g. The partial pressure of CO_2 is decreased.

6. Silver bicarbonate dissolves in water according to the equilibrium equation:

$$AgHCO_3(s) + H_2O(l) \leftrightarrows Ag^+(aq) + HCO_3^-(aq)$$

The bicarbonate ion can further dissociate:

$$HCO_3^-(aq) + H_2O(l) \leftrightarrows H_3O^+(aq) + CO_3^{2-}(aq)$$

a. List all of the species that are present at equilibrium when $AgHCO_3$ is dissolved in water.

b. How would the addition of hydrochloric acid affect the amount of silver bicarbonate present in the reaction mixture?

c. Adding which of the following compounds would result in a decrease in the concentration of H_3O^+ in the equilibrium solution: NaOH, $NaHCO_3$, or NaCl?

Concept Questions:

1. In this chapter, we have learned that a completed chemical reaction is actually still occurring. Since concentrations and reaction rates do not change after equilibrium is established, this can be hard to believe. Suggest an experiment that would allow us to confirm that reactions at equilibrium are, in fact, dynamic processes.

2. A reaction A + B $\leftrightarrows$ C is studied by a group of students in the laboratory. The first experiment shows that K=1. It is suggested that no more experiments need to be done because the concentrations will always be equal. What is wrong with this suggestion?

3. The definition of equilibrium states that the rates of the forward and reverse reactions are equal at equilibrium. For the general reaction:

   $$aA(aq) + bB(aq) \leftrightarrows cC(aq)$$

   Show that $K=k_1/k_{-1}$ using the definition of equilibrium in terms of reaction rates. Hint: start by assuming that this is a one-step reaction and that the rate constant for the forward reaction is k_1 and the rate of the reverse reaction is k_{-1}.

4. Le Châtelier's principle predicts how changes in reactant and product concentrations will cause the equilibrium position to shift. Explain how concentration changes affect reaction rates and use the expression $K=k_1/k_{-1}$ to verify your predictions using Le Châtelier's principle.

5. The discussion of Le Châtelier's principle in this chapter considers changes in temperature as if heat were a product or reactant in the reaction. In fact, the effects of temperature changes are more complicated than the effects of concentration changes because the value of the equilibrium constant is temperature dependent.

   Sketch a potential energy diagram for an exothermic reaction and then use an Arrhenius plot to examine how temperature affects the rates of the forward and reverse reactions. Using this analysis and your expression for K from the previous problem, explain the way that temperature affects the position of the equilibrium for an exothermic reaction.

6. Analyze the effects of temperature changes on equilibrium positions using the Boltzmann distribution of molecular speeds. First, look at the fraction of reactant species that will have sufficient energy to overcome a reaction barrier at two different temperatures. Estimate the ratio of reactive species at the two temperatures:

   $$\frac{\text{fraction of molecules with energy above } E_a \text{ at } T_1}{\text{fraction of molecules with energy above } E_a \text{ at } T_2}$$

   Finally, determine how this ratio will change when the activation energy is changed.

Chapter 15: Acids and Bases

Learning Goals:

- Identify acids and bases using the Arrhenius, Brønsted-Lowry, and Lewis definitions.
- Write acid and base ionization equations and their corresponding equilibrium constants.
- Use acid/base ionization constants to calculate equilibrium concentrations and estimate acid and base strengths.
- Utilize the autoionization of water to determine the pH and pOH of aqueous solutions.
- Understand the acid/base properties of salt solutions.
- Calculate the base ionization constants for solutions containing the conjugate base of a weak acid and acid ionization constants for solutions containing the conjugate acid of a weak base.
- Understand the behavior of polyprotic acids and calculate equilibrium concentrations for their solutions.
- Predict relative acid strengths using structural elements.

Chapter Summary:

In this chapter, a specific type of equilibrium, acid and base ionization, will be explored in detail. First you will learn to identify acids and bases using the Arrhenius and Brønsted-Lowry definitions. Using the reactions that describe the behavior of acids and bases in water, you will learn how to write acid and base dissociation constants; these will then be used to calculate equilibrium concentrations of all species in acid and base solutions. You will learn about the pH scale and how the autoionization of water relates to all acid and base solutions. Once you are comfortable with equilibrium calculations for acids and bases, you will learn how to carry out similar calculations on the conjugate pairs of acids and bases; both a quantitative and qualitative treatment of salt solutions will then be considered. Acids that are able to donate more than one proton will next be explored. After a thorough quantitative treatment, you will learn how to compare acid strengths based on molecular structure. Finally, Lewis acids and bases will be explained.

Chapter Outline:

I. The Nature of Acids and Bases

   a. Acids have a sour taste, dissolve many metals, turn litmus red, and neutralize bases.

      i. Common acids include HCl, H_2SO_4, HNO_3, and $HC_2H_3O_2$.

      ii. Many naturally occurring acids are carboxylic acids which contain the $-CO_2H$ group.

   b. Bases have a bitter taste, feel slippery, turn litmus blue, and neutralize acids.

      i. Common bases include NaOH, KOH, $NaHCO_3$, and NH_3.

II. Definitions of Acids and Bases

   a. According to the Arrhenius definition, acids produce H^+ when dissolved in water and bases produce OH^- when dissolved in water.

      i. When acids dissolve in water, we don't write H^+, instead we write the hydronium ion, H_3O^+, to indicate that a free proton is too attracted to polar water molecules to exist free in water.

ii. Acids and bases react together to produce water (neutral solution). This is what we are referring to when we say that acids neutralize bases and vice versa.

b. According to the Brønsted-Lowry definition, acids donate protons while bases accept protons.

i. Acids and bases always appear together in a reaction.

ii. When an acid donates a proton, it becomes a conjugate base; when a base accepts a proton, it becomes a conjugate acid.

iii. An amphoteric species is one that can serve as an acid or base in a chemical reaction.

EXAMPLE:

Write the acid base reaction that will occur between each pair of species. Identify the conjugate acid base pairs in each equation.

a. Ammonia with water

$NH_3(aq) + H_2O(l) \leftrightarrows NH_4^+(aq) + OH^-(aq)$

Ammonia is the base and ammonium ion is its conjugate acid. Water is an acid here and the hydroxide ion is its conjugate base.

b. Ammonia with hydrochloric acid

$NH_3(aq) + HCl(aq) \leftrightarrows NH_4^+(aq) + Cl^-(aq)$

In this reaction, ammonia is the base and the ammonium ion is its conjugate acid. Hydrochloric acid is the acid and the chloride ion is its conjugate base.

We can write this reaction in an alternate way that emphasizes the strength of hydrochloric acid. Since hydrochloric acid is a strong acid, it will dissociate completely in water to produce chloride ions and hydronium ions:

$NH_3(aq) + H_3O^+(aq) + Cl^-(aq) \leftrightarrows NH_4^+(aq) + H_2O(l) + Cl^-(aq)$

We see that chloride ions are spectator ions in this reaction. Ammonia is again the base while the ammonium ion is its conjugate acid. H_3O^+ is the acid while H_2O is its conjugate base.

c. Carbonic acid with dimethylamine, $(CH_3)_2NH$

$H_2CO_3(aq) + (CH_3)_2NH(aq) \leftrightarrows HCO_3^-(aq) + (CH_3)_2NH_2^+(aq)$

Carbonic acid is the acid while the bicarbonate ion is its conjugate base. Dimethylamine is the base and the dimethyl ammonium ion is its conjugate acid.

d. Sodium acetate with ammonium chloride

$NaC_2H_3O_2(aq) + NH_4Cl(aq) \leftrightarrows HC_2H_3O_2(aq) + NH_3(aq) + NaCl(aq)$

It is difficult to see what the conjugate acid base pairs are here, so we will rewrite the equation without the spectator ions:

$C_2H_3O_2^-(aq) + NH_4^+(aq) \leftrightarrows HC_2H_3O_2(aq) + NH_3(aq)$

We can see that the acetate ion is the base while acetic acid is its conjugate base and the ammonium ion is the acid with ammonia as its conjugate base.

III. Acid Strength and the Acid Dissociation Constant

a. The strength of an acid depends on the equilibrium position of its dissociation in water:

$$HA(aq) + H_2O(l) \leftrightarrows H_3O^+(aq) + A^-(aq)$$

b. The equilibrium constant for this reaction is called the acid dissociation or acid ionization constant and has the form:

$$K_a = \frac{[H_3O^+][A^-]}{[HA]}$$

EXAMPLE:

Write equations for dissociation and the acid ionization constants for the following weak acids:

The acid dissociation reaction will be the reaction of the weak acid with water serving as a base. The acid ionization constant will be the concentrations of the products over the concentrations of reactants; water is not included in the acid ionization constants because it is a pure liquid.

a. Hydrofluoric acid, HF

$HF(aq) + H_2O(l) \leftrightarrows H_3O^+(aq) + F^-(aq)$ $\quad K_a = \frac{[H_3O^+][F^-]}{[HF]}$

b. Oxalic acid, $H_2C_2O_4$

$H_2C_2O_4(aq) + H_2O(l) \leftrightarrows H_3O^+(aq) + HC_2O_4^-(aq)$ $\quad K_a = \frac{[H_3O^+][HC_2O_4^-]}{[H_2C_2O_4]}$

c. Benzoic acid, $C_6H_5CO_2H$

$C_6H_5CO_2H(aq) + H_2O(l) \leftrightarrows H_3O^+(aq) + C_6H_5CO_2^-(aq)$ $\quad K_a = \frac{[H_3O^+][C_6H_5CO_2^-]}{[C_6H_5CO_2H]}$

---

c. The strength of an acid depends on the value of K_a, which is an indication of the position of the equilibrium.

i. Strong acids are species that completely ionize in water; the equilibrium position of a strong acid lies far to the right.

1. Strong acids have such large equilibrium constants that we write their dissociation using a single forward arrow:

$$HA(aq) + H_2O(l) \rightarrow H_3O^+(aq) + A^-(aq)$$

2. A strong acid has a very weak conjugate base.

ii. Weak acids only partially ionize in water so that both the ionized and nonionized forms of the acid exist in water.

1. The extent to which an acid dissociates depends on the attraction of the anion to the H^+. The weaker the attraction, the less the reaction will proceed in the reverse direction and the stronger the acid is.

d. Monoprotic acids have only one ionizable proton, diprotic acids have two ionizable protons, and triproptic acids have three ionizable protons.

IV. Auto-Ionization of Water and pH

a. Water is amphoteric, which means that it can serve as an acid or as a base.

b. The auto-ionization of water is the reaction of water with itself in an acid/base manner:

$$H_2O(l) + H_2O(l) \leftrightharpoons H_3O^+(aq) + OH^-(aq)$$

i. The equilibrium constant for this reaction is called the ion product constant of water:

$$K_w = [OH^-][H_3O^+] = 1\times10^{-14} \text{ at } 25°C$$

c. In pure water $[H_3O^+]=[OH^-]=10^{-7}$ M at 25°C.

d. In an aqueous solution, the equilibrium constant for the auto-ionization of water remains constant (equilibrium constants only change when the temperature changes); this is true even when an acid or base is dissolved in water.

i. In an acidic solution $[H_3O^+]>[OH^-]$

ii. In a basic solution $[H_3O^+]<[OH^-]$

iii. In a neutral solution $[H_3O^+]=[OH^-]$

e. The pH scale is used to compare the acidity of solutions:

$$pH = -\log[H_3O^+]$$

f. The pOH scale is used to compare the basicity of solutions:

$$pOH = -\log[OH^-]$$

g. Using the equilibrium constant of water, we can derive a relationship between pH and pOH:

$$pH + pOH = 14 \text{ at } 25°C$$

EXAMPLE:

Calculate the hydronium ion concentration and the pH of the following solutions and determine whether the solutions are acidic, basic, or neutral.

a. $[OH^-] = 4.63\times10^{-3}$

The hydroxide ion concentration and the hydronium ion concentration are related by the ion product constant of water:

$$10^{-14} = (4.63\times10^{-3})\times[H_3O^+]$$

$$[H_3O^+]= 2.16\times10^{-12}$$

The pH can then be calculated by taking the negative log of the hydronium ion concentration:

$$pH = -\log(2.16\times10^{-12}) = 11.666$$

which is a pH that is greater than 7 meaning that the solution is basic.

b. $[OH^-] = 1.89\times10^{-8}$

Again, we will calculate the hydronium ion concentration using the ion product constant of water:

$$10^{-14} = (1.89\times10^{-8})\times[H_3O^+]$$

$$[H_3O^+]= 5.291\times10^{-7}$$

The pH can then be calculated by taking the negative log of the hydronium ion concentration:

$$pH = -\log(5.291\times10^{-7}) = 6.276$$

A pH that is less than 7 means that the solution is acidic.

V. Finding the Hydronium Ion Concentration and pH of Strong and Weak Acid Solutions

a. In most acid solutions, the concentration of hydronium ions from the auto-ionization of water is negligible. Note that this is not true when the acid is sufficiently dilute or sufficiently weak.

b. The concentration of hydronium ions in a strong acid solution is equal to the concentration of the strong acid.

EXAMPLE:

What is the pH of the following solutions?

a. 1.0×10^{-3} M HCl

HCl is a strong acid so it will completely dissociate in water; the concentration of hydronium ions in water will therefore be 1.0×10^{-3} M. We can easily see that the pH of this solution is 3.00 because the log function is the inverse of raising ten to the power of a number (i.e., $\log 10^x = x$).

b. 1.25×10^{-2} M H_2SO_4

H_2SO_4 is a strong acid and will dissociate completely in water to give a hydronium ion concentration of 1.25×10^{-2} M. This is a number between 10^{-1} M and 10^{-2} M, so the pH should be between 1 and 2. Using a calculator to calculate we find:

$$pH = -\log(1.25\times10^{-2}) = 1.90$$

which is, in fact, between 1 and 2.

c. 6×10^{-9} M HNO_3

HNO_3 is another strong acid so it will dissociate completely in water to give a hydronium ion concentration of 6×10^{-9} M. We can calculate the pH of the solution from this value (pH = 8.22), but this cannot be the pH of the solution since it is basic! Because the nitric acid is so dilute, the hydronium ion contribution from the autoionization of water is much larger than that from the acid. This solution will actually have a hydronium ion concentration of:

$$1.0\times10^{-7}\text{ M (from water)} + 6\times10^{-9}\text{ M (from HNO}_3\text{)} = 1.06\times10^{-7}\text{ M}$$

which gives a pH of:

$$pH = -\log(1.06\times10^{-7}) = 6.97$$

d. 4.51×10^{-2} M NaOH

Since NaOH is a strong base, it will completely dissociate in water to give a hydroxide ion concentration of 4.51×10^{-2} M. We can use this concentration to calculate the pOH:

$$pOH = -\log(4.51\times10^{-2}) = 1.3458$$

This is reasonable since 4.51×10^{-2} is a number between 10^{-2} and 10^{-1}.

The pH can then be calculated by subtracting the pOH from 14.000:

$$pH = 14.000 - 1.346 = 12.6542$$

Our answer should have three significant figures after the decimal point, so the pH is 12.654.

---

c. In a weak acid solution, the concentration of hydronium ions must be solved according to the equilibrium equation using an ICE table as in the previous chapter.

EXAMPLE:

Calculate the pH of a 0.425 M solution of boric acid, H_3BO_3. The K_a of boric acid is 5.4×10^{-10}.

We first need to write out the ionization equation for boric acid.

$$H_3BO_3(aq) + H_2O(l) \leftrightarrows H_3O^+(aq) + H_2BO_3^-(aq) \qquad K_a = \frac{[H_3O^+][H_2BO_3^-]}{[H_3BO_3]}$$

We can use an ICE table to solve for the equilibrium concentrations:

Reaction Condition	$[H_3BO_3]$ (M)	H_2O	$[H_2BO_3^-]$ (M)	$[H_3O^+]$ (M)
Initial	0.425	-	0	0
Change	-x	-	+x	+x
Equilibrium	0.425-x	-	x	x

We can now put values into the acid ionization constant expression:

$$5.4\times10^{-10} = \frac{x \cdot x}{(0.425 - x)}$$

We can compare the acid ionization constant of boric acid with the initial concentration of boric acid to determine if we can approximate $0.425\text{-}x \approx 0.425$. Since the K_a value is nine orders of magnitude smaller than the initial acid concentration, we are safe in making the approximation.

$$5.4\times10^{-10} \approx \frac{x \cdot x}{0.425}$$

$$x = \sqrt{2.295\times10^{-10}}$$

$$x = 1.515\times10^{-5}$$

Since x is equal to the hydronium ion concentration at equilibrium, we simply need to calculate the pH:

$$pH = -\log(1.515\times10^{-5}) = 4.80$$

This value makes sense because we have a very weak acid.

---

d. The percent ionization of an acid is equal to the concentration of the conjugate base at equilibrium or the hydronium ion concentration at equilibrium over the initial acid concentration expressed as a percentage:

$$\frac{[A^-]_{eq}}{[HA]_{initial}}\times100\% = \frac{[H_3O^+]_{eq}}{[HA]_{initial}}\times100\% = \text{Percent Ionization}$$

EXAMPLE:

Calculate the percent ionization for the boric acid solution in the previous example.

We calculated the conjugate base concentration in the boric acid solution as 1.515×10^{-5} M. The percent ionization is then:

$$\frac{1.515\times10^{-5}\ \mathrm{M\ H_2BO_3^-}}{0.425\ \mathrm{M\ H_3BO_3}}\times100\% = 0.0036\,\%\ \mathrm{Ionized}$$

---

e. When a strong acid and a weak acid are mixed, the weak acid can usually be neglected in the equilibrium expression.

- ► This is similar to ignoring the auto-ionization of water in an acid-ionization problem. The idea is that the weak acid contributes an amount of hydronium ion that is negligible compared with the amount contributed by the strong acid.

f. When two weak acids are mixed, the weaker acid can be neglected if the K_a differs by at least two orders of magnitude.

- ► This is similar to the criteria for using the "x is small" approximation.

EXAMPLE:

Calculate the hydronium ion concentration in a solution containing 5.45 mL of a 0.145 M HCl solution and 5.32 mL of a 0.029 M hypiodous acid (K_a=2.3×10^{-11}).

We can see from the acid ionization constant of hypoiodous acid that it is very weak. Since the concentration of hypoiodous acid is very dilute, the contribution of hydronium ions from the weak acid is negligible compared to the strong acid contribution. Since HCl is a strong acid, it will completely ionize and we can calculate the concentration of hydronium ions that will be produced from the HCl:

$$5.45\ \mathrm{mL}\times\frac{1\ \mathrm{L}}{1000\ \mathrm{mL}}\times\frac{0.145\ \mathrm{mol\ HCl}}{1\ \mathrm{L}}\times\frac{1\ \mathrm{mol\ H_3O^+}}{1\ \mathrm{mol\ HCl}} = 7.903\times10^{-4}\ \mathrm{mol\ H_3O^+}$$

The concentration of hydronium ions in the solution will be the number of moles of hydronium ions (calculated above) over the total volume:

$$\frac{7.903\times10^{-4}\ \mathrm{mol\ H_3O^+}}{(5.45+5.32)\ \mathrm{mL}}\times\frac{1000\ \mathrm{mL}}{1\ \mathrm{L}} = 0.07338\ \mathrm{M\ H_3O^+}$$

Our answer should have three significant figures, so the final hydronium concentration is 0.0734 M.

VI. Base Solutions

a. When bases are dissolved in solution, they react with water according to the equilibrium expression:

$$\mathrm{B(aq) + H_2O(l) \leftrightarrows HB^+(aq) + OH^-(aq)}$$

b. The equilibrium constant for this reaction is called the base dissociation or base ionization constant and has the form:

$$K_b = \frac{[HB^+][OH^-]}{[B]}$$

c. The strength of a base depends on the value of K_b, which is an indication of the position of the equilibrium.

  i. Strong bases have such large K_b values that we write their dissociation using a single forward arrow:

$$B(aq) + H_2O(l) \rightarrow HB^+(aq) + OH^-(aq)$$

    1. Most strong bases are metals combined with hydroxide ions.

  ii. Weak bases only partially ionize in water so that both the neutral and protonated form of the base exist in water.

    1. The most common weak bases are ammonia and derivatives of ammonia.

d. The process of calculating the concentration of hydroxide ions and the pH of basic solutions is exactly the same as for acid solutions.

EXAMPLE:

Calculate the pH of a solution prepared by dissolving 9.7 g of sodium hydroxide in enough water to prepare 250 mL of solution.

This is a simple conversion factor problem because sodium hydroxide is a strong base that will completely dissociate into sodium ions and hydroxide ions. First we find the moles of hydroxide ions in the solution:

$$9.7\,g \times \frac{1\,mol\,NaOH}{39.998\,g} \times \frac{1\,mol\,OH^-}{1\,mol\,NaOH} = 0.2425\,mol\,OH^-$$

This is the number of moles in 250 mL, so the concentration of hydroxide ions is:

$$\frac{0.2425\,mol\,OH^-}{250\,mL} \times \frac{1000\,mL}{1\,L} = 0.97007\,M\,OH^-$$

The hydronium ion concentration can be found by dividing the ion product constant of water by the hydroxide ion concentration and then taking the negative log of that value:

$$[H_3O^+] = \frac{1.0 \times 10^{-14}}{0.97007\,M\,OH^-} = 1.0309 \times 10^{-14}$$

$$pH = -\log [H_3O^+] = 13.987$$

Our answer should have two significant figures, so the pH of this solution is 13.99.

EXAMPLE:

Calculate the pH of a 4.5 M nicotine, $C_{10}H_{14}N_2$, solution. The K_b for nicotine is 1.0×10^{-6}.

First we need to write the base ionization equation:

$$C_{10}H_{14}N_2(aq) + H_2O(l) \leftrightarrows C_{10}H_{14}N_2H^+(aq) + OH^-(aq)$$

Now we can use an ICE table to determine the equilibrium concentrations of all species:

Reaction Condition	$[C_{10}H_{14}N_2]$ (M)	$[H_2O]$ (M)	$[C_{10}H_{14}N_2H^+]$ (M)	$[OH^-]$ (M)

Initial	4.5	-	0	0
Change	-x	-	+x	+x
Equilibrium	4.5-x	-	x	x

Now we can write the base ionization constant for the reaction and then include the values given:

$$K_b = \frac{[C_{10}H_{14}N_2H^+][OH^-]}{[C_{10}H_{14}N_2]}$$

$$1.0\times10^{-6} = \frac{(x)(x)}{(4.5-x)}$$

As the concentration of nicotine and the base ionization constant differ by four orders of magnitude, we can approximate 4.5-x ≈4.5:

$$1.0\times10^{-6} = \frac{(x)(x)}{4.5}$$

$$x = \sqrt{4.5\times10^{-6}} = 0.002121$$

This is the concentration of hydroxide ions in the solution. In order to calculate the pH, we need to use the ion product constant for water to find the hydroxide ions first.

$$[H_3O^+] = \frac{1.0\times10^{-14}}{0.0.002121\,M\,OH^-} = 4.714\times10^{-12}$$

The pH is then:

$$pH = -\log(4.714\times10^{-12}) = 11.3266$$

Our answer should have two significant figures, so the pH of the solution is 11.33.

VII. The Acid-Base Properties of Ions and Salts

a. Salts are ionic substances that dissociate in water to produce anions and cations.

b. If the anion of the salt is a weak base (conjugate base of a weak acid), the pH of the solution will be greater than 7.

i. The anion can interact with water as a base:

$$A^-(aq) + H_2O(l) \rightarrow HA(aq) + OH^-(aq)$$

ii. The equilibrium constant for this reaction will be the K_b value of the acid:

$$K_b = \frac{[HA][OH^-]}{[A^-]}$$

1. The value of K_b for a base is related to K_a of its conjugate acid through the water ionization constant:

$$K_b \times K_a = K_w$$

EXAMPLE:

Calculate the pH of a solution of a 1.05 M NaCN given that the K_a of HCN is 4.9×10^{-10}.

We first need to calculate the K_b value of CN^-:

$$K_b = \frac{K_w}{K_a}$$

$$K_b = \frac{1.0 \times 10^{-14}}{4.9 \times 10^{-10}} = 2.0408 \times 10^{-5}$$

Now we can write the base ionization equation and use an ICE table to find equilibrium concentrations:

$$CN^-(aq) + H_2O(l) \leftrightarrows HCN(aq) + OH^-(aq)$$

Reaction Condition	$[CN^-]$ (M)	$[H_2O]$ (M)	[HCN] (M)	$[OH^-]$ (M)
Initial	1.05	-	0	0
Change	-x	-	+x	+x
Equilibrium	1.05-x	-	x	x

We will now write the base ionization expression for the reaction:

$$K_b = \frac{[CN^-][OH^-]}{[HCN]}$$

$$2.0408 \times 10^{-5} = \frac{(x)(x)}{(1.05 - x)}$$

As the concentration of nicotine and the base ionization constant differ by five orders of magnitude, we can approximate $1.05\text{-}x \approx 1.05$:

$$2.0408 \times 10^{-5} = \frac{(x)(x)}{1.05}$$

$$x = \sqrt{2.143 \times 10^{-5}} = 0.0046291$$

This is the concentration of hydroxide ions in the solution. In order to calculate the pH, we need to use the ion product constant for water to find the hydroxide ions first.

$$[H_3O^+] = \frac{10^{-14}}{0.0046291\,M\ OH^-} = 2.1602 \times 10^{-12}$$

The pH is then:

$$pH = -\log(2.1602 \times 10^{-12}) = 11.6655$$

Our answer should have two significant figures, so the pH of the solution is 11.67.

---

c. If the cation is a weak acid (conjugate acid of a weak base), the pH of the solution will be less than 7.

   i. Weakly acidic solutions also form when small, highly charged metals are dissolved in water.

EXAMPLE:

Will a 1.0 M solution of CH_3NH_3Cl be acidic, basic, or neutral?

We need to consider each component of the salt. Cl^- is a very weak base so its effect on the pH of the solution is negligible. $CH_3NH_3^+$ is a moderately weak acid, meaning that it will interact with water to produce CH_3NH_2 and H_3O^+. The presence of H_3O^+ will result in an acidic solution.

---

d. When the cation of a salt serves as a weak acid and the anion of the salt behaves as a weak base, the pH of the solution will depend on the strength of the weak acid and base.

EXAMPLE:

Will a solution containing 10.0 g of ammonium lactate, $NH_4C_3H_5O_3$, be acidic, basic, or neutral?

The K_b of ammonia, NH_3, is 1.76×10^{-5} and the K_a of lactic acid, $HC_3H_5O_3$, is 1.4×10^{-5}.

First we need to calculate the K_a of ammonium and the K_b of the lactate ion because these will be the species of interest in water:

The K_a of ammonium, NH_4^+, is:

$$K_a = \frac{K_w}{K_b}$$

$$K_a = \frac{1.0\times10^{-14}}{1.76\times10^{-5}} = 5.7\times10^{-10}$$

and the K_b of the lactate ion, $C_3H_5O_3^-$, is:

$$K_b = \frac{K_w}{K_a}$$

$$K_b = \frac{1.0\times10^{-14}}{1.4\times10^{-5}} = 7.1\times10^{-10}$$

We can see from these values that because the K_b value of the lactate ion is slightly larger than the K_a value of ammonium, the solution will be slightly basic. Note that this solution will be only very slightly basic because the values of the equilibrium constants are so similar.

VIII. Polyprotic Acids

a. Acids with more than one ionizable proton will lose protons in successive steps; each step (or reaction) will have its own equilibrium constant, K_a.

b. The pH of a polyprotic acid solution is calculated like any other acid solution.

EXAMPLE:

What is the pH of a 0.052 M sulfuric acid solution?

Since sulfuric acid is a strong acid, the dissociation of the first ionizable proton will give a hydronium ion concentration that is equal to the sulfuric acid concentration, or 0.052 M H_3O^+. The acid ionization constant for the second proton is 1.2×10^{-2} which is large enough compared to the concentration of sulfuric acid that we should not neglect its contribution.

We will write the acid ionization equation for the bisulfate ion and set up an ICE table:

$$HSO_4^-(aq) + H_2O(l) \leftrightarrows SO_4^{2-}(aq) + H_3O^+(aq)$$

Reaction Condition	$[HSO_4^-]$ (M)	$[H_2O]$ (M)	$[SO_4^{2-}]$ (M)	$[H_3O^+]$ (M)
Initial	0.052	-	0	0.052
Change	-x	-	+x	+x
Equilibrium	0.052-x	-	x	0.052+x

Note that the concentration of bisulfate and hydronium ions initially present was due to the dissociation of the original sulfuric acid solution.

The reaction will proceed in the forward direction because there is no sulfate ion present initially.

Now we will write the acid ionization constant for the reaction:

$$K_a = \frac{[SO_4^{2-}][H_3O^+]}{[HSO_4^+]}$$

$$1.2\times10^{-2} = \frac{(x)(0.052+x)}{(0.052-x)}$$

We need to use the quadratic formula, so we will first expand this expression to get the proper form:

$$0.000624 - 0.012x = 0.052x + x^2$$

$$x^2 + 0.064x - 0.000624 = 0$$

We see that a=1, b=0.064, and c=-0.000624 for use in the quadratic formula:

$$x = \frac{-0.064 \pm \sqrt{(0.064)^2 - 4(1)(-0.000624)}}{2(1)}$$

$$x = 0.0086 \text{ or } x = -0.0726$$

We cannot have a negative concentration, so we see that the concentration of hydronium ions at equilibrium is 0.052+0.0086=0.0606 M.

The pH is then:

$$pH = -\log(0.0606) = 1.2175$$

Our answer should have two significant figures, so the pH of the solution is 1.22.

EXAMPLE:

Calculate the pH of a 0.0451 M solution of oxalic acid, $H_2C_2O_4$, given that $K_{a1}=5.9\times10^{-2}$ and $K_{a2}=6.4\times10^{-5}$.

We will start by writing the acid ionization equation for oxalic acid:

$$H_2C_2O_4(aq) + H_2O(l) \leftrightarrows HC_2O_4^-(aq) + H_3O^+(aq)$$

Using an ICE table, we find the equilibrium concentrations:

Reaction Condition	$[H_2C_2O_4]$ (M)	$[H_2O]$ (M)	$[HC_2O_4^-]$ (M)	$[H_3O^+]$ (M)

Initial	0.0451	-	0	0
Change	-x	-	+x	+x
Equilibrium	0.0451-x	-	x	x

We will now write the base ionization expression for the reaction:

$$K_a = \frac{[HC_2O_4^-][H_3O^+]}{[H_2C_2O_4]}$$

$$5.9 \times 10^{-2} = \frac{(x)(x)}{(0.0451 - x)}$$

We will use the quadratic formula to solve this problem since the original concentration and first acid ionization constant are so similar:

$$0.0026609 - 0.059x = x^2.$$

$$x^2 + 0.059x - 0.0026609 = 0$$

We see that a=1, b=0.059, and c=-0.0026609 for use in the quadratic formula:

$$x = \frac{-0.059 \pm \sqrt{(0.059)^2 - 4(1)(-0.0026609)}}{2(1)}$$

$$x = 0.0299 \text{ or } x = -0.0889$$

We cannot have a negative concentration, so we see that the concentration of hydronium ions at equilibrium is 0.0299 M.

We do not need to consider the ionization of the second proton because we already have a large hydronium ion concentration (limiting the extent to which the second ionization will occur) and the acid ionization constant for the second proton is very small.

The pH is therefore:

$$pH = -\log(0.0299) = 1.524$$

Our answer should have two significant figures, so the pH of the solution is 1.52.

IX. Acid Strength and Molecular Structure

a. Binary acids consist of hydrogen and one other element and have the general form of H-X. Two factors affect the strength of binary acids.

   i. The strength of a binary acid depends on bond polarity; the more polar the H-X bond, the stronger the acid will be.

   ii. The strength of a binary acid depends on the H-X bond strength; the weaker the H-X bond is, the stronger the acid will be.

b. Oxyacids contain a hydrogen atom bonded to hydrogen and have the general form of H-O-Y where Y is a generic group.

   i. The more electronegative the Y group is, the more polarized the H-O bond will be and the stronger the acid.

   ii. The more oxygen atoms attached to Y, the more electron density will be pulled from the hydrogen atom and the stronger the acid will be.

X. Lewis Acids and Bases

a. A Lewis acid is an electron pair acceptor and a Lewis base is an electron pair donor.

i. Lewis acids must have an empty orbital (or rearrange to provide an empty orbital) in order to accept the Lewis base electron pair.

ii. Small, highly charged metals form acidic solutions in water because they act as Lewis acids.

Fill in the Blank:

1. A ________________ is an acid that ionizes in successive steps.
2. The ________________ is the ratio of the ionized acid at equilibrium over the initial acid concentration expressed as a percentage.
3. Acids are electron pair acceptors according to the ________________ definition.
4. A Brønsted-Lowry base is a(n) ________________.
5. In pure water, the hydronium ion concentration is ________________ and the pH is ________________.
6. A strong acid is a species that completely ________________ when dissolved in water.
7. In general, the stronger the acid, the ________________ the conjugate base.
8. The equilibrium constant for the ionization reaction of a weak acid is called the ________________.
9. The more electronegative the atom attached to hydrogen in a binary acid, the ________________ the acid (other factors being equal).
10. When a salt containing an anion that is the conjugate base of a weak acid is dissolved in water, the solution will be ________________.
11. Accordin g to the Arrhenius definition an acid is a(n) ________________.
12. NH_4^+ is the conjugate ________________ of NH_3.
13. S mall, highly charged metal cations form ________________ solutions.
14. The strengt h of an oxyacid increases with a(n) ________________ number of oxygen atoms.
15. A(n) ________________ substance is one that can act as an acid or a base.

Problems:

1. Formic acid, $HCHO_2$, has a K_a value of 1.8×10^{-4}. 100.4 g of formic acid are used to prepare 10. L of a stock acid solution.

   a. What is the acid ionization equation for formic acid?

   b. Identify the conjugate acid/base pairs in the equation from part a.

   c. What is the pH of the stock solution?

   d. What is the pOH of the stock solution?

   e. What is the percent ionization of formic acid in the stock solution?

   f. A solution for use in a laboratory is made by taking a 150.0-mL aliquot of the stock solution and diluting it to 1.0 L.

      i. What is the pH of the new solution?

      ii. What is the percent ionization of the new solution?

g. Considering your answers in parts c–f, is the preparation of a stock solution a wise use of time when the acid is weak?

2. A 0.529 M solution of an unknown acid is prepared and the pH is measured to be 4.82. What is the acid ionization constant value of the unknown acid?

3. A solution of ethylamine, $C_2H_5NH_2$, of an unknown concentration is found on a laboratory shelf. The pH of the solution is found to be 8.53. Given that the K_b of ethylamine is 5.6×10^{-4}, calculate the concentration of the solution.

4. What mass of nitrous acid must be dissolved in 145 mL of water in order to obtain a solution with a pH of 5.2? The K_a value for nitrous acid is 4.6×10^{-4}.

5. Hypochlorous acid has an acid ionization constant of 3.5×10^{-8}. What are the concentrations of all species at equilibrium in a 0.0015 M solution of hypochlorous acid, and what is the pH of the solution?

6. Calculate the pH of a 1.45 M solution of hydrosulfuric acid given that the K_{a1} value is 8.9×10^{-8} and the K_{a2} value is 1×10^{-19}.

7. Acetic acid and sodium bicarbonate are combined in equal amounts in 1.0 L of water.

   a. Write the balanced chemical reaction for the combination of these in water.

   b. Label the conjugate acid-base pairs in the reaction from part a.

   c. Given that the acid ionization constant for acetic acid is 1.8×10^{-5} and the acid ionization constant for carbonic acid is 4.2×10^{-7}, predict whether or not a reaction will occur when the acetic acid and sodium bicarbonate are mixed.

8. The bicarbonate ion is amphoteric. The K_a value is 5.6×10^{-11} and the K_b value is 1.7×10^{-9}.

   a. Will a solution of sodium bicarbonate be acidic or basic? (You should be able to answer this without doing any calculations.)

   b. Calculate the pH and percent ionization of a 1.0 M bicarbonate ion solution assuming that it behaves as an acid.

   c. Calculate the pH and percent ionization of a 1.0 M bicarbonate ion solution assuming that it behaves as a base.

   d. Use your results in parts b and c to justify or refute the answer that you gave in part a.

   e. Sodium bicarbonate will "fizz" when combined with vinegar. Write the chemical reaction for this process and explain the observation in the context of your answers in the previous parts of this question.

9. A 0.025 M solution of an unknown base has a pH of 10.09.

   a. What are the hydronium ion and hydroxide ion concentrations of the solution?

   b. Calculate the base ionization constant for the unknown base.

   c. Is the base strong, moderately weak, or very weak? Justify your answer based on the base ionization constant.

   d. What is the percent ionization of the base?

   e. Is your answer in part c consistent with your answer in part d? Explain.

10. A scorbic acid is a diprotic acid with a $K_{a1}=6.8\times10^{-5}$ and a $K_{a2}=2.7\times10^{-12}$.

    a. List all species that are present in a solution of ascorbic acid at equilibrium.

    b. What is the pH of a solution that contains 5.0 mg of ascorbic acid per milliliter of solution?

11. Ho w many milliliters of a strong monoprotic acid solution at a pH of 4.12 must be added to 528 mL of a solution of the same acid at a pH of 5.76 in order to prepare a solution with a pH of 5.34? Assume that the volumes are additive.

12. When Cr^{3+} is dissolved in water, a slightly acidic solution results.

    a. Write the balanced acid ionization equation for Cr^{3+} in water.

    b. What is the pH of a solution that has 16.52 g of $Cr(NO_3)_3$ dissolved to give 175 mL of solution? The K_a value of Cr^{3+} is 1.6×10^{-4}.

13. Explain why H-O-CN is a stronger acid than H-CN.

14. Ethene, CH_2CH_2, contains two carbons with a double bond between them. Explain how and why ethene is a Lewis acid.

Concept Questions:

1. Explain the auto-ionization of water using the Arrhenius, Brønsted-Lowry, and Lewis definitions of acids and bases.

2. Use Le Châtelier's principle to explain why the percent ionization of a weak acid is larger for a dilute solution of the acid while the hydronium ion concentration is larger for a more concentrated solution.

3. Explain, using resonance structures, why you think that so many naturally occurring acids contain the carboxylic acid ($-CO_2H$) group.

4. Which of the following do you expect to have the largest K_b value? Explain your answer in 1–2 sentences.

    a. H_3PO_4 b. $H_2PO_4^-$ c. HPO_4^{2-} d. PO_4^{2-}

5. Which of the following will result in the largest hydronium ion concentration when dissolved in water?

    a. CO_2 b. SO_3 c. Al_2O_3 d. Na_2O

6. Explain why the hydronium ion is the strongest acid that can exist in water.

7. The strong acids sulfuric, nitric, perchloric, hydrochloric, hydrobromic, and hydroiodic acids actually have varying strengths. For example, hydroiodic acid is stronger than hydrochloric acid.

    a. The strength of these acids cannot be compared when they are dissolved in water. Explain why?

    b. Suggest an experiment that would allow you to test the strengths of these acids.

    c. Using what you have learned about acid strengths, predict the relative strengths of these acids.

8. Explain why the concentration of hydronium ions from the autoionization of water can be neglected for a 0.0010 M HCl solution using the rules for significant figures.

9. Explain why the pH scale is used rather than the hydronium ion concentration to report acidity. Hint: most acids and bases are relatively weak.

10. Most weak bases are derivatives of ammonia called amines. Explain why you think that this observation is true based on the structure of an amine.

11. Pure, elemental boron is very rare; it is more commonly found as borax. Look up the structure and chemistry of borax and explain why you think that this is, using concepts of Lewis acid/base chemistry.

Chapter 16: Aqueous Ionic Equilibrium

Learning Goals:

- Understand how and why buffers resist pH changes.
- Learn how to prepare a buffer system with a particular pH value and an adequate buffering capacity.
- Calculate pH values, sketch titration curves, and identify an appropriate indicator for a titration experiment.
- Write equilibrium reactions, solubility product constants, and calculate the solubility for a sparingly soluble salt.
- Determine whether a precipitate will form in a given solution and use precipitation reactions to separate cations in a solution.
- Write formation reactions and formation constants for complex ion equilibria.
- Predict the manner in which complex ion formation affects solubility.
- Understand how pH affects solubility and influences the formation of amphoteric metal hydroxides.

Chapter Summary:

Now that you are familiar with general equilibrium reactions and acid/base equilibrium reactions in particular, we will explore more complex equilibrium reactions. First, you will learn about buffer solutions, which are weak acid/conjugate base or weak base/conjugate acid solutions that resist pH changes. You will learn how and why buffer systems resist pH changes, how to calculate the pH of a buffer, and how to determine the buffer capacity. Your understanding of solutions containing acid/base conjugate pairs will allow you to carry out a titration experiment, sketch the resulting titration curve, and identify important regions of that curve. The dissolution of an "insoluble" salt will then be explored; you will learn how to calculate the solubility of a salt and how that solubility changes upon the addition of acids and/or bases. With your understanding of solubility, you will learn how to qualitatively identify the presence of certain ions in solution. Finally, you will look at complex ion equilibria with an emphasis on how complex ion formation affects solubility and how pH affects complex ion formation.

Chapter Outline:

I. Buffers: Solutions That Resist pH Change

  a. Buffers are solutions that resist pH changes upon the addition of small amounts of acid or base.

    i. Buffers contain significant amounts of a weak acid and a salt of its conjugate base (or a weak base and a salt of its conjugate acid).

    ii. Buffers resist pH changes by converting strong acids into weak acids and strong bases into weak bases:

$$H_3O^+(aq) + A^-(aq) \leftrightarrows H_2O(l) + HA(aq)$$

$$OH^-(aq) + HA(aq) \leftrightarrows H_2O(l) + A^-(aq)$$

    iii. Weak acids and bases do not ionize to a significant extent in water so that the hydronium ion concentration is not changed dramatically even after a strong acid or base has been added.

iv. The pH of a buffer solution can be calculated in the same way as in the acid/base ionization problems from the last chapter.

1. This is an equilibrium problem in which a common ion is present.

EXAMPLE:

A buffer solution is prepared by adding 10.4 g of sodium hypochlorite to 1.23 L of a 0.105 M hypochlorous acid solution. What is the pH of the buffer? The K_a of HClO is 2.9×10^{-8}.

First we will determine the concentration of sodium hypochlorite that will initially be present:

$$10.4 \text{ g NaClO} \times \frac{1 \text{ mol NaClO}}{74.44 \text{ g NaClO}} = 0.1397 \text{ mol NaClO}$$

This is dissolved in 1.23 L of solution, so the concentration of NaClO is 0.1136 M.

Now we will write the acid ionization equation and use an ICE table to determine the equilibrium concentrations of all species:

$$HClO(aq) + H_2O(l) \leftrightarrows ClO^-(aq) + H_3O^+(aq)$$

Reaction Condition	[HClO] (M)	$[H_2O]$ (M)	$[ClO^-]$ (M)	$[H_3O^+]$ (M)
Initial	0.105	-	0.1136	0
Change	-x	-	+x	+x
Equilibrium	0.105-x	-	0.1136+x	x

We will now write the base ionization expression for the reaction:

$$K_a = \frac{[ClO^-][H_3O^+]}{[HClO]}$$

$$2.9\times10^{-8} = \frac{(0.1136+x)(x)}{(0.105-x)}$$

Because the acid ionization constant is very small (10^{-8}) and the presence of hypochlorite ions is initially high, the "x is small" approximation should be valid:

$$2.9\times10^{-8} = \frac{(0.1136)(x)}{(0.105)}$$

We find that $x = 2.695\times10^{-8}$ which is equal to the hydronium ion concentration at equilibrium. The pH is then equal to 7.569.

EXAMPLE:

Calculate the change in pH that occurs in the hypochlorous acid buffer from the previous example upon the addition of 0.0012 mol of HCl.

We will start by doing a stoichiometry problem in which the added acid reacts completely with the hypochlorite ions that are initially present:

$$0.0012\,\text{mol HCl}\times\frac{1\,\text{mol ClO}^-}{1\,\text{mol HCl}}=0.0012\,\text{mol ClO}^-$$

The initial concentration of hypochlorite ions is the difference between the initial amount and the amount that reacted over the solution volume:

$$\frac{0.1397\,\text{mol ClO}^- - 0.0012\,\text{mol ClO}^-}{1.23\,\text{L solution}}=0.1126\,\text{M ClO}^-$$

Now we will use an ICE table to calculate the pH as in the above example:

$$HClO(aq) + H_2O(l) \leftrightarrows ClO^-(aq) + H_3O^+(aq)$$

Reaction Condition	[HClO] (M)	[H_2O] (M)	[ClO^-] (M)	[H_3O^+] (M)
Initial	0.105	-	0.1126	0
Change	-x	-	+x	+x
Equilibrium	0.105-x	-	0.1126+x	x

We will now write the base ionization expression for the reaction:

$$K_a=\frac{[ClO^-][H_3O^+]}{[HClO]}$$

$$2.9\times10^{-8}=\frac{(0.1126+x)(x)}{(0.105+x)}$$

Again, we can assume that the "x is small" approximation is valid:

$$2.9\times10^{-8}=\frac{(0.1126)(x)}{(0.105)}$$

We find that $x = 2.704\times10^{-8}$ which is equal to the hydronium ion concentration at equilibrium. The pH is then equal to 7.57.

Notice that the pH of the solution with the added HCl is the same as in the original solution.

---

v. A buffer calculation can be simplified using the K_a expression:

$$K_a=\frac{[H_3O^+][A^-]}{[HA]}$$

Taking the negative log of both sides:

$$-\log K_a=-\log[H_3O^+]+-\log\frac{[A^-]}{[HA]}$$

We recognize that $-\log K_a$ is the pK_a and the $-\log[H_3O^+]$ is the pH:

$$pK_a=pH+-\log\frac{[A^-]}{[HA]}$$

Rearranging to solve for pH:

$$pH = pK_a + \log\frac{[A^-]}{[HA]}$$

This is called the Henderson-Hasselbalch equation and can be used to simplify buffer calculations. This expression is valid for use when the percent ionization of a weak acid is less than 5% (i.e., when the "x is small" approximation is valid).

EXAMPLE:

Calculate the pH of the buffer system from the previous example (before the addition of HCl) using the Henderson-Hasselbalch equation.

In order to use the Henderson-Hasselbalch equation, we first need to calculate the pK_a of the acid:

$$pK_a = -\log K_a$$

$$pK_a = -\log (2.9\times10^{-8})$$

So the pK_a is 7.5376. Using the initial concentrations of the acid and conjugate base, we have:

$$pH = 7.5376 + \log\frac{(0.1136)}{(0.105)}$$

The pH of the solution is 7.57 which is exactly the same as the pH calculated in the previous problem.

EXAMPLE:

Calculate the pH of a buffer system prepared by dissolving 2.24 g of dimethylamine, $(CH_3)_2NH$, and 5.08 g of $(CH_3)_2NH_2Cl$ to prepare 545 mL of solution. The K_b of dimethylamine is 5.4×10^{-4}.

We will first calculate the concentration of the weak base and its conjugate acid:

$$2.24\,g\,(CH_3)_2\,NH \times \frac{1\,mol\,(CH_3)_2\,NH}{45.086\,g\,(CH_3)_2\,NH} = 0.04968\,mol\,(CH_3)_2\,NH$$

$$5.08\,g\,(CH_3)_2\,NH_2Cl \times \frac{1\,mol\,(CH_3)_2\,NH_2Cl}{81.544\,g\,(CH_3)_2\,NH_2Cl} = 0.06230\,mol\,(CH_3)_2\,NH_2Cl$$

In 545 mL, the concentration of $(CH_3)_2NH$ is 0.09116 M and of $(CH_3)_2NH_2Cl$ is 0.1143 M.

In order to use the Henderson-Hasselbalch equation, we need the pK_a of the acid:

$$pK_w = pK_a + pK_b$$

$$14.00 = pK_a + 3.2676$$

So the pK_a of $(CH_3)_2NH_2^+$ is 10.7324.

Inserting the pK_a and the concentrations into the Henderson-Hasselbach equation, we have:

$$pH = 10.7324 + \log\frac{(0.09116)}{(0.1143)}$$

The pH of the buffer solution is 10.634. Our final answer should have two significant figures (from the K_b value), so the pH is reported as 10.63.

II. Buffer Effectiveness: Buffer Capacity and Buffer Change

a. A buffer system is most effective when the concentration of the weak acid (or weak base) is approximately equal to the concentration of the conjugate base (or conjugate acid).

i. When the concentrations of the weak acid (or weak base) and conjugate base (or conjugate acid) are exactly equal at equilibrium, the pH of the solution will be equal to the pK_a.

ii. An effective range for a buffer system is a pH = $pK_a \pm 1$; this is where the concentrations of the weak conjugate pairs are not more than a factor of ten different from one another.

EXAMPLE:

Select the most appropriate acid for preparing a buffer solution with a pH of 4.97 using the K_a values listed in Appendix II of your book.

We want to find an acid that has a pK_a value that is very close to the desired pH. We can more easily compare the desired $[H_3O^+]$ to the K_a values listed in the Appendix. The $[H_3O^+]$ value is equal to 10^{-pH} which is 1.07×10^{-5}. Looking at the K_a values in the appendix, we find that propanoic acid has a K_a value of 1.3×10^{-5} which is closest to the value that we are looking for.

---

b. The buffer capacity is the amount of acid or base that can be added without the buffer losing its effectiveness.

i. The concentrations of the weak acid (or base) and weak base (or weak acid) should be large as compared with the amount of acid or base that will be added.

III. Titrations and pH Curves

a. In an acid-base titration, an acid (or base) solution of unknown concentration is reacted with a base (or acid) of known concentration.

i. The equivalence point of an acid-base titration is where the moles of acid are stoichiometrically equal to the moles of base.

1. The equivalence point can be monitored using an indicator or a pH probe.

ii. A titration curve is a plot of pH versus the volume of acid (or base) added as the titrant.

b. The titration of a strong acid with a strong base is essentially a stoichiometry problem.

► This is because the strong acid reacts completely – there is no reverse reaction.

c. The titration of a weak acid (or base) with a strong base (or acid) is a two-part problem in which the stoichiometry is first accounted for and then the equilibrium reaction of the weak acid (or base) is considered.

► This is done for convenience. It is simpler to remove the strong base in the calculation and then allow the equilibrium to be re-established. You should realize that in reality, the equilibrium is established immediately – the reaction does not go all the way in one direction and then go back.

EXAMPLE:

Lactic acid, $HC_3H_5O_3$, has a K_a value of 1.4×10^{-4}. Consider titrating 25.0 mL of a 0.187 M lactic acid solution with a 0.167 M KOH solution. Calculate the initial pH, the pH after the addition of 15.2 mL of KOH, the pH at the equivalence point, and the pH after the addition of 40.5 mL of KOH.

All parts of this problem will require that we have the acid-ionization equation and K_a expression:

$$HC_3H_5O_3(aq) + H_2O(l) \leftrightarrows H_3O^+(aq) + C_3H_5O_3^-(aq)$$

$$K_a = \frac{[H_3O^+][C_3H_5O_3^-]}{[HC_3H_5O_3]}$$

Now we will approach each point in the titration as a separate problem:

1. The initial pH:

   This is an equilibrium problem similar to those from Chapter 15. We will use an ICE table and the acid-ionization constant as before:

Reaction Condition	$[HC_3H_5O_3]$ (M)	$[H_2O]$ (M)	$[C_3H_5O_3^-]$ (M)	$[H_3O^+]$ (M)
Initial	0.187	-	0	0
Change	-x	-	+x	+x
Equilibrium	0.187-x	-	x	x

$$1.4\times10^{-4} = \frac{(x)(x)}{(0.187-x)}$$

   Since the acid ionization constant and the initial acid molarity differ by a factor of 10^3, we can use the "x is small" approximation:

$$1.4\times10^{-4} = \frac{(x)(x)}{(0.187)}$$

$$x = 0.005117$$

   The pH of the solution is then 2.29 before any of the strong base has been added.

2. The pH after the addition of 15.2 mL of KOH:

   The first step is to use the reaction stoichiometry in order to convert the strong base (KOH) into a weak base ($C_3H_5O_3^-$). We will start by calculating the moles of acid and base present in the solution.

$$15.2\text{ mL KOH} \times \frac{1\text{ L}}{1000\text{ mL}} \times \frac{0.167\text{ mol KOH}}{1\text{ L KOH}} = 0.002538\text{ mol KOH}$$

$$25.0\text{ mL } HC_3H_5O_3 \times \frac{1\text{ L}}{1000\text{ mL}} \times \frac{0.187\text{ mol } HC_3H_5O_3}{1\text{ L } HC_3H_5O_3} = 0.004675\text{ mol } HC_3H_5O_3$$

   We see that the base is the limiting reagent. The neutralization reaction will have a 1:1 mole ratio of acid to base:

$$HC_3H_5O_3(aq) + OH^-(aq) \leftrightarrows H_2O(l) + C_3H_5O_3^-(aq)$$

   All of the KOH will react to form a stoichiometric amount of $C_3H_5O_3^-$. The lactic acid that remains will be the difference between the moles that are initially present (0.004675 mol) and those that reacted with the base (0.002538 mol); so 0.002137 mol of lactic acid are present after KOH is added.

In order to use the ICE table to determine the pH of the solution, we need to convert the number of moles present into concentrations. The total volume is the volume of acid plus the volume of base, or 40.2 mL. The new concentrations of acid and conjugate base are:

$$\frac{0.002538 \text{ mol } C_3H_5O_3^-}{0.0402 \text{ L solution}} = 0.06313 \text{ M } C_3H_5O_3^-$$

$$\frac{0.002137 \text{ mol } HC_3H_5O_3}{0.0402 \text{ L solution}} = 0.05316 \text{ M } HC_3H_5O_3$$

Now we will use an ICE table to calculate equilibrium concentrations:

Reaction Condition	$[HC_3H_5O_3]$ (M)	$[H_2O]$ (M)	$[C_3H_5O_3^-]$ (M)	$[H_3O^+]$ (M)
Initial	0.05316	-	0.06313	0
Change	-x	-	+x	+x
Equilibrium	0.05316-x	-	0.06313+x	x

$$1.4 \times 10^{-4} = \frac{(0.06313 + x)(x)}{(0.05316 - x)}$$

We will use the "x is small" approximation:

$$1.4 \times 10^{-4} = \frac{(0.06313)(x)}{(0.05316)}$$

$$x = 0.0001179$$

Since x is less than 5% of the original concentrations of acid and conjugate base, the approximation is valid. The pH of the solution is then 3.93 after the addition of 15.2 mL of KOH.

3. The pH at the equivalence point:

   We need to calculate the volume of the KOH that has been added at the equivalence point. From the previous problem, we know that there are 0.04675 mol of lactic acid initially in the solution. We can use this with the molarity of the KOH solution to calculate the volume:

$$0.004675 \text{ mol } HC_3H_5O_3 \times \frac{1 \text{ mol KOH}}{1 \text{ mol } HC_3H_5O_3} \times \frac{1 \text{ L KOH solution}}{0.167 \text{ mol KOH}} \times \frac{1000 \text{ mL}}{1 \text{ L}} = 27.99 \text{ mL KOH solution}$$

   All of the weak acid and strong base will be consumed to produce an equimolar amount of the conjugate base. The initial concentration of the conjugate base is then:

$$\frac{0.004675 \text{ mol } C_3H_5O_3^-}{0.05299 \text{ L solution}} = 0.08822 \text{ M } C_3H_5O_3^-$$

   In order to calculate the equilibrium concentrations, we need to consider the behavior of this weak base in water:

$$H_2O(l) + C_3H_5O_3^-(aq) \leftrightarrows HC_3H_5O_3(aq) + OH^-(aq)$$

   This is the base-ionization equation, so we can write the equilibrium constant expression:

$$K_b = \frac{[OH^-][HC_3H_5O_3]}{[C_3H_5O_3^-]}$$

   Since we are given the K_a for lactic acid, we can solve for the K_b value using K_w:

$$K_w = K_a K_b$$

The value of K_b for lactate is 7.14×10^{-11}. Using the base-ionization equation and the K_b value, we solve for x:

Reaction Condition	$[C_3H_5O_3^-]$ (M)	$[H_2O]$ (M)	$[HC_3H_5O_3]$ (M)	$[OH^-]$ (M)
Initial	0.08822	-	0	0
Change	-x	-	+x	+x
Equilibrium	0.08822-x	-	x	x

$$7.14\times10^{-11} = \frac{(x)(x)}{(0.08822-x)}$$

Using the "x is small" approximation, x is 2.510×10^{-6} M. This is the concentration of hydroxide ions in the solution, so the pOH is 5.60 and the pH is 8.40 at the equivalence point. This value makes sense because we expect that the presence of a weak base will cause the pH to be greater than 7.

4. The pH after the addition of 40.5 mL of KOH:

At this point in the titration, all of the weak acid has been converted to a weak base and excess strong base is in solution. We will determine the amount of KOH that remains in solution using the stoichiometry of the reaction as in the previous two parts of this reaction.

$$40.5\text{ mL KOH}\times\frac{1\text{ L}}{1000\text{ mL}}\times\frac{0.167\text{ mol KOH}}{1\text{ L KOH}} = 0.006764\text{ mol KOH}$$

From part 3, we know that 0.004675 mol of this will be consumed by the weak acid. This means that after the addition, there are 0.002089 mol of KOH remaining in solution.

The concentration of hydroxide ions from the strong base is one thousand times more than the concentration of hydroxide ions produced when the conjugate base ionizes in water (from part 3). This means that the pOH of the solution will be dominated by the excess KOH.

$$\text{pOH} = -\log\left(\frac{0.002089\text{ mol}}{0.0655\text{ L}}\right)$$

The pOH is then 1.50 and the pH of the solution is 12.50.

---

d. When a titration experiment involves a polyprotic acid, two equivalence points must be considered.

e. A pH indicator is a weak acid that changes color at its equivalence point.

i. In water, the indicator (HInd) dissociates as a weak acid:

$$\text{HInd}(aq) + H_2O(l) \leftrightharpoons H_3O^+(aq) + \text{Ind}^-(aq)$$

ii. A shift in equilibrium results in a change in the solution color.

1. If the solution pH is greater than the pK_a of the indicator, then the color of the solution will be dominated by the color of Ind^-.

2. If the solution pH is less than the pK_a of the indicator, then the color of the solution will be dominated by the color of HInd.

► You can think about Le Chatelier's principle in predicting the dominant species at a given pH. A high pH means that there is a low concentration of H_3O^+ and the reaction shifts forward to make more H_3O^+ and more of the

Ind^-. A low pH means that there is a high concentration of H_3O^+ and the reaction shifts in the reverse direction to make more HInd.

IV. Solubility Equilibria and the Solubility Product Constant

a. The equilibrium expression for a reaction in which an ionic solid dissolves is called a solubility product constant, K_{sp}:

$$aXY(s) \leftrightarrows bX^+(aq) + cY^-(aq)$$

$$K_{sp} = [X^+]^a[Y^-]^b$$

b. The K_{sp} value is a measure of the solubility of a solid in water.

i. The molar solubility is the quantity (in mol) of a solid that will dissolve in 1 L of water.

EXAMPLE:

Calculate the solubility of calcium sulfate given that the K_{sp} value is 7.10×10^{-5}.

The solubility equation for calcium sulfate is:

$$CaSO_4(s) \leftrightarrows Ca^{2+}(aq) + SO_4^{2-}(aq)$$

The solubility product constant expression is:

$$K_{sp} = [Ca^{2+}][SO_4^{2-}]$$

We can set up an ICE table for this reaction using s to represent the solubility:

Reaction Condition	$[CaSO_4]$ (M)	$[Ca^{2+}]$ (M)	$[SO_4^{2-}]$ (M)
Initial	-	0	0
Change	-	+s	+s
Equilibrium	-	s	s

Including the K_{sp} value and the equilibrium concentrations into the K_{sp} expression:

$$7.10\times10^{-5} = s^2$$

$$s = \sqrt{7.10\times10^{-5}}$$

So the solubility of $CaSO_4$ in water is 0.00843 M. Notice that this is the concentration of Ca^{2+} and SO_4^{2+} in solution and also the number of moles of the solid $CaSO_4$ that will dissolve in 1 L of solution.

EXAMPLE:

The solubility of $PbBr_2$ is 0.01058 M. Calculate the solubility product constant value for $PbBr_2$.

We will start by writing the solubility equation, the solubility product constant expression, and then setting up an ICE table:

$$PbBr_2(s) \leftrightarrows Pb^{2+}(aq) + 2Br^-(aq)$$

$$K_{sp} = [Pb^{2+}][Br^-]^2$$

Reaction Condition	$[PbBr_2]$ (M)	$[Pb^{2+}]$ (M)	$[Br^-]$ (M)
Initial	-	0	0
Change	-	+s	+2s
Equilibrium	-	s	2s

In this case, we have two bromide ions produced for each $PbBr_2$ molecule that dissolves in solution. The equilibrium concentration of bromide ions is then twice the solubility:

$$K_{sp} = (s)(2s)^2$$

$$K_{sp} = (0.01058)(0.02116)^2$$

$$K_{sp} = 4.737\times10^{-6}$$

---

c. The common ion affect will cause the solubility of an ionic solid to decrease.

EXAMPLE:

Calculate the solubility of $PbBr_2$ in a 0.100 M NaBr solution.

We will use the same equilibrium equation and the same solubility product constant as in the previous example. This time, however, we will set up the ICE table with an initial concentration of bromide ions:

Reaction Condition	$[PbBr_2]$ (M)	$[Pb^{2+}]$ (M)	$[Br^-]$ (M)
Initial	-	0	0.100
Change	-	+s	0.100+2s
Equilibrium	-	s	0.100+2s

We know the solubility constant expression and the solubility product constant value so:

$$K_{sp} = [Pb^{2+}][Br^-]^2$$

$$4.737\times10^{-6} = (s)(0.100+2s)^2$$

Since the initial concentration of bromide ions and the solubility product constant value are five orders of magnitude different, we can use the "x is small" (or in this case, "s is small") approximation:

$$4.737\times10^{-6} = (s)(0.100)^2$$

$$s = 4.737\times10^{-4}$$

So the solubility of $PbBr_2$ in a 0.100 M NaBr solution is 4.737×10^{-4} M.

---

d. The solubility of an ionic compound changes when the pH of the solution is changed.

   i. The way in which the solubility changes depends on how the ionic compound interacts with the added acid or base.

EXAMPLE:

Which of these will be more soluble in an acidic solution rather as compared with their solubility in a neutral solution: $Al(OH)_3$, $BaSO_4$, or $Hg_2(CN)_2$?

For all of these salts, we need to consider the reaction that would occur upon the addition of an acid:

$$Al(OH)_3(s) + 3H_3O^+(aq) \rightarrow 6H_2O(l) + Al^{3+}(aq)$$

When water is formed, it will not ionize to reform OH^- to any appreciable extent. The solubility of $Al(OH)_3$ will be increased in an acidic solution because little OH^- is available to react with Al^{3+}.

$$BaSO_4(s) + 2H_3O^+(aq) \rightarrow H_2SO_4(aq) + H_2O(l) + Ba^{2+}(aq)$$

The sulfuric acid that forms is a strong acid that will dissociate completely to give hydronium ions and sulfate ions. The barium ions will combine with the sulfate ions to reform barium sulfate. The solubility of $BaSO_4$ will not, therefore, be more soluble in an acidic solution.

$$Hg_2(CN)_2(s) + 2H_3O^+(aq) \rightarrow 2HCN(aq) + 2H_2O(l) + Hg_2^{2+}(aq)$$

A weak acid, HCN, is formed in this reaction. HCN will dissociate partially to give hydronium ions and cyanide ions. The solubility will increase in this example, but it will not increase as much as it did in the first example where water was formed.

V. Precipitation

a. In order to determine whether or not a precipitate will form under a given set of conditions, we compare the reaction quotient to the solubility product constant of a reaction.

i. If $Q<K_{sp}$, then the solution is unsaturated and additional solid will dissolve.

ii. If $Q=K_{sp}$, then the solution is saturated – it is at equilibrium.

iii. If $Q>K_{sp}$, then the solution is supersaturated and a precipitate will form.

EXAMPLE:

Will a solution that is prepared by dissolving 1.04 g of NaOH and 2.34 g of $Pb(NO_3)_2$ in 1.42 L of water be unsaturated, saturated, or supersaturated?

We need to first consider the chemical reaction that will occur between these two substances:

$$2NaOH(aq) + Pb(NO_3)_2(aq) \leftrightarrows Pb(OH)_2(s) + 2NaNO_3(aq)$$

This reaction shows that we are concerned with the solubility of lead(II) hydroxide:

$$Pb(OH)_2(s) \leftrightarrows Pb^{2+}(aq) + 2OH^-(aq) \qquad K_{sp} = 1.43\times10^{-20}$$

We will now calculate the concentrations of lead ions and hydroxide ions in the solution:

$$1.04\text{ g NaOH}\times\frac{1\text{ mol NaOH}}{39.998\text{ g NaOH}}\times\frac{1\text{ mol OH}^-}{1\text{ mol NaOH}} = 0.02600\text{ mol OH}^-$$

This amount of hydroxide ions is dissolved in 1.42 L of water, so the $[OH^-]$=0.01831 M.

$$2.34\text{ g Pb(NO}_3)_2\times\frac{1\text{ mol Pb(NO}_3)_2}{331.22\text{ g Pb(NO}_3)_2}\times\frac{1\text{ mol Pb}^{2+}}{1\text{ mol Pb(NO}_3)_2} = 0.007065\text{ mol Pb}^{2+}$$

This amount of lead ions is dissolved in 1.42 L of water, so the $[Pb^{2+}]$=0.004975 M.

We will use the initial concentrations to calculate Q for this reaction:

$$Q = [Pb^{2+}][OH^-]^2$$

$$Q = (0.004975)(0.01931)^2$$

$$Q = 1.668 \times 10^{-6}$$

Since $Q > K_{sp}$ the solution is supersaturated and a precipitate will form in order to establish equilibrium.

---

b. Selective precipitation is a method for separating out cations in a solution.

EXAMPLE:

A solution contains an equimolar amount of Fe^{2+}, Cu^{2+}, and Co^{2+}. If the total ion concentration in the solution is 4.52 M, what is the minimum number of grams of Na_2CO_3 that must be added to 1.00 L of the solution in order for a precipitate to form? What is the formula of the precipitate that forms?

First we need to find the K_{sp} values for the metal carbonate solids that will form. From Appendix II:

$K_{sp} = 3.07 \times 10^{-11}$ for $FeCO_3$

$K_{sp} = 2.4 \times 10^{-10}$ for $CuCO_3$

$K_{sp} = 1.0 \times 10^{-10}$ for $CoCO_3$

The smaller the K_{sp} value is, the less the solid will dissolve. Comparing the K_{sp} values for the carbonate salts, we see that iron(II) carbonate is the least soluble so $FeCO_3$ will precipitate out of solution first.

Next we need to determine the concentration of iron ions in the solution. Since there are three ions present in equimolar amounts and the total ion concentration is 4.52 M, each metal cation will have a concentration of 1.5067 M. We will now calculate the concentration of carbonate ions that must be added to the solution in order for the iron(II) carbonate to precipitate out:

$$K_{sp} = [Fe^{2+}][CO_3^{2-}]$$

$$3.07 \times 10^{-11} = (1.5067)[CO_3^{2+}]$$

$$[CO_3^{2+}] = 2.038 \times 10^{-11} \text{ M}$$

Now we will convert this into grams of sodium carbonate:

$$\frac{2.038 \times 10^{-11} \text{ mol } CO_3^{2-}}{1 \text{ L solution}} \times 1.00 \text{ L solution} \times \frac{1 \text{ mol } Na_2CO_3}{1 \text{ mol } CO_3^{2-}} \times \frac{105.99 \text{ g } Na_2CO_3}{1 \text{ mol } Na_2CO_3} = 2.16 \times 10^{-9} \text{ g } Na_2CO_3$$

VI. Qualitative Chemical Analysis

a. We can use selective precipitation to determine the type of ions that are in a given solution.

i. The ions will be separated from the solution in a sequential manner.

1. Insoluble chlorides will be separated out by adding a chloride salt.
2. Hydrosulfuric acid will be added to separate out very insoluble sulfides.
3. A strong base will then be added to identify sparingly soluble sulfides and insoluble hydroxide salts.
4. A phosphate salt is then added to remove insoluble phosphates.
5. Finally, a flame test can be used to test for alkali metals.

VII. Complex Ion Equilibria

a. In solution, metal cations, acting as Lewis acids, are hydrated by water molecules, acting as Lewis bases.

b. A complex ion is a compound that contains a central metal ion bound to one or more ligands.

i. Ligands are Lewis bases that can be charged or neutral.

ii. Hydrated metal ions, $[M(H_2O)_x]^{y+}$, are usually written without water ligands, $M^{y+}(aq)$, for simplicity.

c. In a solution, another ligand can displace water in a complex ion formation equilibrium:

$$M^+(aq) + X\,L(aq) \leftrightarrows [ML_x]^+(aq)$$

$$K_f = \frac{[ML_x{}^+]}{[M^+][L]^X}$$

where K_f is called the formation constant.

EXAMPLE:

What is the concentration of $CdBr_4{}^{2-}$ that forms when a 452 mL solution of 0.143 M $Cd(C_2H_3O_2)_2$ is combined with 596 mL of a 0.218 M solution of KBr? The formation constant for $CdBr_4{}^{2-}$ is 5.3×10^3.

We will first calculate the number of moles of Cd^{2+} and Br^- initially in the solution:

$$452\text{ mL solution} \times \frac{1\text{ L}}{1000\text{ mL}} \times \frac{0.143\text{ mol Cd}(C_2H_3O_2)_2}{1\text{ L solution}} \times \frac{1\text{ mol Cd}^{2+}}{1\text{ mol Cd}(C_2H_3O_2)_2} = 0.06464\text{ mol Cd}^{2+}$$

$$596\text{ mL solution} \times \frac{1\text{ L}}{1000\text{ mL}} \times \frac{0.218\text{ mol KBr}}{1\text{ L solution}} \times \frac{1\text{ mol Br}^-}{1\text{ mol KBr}} = 0.1299\text{ mol Br}^-$$

We will assume that the reaction goes to completion because the formation constant is so large. We note that bromide ions will limit the formation of the complex ion (Br^- is the limiting reagent), so:

$$0.1299\text{ mol Br}^- \times \frac{1\text{ mol Cd}^{2+}}{4\text{ mol Br}^-} = 0.03248\text{ mol Cd}^{2+}$$

The reaction stoichiometry shows us that the number of moles of Cd^{2+} used equals the moles of $CdBr_4{}^{2-}$ produced, so we are able to calculate the concentrations of all species in solution after the reaction occurs:

$$[Cd^{2+}] = \frac{(0.06464 - 0.003248)\text{ mol Cd}^{2+}}{1.048\text{ L}} = 0.003069\text{ M Cd}^{2+}$$

$$[Br^-] = \frac{0\text{ mol Br}^-}{1.048\text{ L}} = 0\text{ M Br}^-$$

$$[CdBr_4{}^{2-}] = \frac{0.03248\text{ mol CdBr}_4{}^{2-}}{1.048\text{ L}} = 0.03099\text{ M CdBr}_4{}^{2-}$$

We will now allow the reaction to proceed in the backward direction in order to reach equilibrium. We will write the balanced equilibrium reaction, the equilibrium constant expression (which is the inverse of the formation constant expression), and an ICE table:

$$CdBr_4{}^{2-}(aq) \leftrightarrows Cd^{2+}(aq) + 4Br^-(aq)$$

$$K_c = \frac{1}{K_f} = \frac{[Cd^{2+}][Br^-]^4}{[CdBr_4^{2-}]}$$

Reaction Condition	$[CdBr_4^{2-}]$ (M)	$[Cd^{2+}]$ (M)	$[Br^-]$ (M)
Initial	0.03099	0.03069	0
Change	-x	+x	+4x
Equilibrium	0.03099-x	0.03069+x	4x

$$1.887 \times 10^{-4} = \frac{(0.03069 + x)(4x)^4}{(0.03099 - x)}$$

We will use the "x is small" approximation to solve for x:

$$1.887 \times 10^{-4} = \frac{(0.03069)(4x)^4}{(0.03099)}$$

$$x = 0.02937$$

$$1.887 \times 10^{-4} = \frac{(0.06006)(4x)^4}{(0.00162)}$$

$$x = 0.01187$$

$$1.887 \times 10^{-4} = \frac{(0.04256)(4x)^4}{(0.01912)}$$

$$x = 0.02399$$

$$1.887 \times 10^{-4} = \frac{(0.05468)(4x)^4}{(0.00700)}$$

$$x = 0.01753$$

$$1.887 \times 10^{-4} = \frac{(0.04822)(4x)^4}{(0.01346)}$$

$$x = 0.02130$$

$$1.887 \times 10^{-4} = \frac{(0.05199)(4x)^4}{(0.00969)}$$

$$x = 0.01925$$

$$1.887 \times 10^{-4} = \frac{(0.04994)(4x)^4}{(0.01174)}$$

$$x = 0.02040$$

$$1.887 \times 10^{-4} = \frac{(0.05109)(4x)^4}{(0.01059)}$$

$$x = 0.01977$$

Which agrees with the previous solution (to two sig. figs.). We have the amount of $CdBr_4^{2-}$ that dissociates, but we are asked to solve for the equilibrium concentration of $CdBr_4^{2-}$:

$$[CdBr_4^{2-}] = (0.03099 - 0.01977)\ M = 0.0112\ M$$

Our answer should have two significant figures, so the equilibrium concentration is reported as 0.011 M.

---

d. The solubility of a salt increases when complex ion formation is favorable.

EXAMPLE:

Which of the following insoluble salts will become more soluble upon the addition of NaCN: AgBr, $Al(OH)_3$, $FeBr_2$, and $PbBr_2$?

The salts that will become more soluble upon the addition of cyanide ions are those that form a complex ion with it. The formation of the complex ion increases solubility by removing the metal cation from the solution.

Looking in Appendix II, we see that Ag^+ and Fe^{2+} form complex ions with cyanide ions. This means that the solubilities of $FeBr_2$ and AgBr will both increase when NaCN is added to solutions of them.

---

e. Some metal hydroxide complex ions are amphoteric; these complex ions will be soluble at high and low pH, but will be insoluble at neutral pH.

i. A metal hydroxide that is soluble in an acidic solution reacts according to a neutralization reaction that gives hydronium ions and the metal cation:

$$M(OH)_x(aq) + X\ H_2O(l) \leftrightarrows X\ H_3O^+(aq) + M^{x+}(aq)$$

ii. A metal hydroxide that is soluble in a basic solution reacts with the hydroxide ion to give a complex ion:

$$M(OH)_x(aq) + Y\ OH^-(aq) \leftrightarrows M(OH)_{x+y}^{y-}(aq)$$

iii. Only hydroxides of Al^{3+}, Cr^{3+}, Zn^{2+}, Pb^{2+}, and Sn^{2+} are amphoteric; all other metal hydroxides will only be soluble in acidic solutions.

Fill in the Blank:

1. A titration curve is a plot of ________________ vs. ________________ in a titration experiment.
2. A complex ion contains a central metal atom that acts as a ________________ bound to one or more ligands that act as ________________.
3. Buffer solutions contain significant amounts of a weak acid and ________________.
4. The equilibrium constant expression for a reaction that represents the dissolution of a sparingly soluble salt is called the ________________.
5. ________________ can be used to determine the metal ions present in an unknown solution.
6. At the ________________ of a titration, the moles of acid and base are stoichiometrically equal.
7. The solubility of an ionic compound ________________ in the presence of a Lewis base that can complex to the cation.
8. The pH of a buffer system is approximately equal to the acid ________________ value.
9. The ________________ of a salt is the number of moles of the salt that dissolve in 1 L of solution.
10. When the change in concentration to achieve equilibrium concentrations is small, the pH of a buffer solution can be calculated using the ________________ equation.
11. The effecti ve range for a buffering system is ________________.

12. The solubility of a(n) _______________ metal hydroxide increases when either acids and bases are added.

13. A _______ __________ will have multiple equivalence points when titrated with a strong base.

14. When two cations both form salts with a given anion, the cation with the _______________ value of K_{sp} will precipitate out of solution first upon the addition of the anion.

15. The _______________ is the amount of acid or base that can be added to a buffer without destroying its effectiveness.

16. A sol ution is supersaturated when the reaction quotient is _______________ the K_{sp} value.

17. The __ _______________ of a titration is where the indicator color changes.

18. The relative concentrations of the acid and conjugate base in a buffer system should not differ by more than _______________.

19. The equilibriu m constant for the formation of a complex ion is called the _______________.

20. The solubility of an ionic substance _______________ when dissolved in a solution containing a common ion.

Problems:

1. 1.0 L of a buffer is prepared using 24.2 g of acetic acid.

   a. What quantity (in g) of sodium acetate must be used in order for the concentrations of the acid and conjugate base to be equal at equilibrium?

   b. What will the pH of the buffer system be when 24.2 g of sodium acetate are added to the original solution? Assume that no volume change occurs upon the addition.

   c. What will the pH of the buffer system be when the amount of sodium acetate added to the acetic acid solution is equal to the number of moles of acetic acid originally in the solution? Assume that no volume change occurs upon the addition.

   d. Explain the difference between your answers in parts b and c and determine which solution is a better buffer.

   e. Calculate the pH of the buffer system that you prepared in part a, upon the addition of 50.0 mL of a 0.024 M NaOH solution.

   f. What volume of the 0.024 M NaOH solution is required to exceed the buffering capacity of the solution that you prepared in part a? Hint: consider how you would make the concentrations of the buffer components differ by a factor of 10 or more.

2. Consider the titration of 25.0 mL of a 0.143 M solution of NaOH with a 0.143 M solution of HCl.

   a. Without doing any calculations, what volume of HCl will be required to reach the equivalence point?

   b. Will this titration experiment have a buffer region? Why or why not?

   c. Calculate the pH of the solution at the beginning of the experiment (before the addition of any HCl).

   d. Calculate the pH of the solution after 15.0 mL of the HCl solution has been added.

   e. Calculate the pH of the solution at the equivalence point.

   f. Calculate the pH of the solution after 45.0 mL of the HCl solution has been added.

3. Consider titrating 15.0 mL of a 0.245 M solution of hydrazine, H_2NNH_2, with a 0.453 M HCl solution. The K_b value for hydrazine is 1.3×10^{-6}.

a. Without doing any calculations, determine whether the pH of the solution at the equivalence point is greater than 7, equal to 7, or less than 7.

b. What volume of HCl needs to be added to the solution in order to have a buffer system with the maximum capacity for added acid or base?

c. What is the pH of the solution before any HCl has been added?

d. What is the pH of the solution after the addition of 3.0 mL of HCl?

e. What is the pH of the solution at the equivalence point?

f. What is the pH of the solution when 5.0 mL more of the HCl solution is added after the equivalence point has been reached?

g. What is the pH of the solution when 10.0 mL more of the HCl solution is added after the equivalence point has been reached?

h. What is the pH of the solution when 15.0 mL more of the HCl solution is added after the equivalence point has been reached?

4. The pH at half equivalence is an important reference point for an acid.

   a. Calculate the pH at half equivalence when 50.0 mL of 0.0252 M NH_3 is titrated with a 0.0385 M HCl solution.

   b. Based on your answer in part a, what do you think the significance of the pH at half equivalence is for chemists?

5. Consider titrating 20.0 mL of a 0.150 M solution of iodoacetic acid (ICH_2CO_2H) with a 0.100 M solution of sodium hydroxide. The pK_a of iodoacetic acid is 3.18.

   a. What is the pH of the solution after 10.0 mL of NaOH has been added to the acid?

   b. How will the pH of the solution that you calculated in part a change upon the addition of 5.0 mL of a 0.100 M solution of HCl?

   c. How will the pH of the solution that you calculated in part a change upon the addition of 10.0 mL of a 0.100 M solution of HCl?

   d. Comment on any differences between the values that you calculated in parts b and c.

   e. What is the pH of the solution at the equivalence point?

   f. Which of the following indicators is best for the titration of iodoacetic acid with sodium hydroxide?

      i. Thymol blue ($pK_a \approx 1.9$)

      ii. Methyl red ($pK_a \approx 5.1$)

      iii. Bromothymol blue ($pK_a \approx 6.8$)

      iv. Phenol red ($pK_a \approx 7.9$)

      v. Phenolphthalein ($pK_a \approx 9.1$)

   g. Sketch the titration curve for this experiment, being sure to indicate the buffer region(s), the initial pH, the pH at the equivalence point, the pH at the end of the experiment, and appropriate labels for the axes.

6. 14.3 g of K_2CO_3 are dissolved in enough water to make 1.0 L of solution.

   a. What is the solubility of $Mg(NO_3)_2$ in this solution?

   b. What is the maximum amount of $Mg(NO_3)_2$ that you can dissolve in this solution without forming a precipitate?

c. Will a precipitate form upon the addition of HCl (after the addition of the $Mg(NO_3)_2$)? If so, what is the precipitate? If not, why not?

d. Will a precipitate form upon the addition of NaOH? If so, what is the precipitate? If not, why not?

7. Na_2CO_3 is added to a solution containing Ca^{2+}, Mg^{2+}, and Fe^{2+}. In what order will these cations precipitate out of solution? Explain your answer.

8. Silver bromide is a sparingly soluble salt.

   a. What is the solubility of silver bromide in water?

   b. What is the solubility of silver bromide in a 0.0015 M solution of HCl?

   c. What is the solubility of silver bromide in a 0.0150 M solution of HBr?

   d. What is the solubility of silver bromide in a 0.250 M solution of NaCN? (Hint: Ag forms a complex ion with the cyanide ion)

9. Describe an experiment that you could use to separate out Fe^{2+}, Ca^{2+}, Mg^{2+}, Na^{+}, Pb^{2+}, and Cr^{3+}.

<u>Concept Questions:</u>

1. pH indicators are weak acids as discussed in this chapter. Explain why their influence on the pH of a solution is never accounted for in a titration experiment.

2. Which species (HInd or Ind^{-}) will be the dominant form of the indicator just before and just after the equivalence point when titrating a weak acid with a strong base? Which species will dominate when titrating a weak base with a strong acid?

3. The equivalence point for a titration can be estimated based on the pK_a of the weak acid used. Explain how and why this can be done. Estimate the pH at the equivalence point for the titration of pyruvic acid with sodium hydroxide.

4. In many of the calculations described in this chapter, we suspend the universe. For example, when calculating the pH of a titration experiment at equivalence, we split the problem into two parts: a reaction between the strong base and weak acid that goes to completion and then an equilibrium problem; in reality, these reactions are constantly occurring in both directions.

   a. Explain why this is a reasonable approach to the problem?

   b. Is this actually what occurs (on a molecular level) in the experiment?

   c. If this is NOT what actually occurs, explain why the problem isn't solved according to the physical reality of the situation.

5. The region of a titration curve in which a weak acid has been titrated with an amount of strong base that is less than the amount needed to reach the equivalence point is called the buffer region. On a titration curve, this region looks very flat, i.e., the pH doesn't change much as base is continually added. The region after the equilibrium point also looks flat on a titration curve, but this is not a buffer region. Explain why this region is not a buffer and why, despite this, it looks flat.

Chapter 17: Free-energy and Thermodynamics

Learning Goals:

- Understand the criteria for a spontaneous chemical process.
- Become familiar with entropy as a measure of energy dispersal and probability.
- Identify processes for which the entropy of a system increases under all conditions.
- Understand Gibbs free-energy as the criteria for spontaneous change, the maximum available work, and the entropy of the universe defined in terms of a thermodynamic system.
- Identify temperatures at which a chemical process is spontaneous.
- Calculate the absolute entropy of a substance using the third law of thermodynamics.
- Calculate the free-energy change of a reaction under standard and nonstandard conditions.
- Understand the relationship between free-energy and equilibrium position.

Chapter Summary:

In this chapter, we will explore the fundamental concepts that dictate whether or not a chemical change will occur spontaneously. We will first define spontaneous change and then use the second law of thermodynamics as its predictor. Entropy will be explained as a measure of a system's ability to disperse or arrange itself energetically. We will use the Boltzmann equation to relate the probability of a given arrangement to a system's entropy and then apply these ideas to common chemical systems. We will identify the entropy change of the universe as the predictor of spontaneous change and then define it in terms of system properties. Gibbs free-energy will then be introduced as a thermodynamic function that determines spontaneity and is equal to the maximum work available from a system. We will learn how to use the third law of thermodynamics in defining the zero of entropy and investigate how to use this information to calculate the entropy change of a chemical reaction. Standard free-energy changes will be calculated using three different methods and then nonstandard free-energy changes will be explored. Finally, you will learn about the relationship between the equilibrium constant value and the thermodynamic properties of a chemical system.

Chapter Outline:

I. Nature's Heat Tax: You Can't Win and You Can't Break Even
   a. Energy is conserved in a chemical process (as discussed in Chapter 6).

II. Spontaneous and Non-spontaneous Processes
   a. A spontaneous process is one that occurs without any ongoing outside intervention.
      i. A spontaneous process occurs in order to minimize a system's chemical potential energy, just as a mechanical system occurs in a direction that minimizes its mechanical potential energy.
   b. Thermodynamics studies the direction and extent of a reaction, not the speed of a chemical reaction.
      i. Thermodynamics is unaffected by catalysts, which increase the rate or speed of a reaction but do not make the reaction occur to any greater (or lesser) extent.
   c. Non-spontaneous processes can be made spontaneous through the application of energy from an external source.

III. Entropy and the Second Law of Thermodynamics

a. Enthalpy is not the criteria for spontaneous change – some endothermic processes occur spontaneously.

b. Entropy is a thermodynamic function that is related to the number of energetically equivalent ways to arrange a system:

$$S = k \ln W$$

where k is the Boltzmann constant (1.38×10^{-23} J/K) and W is the multiplicity or number of energetically equivalent ways to arrange a system.

i. As the number of ways to arrange a system increases, the entropy of the system increases.

c. Entropy has units of J/K and is a state function:

$$\Delta S = S_{final} - S_{initial}$$

d. The multiplicity (W) is a measure of probability; the more ways that a system can be arranged in a given state, the more likely the system is to be found in that state.

e. Entropy is maximized when energy dispersion is maximized.

i. Heat flows from a hot body to a cold body as a result of energy dispersion.

f. The second law of thermodynamics states that a chemical system will proceed in a way that increases the entropy of the universe:

$$\Delta S_{universe} > 0 \text{ for a spontaneous process}$$

g. The entropy of a system always increases when:

i. A solid changes into a liquid.

ii. A solid changes into a gas.

iii. A liquid changes into a gas.

iv. A chemical reaction occurs in which the moles of gas increases ($\Delta n>0$).

EXAMPLE:

Determine the sign of the entropy change in the following processes:

a. $PCl_5(s) \rightarrow PCl_3(s) + Cl_2(g)$

The formation of a gas upon the decomposition of a solid results in a positive entropy change.

b. $Hg(l) \rightarrow Hg(g)$

A phase transition in which a liquid becomes a gas has a positive entropy change.

c. $C_2H_5OH(g) \rightarrow C_2H_5OH(l)$

The condensation of gaseous ethanol to liquid ethanol has a negative entropy change.

IV. Heat Transfer and Changes in the Entropy of the Surroundings

a. The entropy change of the universe is the sum of the entropy changes of the system (ΔS_{sys}) and its surroundings (ΔS_{surr}):

$$\Delta S_{universe} = \Delta S_{surr} + \Delta S_{sys}$$

b. Since the entropy change of the universe must be positive for a spontaneous process:

$$\Delta S_{surr} + \Delta S_{sys} > 0$$

c. This means that the change in entropy of the surroundings must be greater than the negative change in entropy of the system for a spontaneous process:

$$\Delta S_{surr} > -\Delta S_{sys}$$

i. For an exothermic process ($\Delta S_{surr} > 0$), the entropy of the surroundings must increase more than the entropy of the system decreases (if it decreases).

ii. For an endothermic process ($\Delta S_{surr} < 0$), the entropy of the system must increase more than the entropy of the surroundings decreases.

d. The spontaneity of a process is temperature dependent since entropy measures heat dispersion per unit temperature.

e. Since entropy is temperature dependent and related to the enthalpy change of a chemical reaction, we can relate the entropy change of the surroundings to the temperature and enthalpy change:

$$\Delta S_{surr} = \frac{-\Delta H_{sys}}{T}$$

EXAMPLE:

Calculate the entropy change of the surroundings and the sign of ΔS_{sys} when the following reactions are carried out.

a. $4Fe(s) + 3O_2(g) \rightarrow 2Fe_2O_3(s)$ $\Delta H = -1648$ kJ/mol at 25°C

The change in entropy of the surroundings is:

$$\Delta S_{surr} = \frac{-(-1648\text{ kJ/mol})}{(298\text{ K})} = +5.53\text{ kJ/mol}\cdot\text{K}$$

The entropy change of the system is negative since the change in gas molecules is negative.

b. $2H_2O(l) + O_2(g) \rightarrow 2H_2O_2(l)$ $\Delta H = +196$ kJ/mol at 25°C

$$\Delta S_{surr} = \frac{-(+196\text{ kJ/mol})}{(298\text{ K})} = -0.658\text{ kJ/(mol}\cdot\text{K)} = -658\text{ J/(mol}\cdot\text{K)}$$

The entropy change of the system is negative since the change in gas molecules is negative.

c. $2F_2(g) + Si(s) \rightarrow SiF_4(g)$ $\Delta H = -1615$ kJ/mol at 25°C

$$\Delta S_{surr} = \frac{-(-1615\text{ kJ/mol})}{(298\text{ K})} = +5.42\text{ kJ/mol}\cdot\text{K}$$

The entropy change of the system is negative since the change in gas molecules is negative.

V. Gibbs Free-energy

a. We can rewrite the change in entropy of the universe using system terms by replacing the entropy change of the surroundings with the enthalpy change divided by the temperature:

$$\Delta S_{universe} = \frac{-\Delta H_{sys}}{T} + \Delta S_{sys}$$

b. Rearranging, we get:

$$T\Delta S_{universe} = -\Delta H_{sys} + T\Delta S_{sys}$$

c. We introduce a new thermodynamic function, Gibbs free-energy (G):

$$G = H - TS$$

d. The Gibbs free-energy change at constant temperature is:

$$\Delta G = \Delta H - T\Delta S$$

e. We can see that the expressions for the entropy change of the universe and the Gibbs free-energy are the same. This means that if the entropy change of the universe is a predictor of spontaneity, the Gibbs free-energy is as well.

f. The free-energy is often referred to as the chemical potential and is used to predict if a chemical process is spontaneous.

g. When $\Delta G<0$, the process occurs spontaneously in the forward direction.

   i. When the entropy change is positive and the enthalpy change is negative, the free-energy change will always be negative.

   ii. When the entropy change is positive and the enthalpy change is positive, the free-energy change will be negative at high temperatures.

      1. This is called an entropically driven reaction.

   iii. When the entropy change is negative and the enthalpy change is negative, the free-energy change will be negative at low temperatures.

      1. This is called an enthalpically driven reaction.

   iv. When the entropy change is negative and the enthalpy change is positive, the free-energy change will never be negative.

EXAMPLE:

Calculate the free-energy change for the following processes and determine whether or not the process occurs spontaneously at 25°C. If the reaction is non-spontaneous at 25°C, at what temperature, if any, will it become spontaneous?

a. $HCl(s) + NH_3(g) \rightarrow NH_4Cl(s)$ ΔH=-176.2 kJ/mol, ΔS= -285.1 J/(mol·K)

In order to add the enthalpy and entropy terms, both need to be in units of kJ or J. We will convert the entropy into units of kJ/(mol·K) and then calculate the free-energy change at 298 K:

$$\Delta G = -176.2 \text{ kJ/mol} - (298 \text{ K})(-0.2851 \text{ kJ/(mol·K)}) = -91.2 \text{ kJ/mol}$$

So the reaction is spontaneous at 25°C.

b. $2C_4H_{10}(l) + 13O_2(g) \rightarrow 8CO_2(g) + 10H_2O(g)$ ΔH=-5271 kJ/mol, ΔS= +468.8 J/(mol·K)

Again, we need to have the same energy units, so we will use kJ/(mol·K) for the entropy change:

$$\Delta G = -5271 \text{ kJ/mol} - (298 \text{ K})(+0.4688 \text{ kJ/(mol·K)}) = -5411 \text{ kJ/mol}$$

This reaction is also spontaneous at 25°C.

VI. Entropy Changes in Chemical Reactions: Calculating ΔS°_{rxn}

a. The standard entropy change of a reaction (ΔS°_{rxn}) is the change in entropy for a process in which all species are in their standard states.

b. The third law of thermodynamics states that the entropy of a perfect crystal at 0 K is zero.

   i. A perfect crystal has only one configuration, so W=1.

   ii. The third law provides a zero of entropy so that we can calculate absolute entropy values of all substances.

c. Entropy is an extensive property, so entropy values are given as molar quantities.

d. The relative entropy values of substances can be compared in a number of different cases.

   i. More disordered phases of matter have higher entropy.

   ii. The more massive a substance is, the higher its entropy.

   iii. Entropy increases with increasing molecular complexity.

   iv. The dissolution of a salt in water increases the system entropy.

e. The standard entropy of a reaction can be calculated using the absolute entropy values of reactants and products multiplied by their stoichiometric coefficients:

$$\Delta S^\circ_{rxn} = \sum n_p S^\circ(\text{products}) - \sum n_r S^\circ(\text{reactants})$$

► Note that this looks just like Hess's Law from Chapter 6.

EXAMPLE:

Calculate the standard entropy change for the following reactions using the values given in Appendix II of your textbook.

a. $C(s, \text{graphite}) + O_2(g) \rightarrow CO_2(g)$

The standard entropy change for the reaction is given by the expression:

$$\Delta S^\circ_{rxn} = \sum n_p S^\circ(\text{products}) - \sum n_r S^\circ(\text{reactants})$$

From Appendix II:

C(s, graphite)	$S^\circ = 5.7$ J/(mol·K)
$O_2(g)$	$S^\circ = 205.2$ J/(mol·K)
$CO_2(g)$	$S^\circ = 213.8$ J/(mol·K)

So the enthalpy change for the reaction is:

$$\Delta S^\circ_{rxn} = S^\circ(CO_2(g)) - \{S^\circ(O_2(g)) + S^\circ(C(s, \text{graphite}))\}$$

$$\Delta S^\circ_{rxn} = 213.8 \text{ J/(mol·K)} - \{205.2 \text{ J/(mol·K)} + 5.7 \text{ J/(mol·K)}\} = +2.9 \text{ J/(mol·K)}$$

The entropy change is very small, which makes sense because the change in moles of gas is zero.

b. $NaCl(s) \xrightarrow{H_2O} Na^+(aq) + Cl^-(aq)$

From Appendix II:

$Cl^-(aq)$	$S^\circ = 56.6$ J/(mol·K)
$Na^+(aq)$	$S^\circ = 58.45$ J/(mol·K)

$NaCl(s)$ $S^o = 72.1$ J/(mol·K)

So the enthalpy change for the reaction is:

$$\Delta S^o_{rxn} = \{S^o(Cl^-(aq)) + S^o(Na^+(aq))\} - S^o(NaCl(s))$$

$$\Delta S^o_{rxn} = \{56.6 \text{ J/(mol·K)} + 58.45 \text{ J/(mol·K)}\} - 72.1 \text{ J/(mol·K)} = +43.0 \text{ J/(mol·K)}$$

This reaction describes the dissolution of salt in water, which should have a positive entropy change in agreement with our calculated value.

c. $2H_2(g) + O_2(g) \rightarrow 2H_2O(g)$

From Appendix II:

$H_2O(g)$ $S^o = 188.8$ J/(mol·K)

$H_2(g)$ $S^o = 130.7$ J/(mol·K)

$O_2(g)$ $S^o = 205.2$ J/(mol·K)

So the enthalpy change for the reaction is:

$$\Delta S^o_{rxn} = \{2 \times S^o(H_2O(g))\} - \{2 \times S^o(H_2(g)) + S^o(O_2(g))\}$$

$$\Delta S^o_{rxn} = \{2 \times 188.8 \text{ J/(mol·K)}\} - \{2 \times 130.7 \text{ J/(mol·K)} + 205.2 \text{ J/(mol·K)}\} = -89.0 \text{ J/(mol·K)}$$

In this reaction the number of moles of gas decreases, so it is reasonable to expect the entropy change to be negative.

VII. Free-energy Changes in Chemical Reactions: Calculating ΔG^o_{rxn}

a. There are three ways to calculate a ΔG^o_{rxn} value:

i. $\Delta G^o_{rxn} = \Delta H^o_{rxn} - T\Delta S^o_{rxn}$

EXAMPLE:

Calculate the standard free-energy change for the reaction of aluminum oxide with hydrogen to produce solid aluminum and water vapor at 45°C:

$$Al_2O_3(s) + 3H_2(g) \rightarrow 2Al(s) + 3H_2O(g)$$

We can look up the standard enthalpy and entropy values in Appendix II:

$Al_2O_3(s)$	ΔH^o_f = -1675.7 kJ/mol	ΔS^o_f = 50.9 J/(mol·K)
$H_2(g)$	ΔH^o_f = 0 kJ/mol	ΔS^o_f = 130.7 J/(mol·K)
$Al(s)$	ΔH^o_f = 0 kJ/mol	ΔS^o_f = 28.32 J/(mol·K)
$H_2O(g)$	ΔH^o_f = -241.8 kJ/mol	ΔS^o_f = 188.8 J/(mol·K)

We will now calculate the value of ΔH^o_{rxn} and ΔS^o_{rxn} from these values:

$$\Delta H^o_{rxn} = \{2 \times 0 \text{ kJ/mol} + 3 \times (-241.8 \text{ kJ/mol})\} - \{-1675.7 \text{ kJ/mol} + 3 \times 0 \text{ kJ/mol}\} = +950.3 \text{ kJ/mol}$$

$$\Delta S^{o}_{rxn} = \{2\times28.32\ J/(mol\cdot K) + 3\times188.8\ J/(mol\cdot K)\} - \{50.9\ J/(mol\cdot K) + 3\times130.7\ J/(mol\cdot K)\}$$

$$= +180.0\ J/(mol\cdot K) = +0.1800\ kJ/(mol\cdot K)$$

Using these values and the temperature, 318 K, we can calculate the standard free-energy change:

$$\Delta G^{o}_{rxn} = +950.3\ kJ/mol - (318\ K)(+0.1800\ kJ/(mol\cdot K))$$

$$\Delta G^{o}_{rxn} = +893.1\ kJ/mol$$

This reaction is non-spontaneous at room temperature because the value of the free-energy change is positive.

---

ii. $\Delta G^{o}_{rxn} = \sum n_p G^{o}(\text{products}) - \sum n_r G^{o}(\text{reactants})$

► Again, note that this looks just like the calculation of a reaction enthalpy in Chapter 6.

EXAMPLE:

Using the free-energy of formation values in Appendix II, calculate the standard free-energy change for the formation of rust at 25°C:

$$4Fe(s) + 3O_2(g) \rightarrow 2Fe_2O_3(s)$$

From Appendix II:

$Fe_2O_3(s)$ $\Delta G^{o}_{f} = -742.2$ kJ/mol

$O_2(g)$ $\Delta G^{o}_{f} = 0$ kJ/mol

$Fe(s)$ $\Delta G^{o}_{f} = 0$ kJ/mol

The standard free-energy of this reaction is:

$$\Delta G^{o}_{f} = \{2\times\text{-}742.2\ kJ/mol\} - \{4\times0\ kJ/mol + 3\times0\ kJ/mol\} = -1484.4\ kJ/mol$$

---

iii. Using a procedure similar to Hess's law for calculating ΔH^{o}_{rxn}; we can use a series of reactions that add up to the overall reaction to calculate ΔG^{o}_{rxn}.

EXAMPLE:

Calculate the value of ΔG^{o}_{rxn} for the synthesis of boron oxide from its elements:

$$4B(s) + 3O_2(g) \rightarrow 2B_2O_3(s)$$

Given the following ΔG^{o}_{rxn} values:

1. $B_2O_3(s) + 3H_2O(g) \rightarrow 3O_2(g) + B_2H_6(g)$ $\Delta G^{o}_{rxn} = +1967.7$ kJ/mol

2. $H_2O(l) \rightarrow H_2O(g)$ $\Delta G^\circ_{rxn} = -108.2$ kJ/mol

3. $2H_2(g) + O_2(g) \rightarrow 2H_2O(l)$ $\Delta G^\circ_{rxn} = -240.8$ kJ/mol

4. $2B(s) + 3H_2(g) \rightarrow 2B_2H_6(g)$ $\Delta G^\circ_{rxn} = +87.6$ kJ/mol

We first need to arrange these equations so that they will add up to give the correct overall equation.

The first reaction will be reversed and multiplied by 2 to give the $2B_2O_3(s)$ molecules as products; this means that the ΔG°_{rxn} of the reaction will be multiplied by 2 and the sign will be reversed:

1. $6O_2(g) + 2B_2H_6(g) \rightarrow 2B_2O_3(s) + 6H_2O(g)$ $\Delta G^\circ_{rxn} = -2\times(+1967.7 \text{ kJ/mol}) = -3935.4$ kJ/mol

The second reaction will be reversed and multiplied by 6 so that the gaseous water in the first reaction is replaced by liquid water:

2. $6H_2O(g) \rightarrow 6H_2O(l)$ $\Delta G^\circ_{rxn} = -6\times(-108.2 \text{ kJ/mol}) = +649.2$ kJ/mol

The third reaction will be multiplied by 3 and reversed to cancel out the formation of liquid water:

3. $6H_2O(l) \rightarrow 6H_2(g) + 3O_2(g)$ $\Delta G^\circ_{rxn} = -3\times(-240.8 \text{ kJ/mol}) = +722.4$ kJ/mol

Finally we multiply the last equation by 2 to give the correct number of boron atoms as a reactant:

4. $4B(s) + 6H_2(g) \rightarrow 4B_2H_6(g)$ $\Delta G^\circ_{rxn} = 2\times(+87.6 \text{ kJ/mol}) = +175.2$ kJ/mol

Adding these reactions and their corresponding ΔG°_{rxn} values gives:

$4B(s) + 3O_2(g) \rightarrow 2B_2O_3(s)$ $\Delta G^\circ_{rxn} = -2388.6$ kJ/mol

---

b. Free-energy is a measure of the maximum amount of work available from a system.

i. Entropy measures the amount of energy that is lost as heat (divided by temperature) while free-energy measures the energy that remains to do work.

ii. Reversible reactions are reactions that occur infinitesimally slowly; in a reversible reaction, the free-energy can be drawn out of the system in infinitesimally small amounts that exactly match the energy produced.

iii. All real reactions are irreversible meaning that the real energy that can be extracted as work is less than the theoretical amount (ΔG).

VIII. Free-energy Changes for Non-standard States: the Relationship Between ΔG°_{rxn} and ΔG_{rxn}

a. Ordinarily, reactions are not carried out at standard conditions.

b. The relationship between the standard free-energy change and the free-energy change for an ordinary system is:

$$\Delta G_{rxn} = \Delta G^\circ_{rxn} + RT\ln Q$$

where R is the gas constant, T is the absolute temperature, and Q is the reaction quotient with all gas phase species expressed as partial pressures and all aqueous species expressed in molarity.

i. At standard conditions, Q=1 and $\Delta G_{rxn} = \Delta G^\circ_{rxn}$.

ii. At equilibrium, Q=K and $\Delta G_{rxn} = 0$.

- Note that at equilibrium ΔG^{o}_{rxn} does NOT equal zero – this is a common mistake.

EXAMPLE:

Calculate the free-energy change for the combustion of methane, $CH_4(g)$, at 25°C given the following initial reaction conditions and the standard free-energy change:

a. $P_{CH_4} = 0.142\text{ atm}, P_{O_2} = 0.715\text{ atm}, P_{H_2O} = 0.574\text{ atm, and } P_{CO_2} = 0.283\text{ atm}$

First we will need to calculate the equilibrium reaction:

$$CH_4(g) + 2O_2(g) \leftrightarrows CO_2(g) + 2H_2O(g) \quad \Delta G^{o}_{rxn} = -801.1\text{ kJ/mol}$$

The reaction quotient expression for this reaction is:

$$Q = \frac{P_{CO_2} P_{H_2O}^{\ 2}}{P_{CH_4} P_{O_2}^{\ 2}}$$

Now we can calculate the value of Q for the conditions given:

$$Q = \frac{(0.283\text{ atm})(0.574\text{ atm})^2}{(0.142\text{ atm})(0.715\text{ atm})^2} = 1.284$$

Calculating the value of ΔG_{rxn}:

$$\Delta G_{rxn} = -801.1\text{ kJ/mol} + (8.3145\text{ J/(mol·K)})(298\text{ K})\ \ln(1.284) \times \frac{1\text{ kJ}}{1000\text{ J}} = -800.5\text{ kJ/mol}$$

Since the value of ΔG_{rxn} is less than zero, the reaction will proceed spontaneously in the forward direction.

b. $P_{CH_4} = 0.00082\text{ atm}, P_{O_2} = 0.000051\text{ atm}, P_{H_2O} = 42.10\text{ atm, and } P_{CO_2} = 29.53\text{ atm}$

We will recalculate Q under the new conditions:

$$Q = \frac{(29.53\text{ atm})(42.10\text{ atm})^2}{(0.00082\text{ atm})(0.000051\text{ atm})^2} = 2.454 \times 10^{16}$$

$$\Delta G_{rxn} = -801.1\text{ kJ/mol} + (8.3145\text{ J/(mol·K)})(298\text{ K})\ \ln(2.454 \times 10^{16}) \times \frac{1\text{ kJ}}{1000\text{ J}} = -707.6\text{ kJ/mol}$$

The reaction is spontaneous in the forward direction under this second set of conditions.

IX. Free-energy and Equilibrium: Relating ΔG^{o}_{rxn} to the Equilibrium Constant (K)

a. The standard free-energy change predicts the direction of spontaneous change when all species (reactants and products) are combined under standard conditions.

b. The equilibrium constant indicates whether a reaction is product-favored or reactant-favored.

c. The standard free-energy change and the equilibrium constant are related:

$$\Delta G^{o}_{rxn} = -RT \ln K$$

i. When the reaction favors products (K>1), the reaction will proceed spontaneously in the forward direction from standard conditions ($\Delta G^{o}_{rxn} < 0$).

ii. When the reaction favors reactants (K<1), the reaction will not proceed spontaneously in the forward direction from standard conditions ($\Delta G^{o}_{rxn} > 0$).

iii. When the reaction reaches equilibrium at a position where the ratio of products to reactants is 1 (K=1), the reaction is at equilibrium under standard conditions ($\Delta G^{o}_{rxn} = 0$).

EXAMPLE:

Calculate the equilibrium constant for the combustion of methane at 25°C and determine if the reaction is product favored or reactant favored at this temperature.

We were given $\Delta G^{o}_{rxn} = -801.1$ kJ/mol in the previous example, so we can calculate the equilibrium constant using the equation:

$$\Delta G^{o}_{rxn} = -RT \ln K$$

We need to rearrange to solve for K:

$$K = e^{-\Delta G^{o}_{rxn}/RT}$$

Putting in the given values:

$$K = e^{-(-801100\ \text{J/mol}/(8.3145\ \text{J/K}\cdot\text{mol})(298\ \text{K})}$$

The equilibrium constant is so large that most calculators will overflow with this input! This is consistent with the calculations that we did in the previous problem where we saw that even at very high partial pressures of the products, the reaction would continue to proceed in the forward direction spontaneously.

---

d. The temperature dependence of the equilibrium constant can be used to determine thermodynamic quantities by combining the expression for ΔG^{o}_{rxn} in terms of the equilibrium constant and the expression for ΔG^{o}_{rxn} in terms of the enthalpy and entropy changes:

$$-RT\ln K = \Delta H^{o}_{rxn} - T\Delta S^{o}_{rxn}$$

e. This expression can be arranged to give the equation of a line, plotting ln K vs. 1/T:

$$\ln K = \frac{-\Delta H^{o}_{rxn}}{R}\left(\frac{1}{T}\right) + \frac{\Delta S^{o}_{rxn}}{R}$$

where the slope is equal to $-\Delta H^{o}_{rxn}/R$ and the y-intercept is equal to $\Delta S^{o}_{rxn}/R$.

Fill in the Blank:

1. The change in ________________ is proportional to the negative of the entropy change of the universe.

2. The entropy of a sample of matter ________________ when it goes from the liquid phase to the solid phase.
3. A(n) ________________ is a process that cannot occur without ongoing outside intervention.
4. The free-energy of formation of pure elements in their standard states is ________________.
5. ________________ is a measure of the number of energetically equivalent ways that a system can be prepared.
6. The entropy of a perfect crystal at 0 K is ________________.
7. The value of the equilibrium constant is ________________ when the standard free-energy change is a positive number.
8. Entropy is a ________________ meaning that the change in entropy can be calculated by finding the difference between the final entropy and the initial entropy of a system.
9. In a reversible chemical reaction, the change in magnitude of the free-energy is equal to the ________________.
10. A reaction that is non-spontaneous at high temperatures but spontaneous at low temperatures has a ________________ entropy change and a ________________ enthalpy change.
11. In a spontaneous process, a chemical system proceeds in a direction that ________________ the entropy of the universe.
12. The entropy of He i s ________________ than the entropy of Ne because of ________________.
13. When a che mical reaction is multiplied by some factor, the ΔG_{rxn} is ________________.
14. ____ ________________ is the study of the extent of a reaction while ________________ is the study of the speed of a reaction.
15. In a spontaneous exothermic reaction, the change in enthalpy of the system is ________________ while the change in entropy of the surroundings is ________________.
16. The free-energy of a reaction under standard conditions can be calculated by ________________.
17. A spon taneous change occurs when the change in free-energy is ________________.

Problems:

1. Which of the following substances would you expect to have the highest entropy at room temperature?

   a. $BF_3(s)$ b. $COF_2(s)$ c. $C(s)$ d. $BF_3(l)$ e. $COF_2(l)$

   Explain the reasoning behind your choice.

2. Without doing any calculations, predict the signs of ΔH°_{rxn}, ΔS°_{rxn}, and ΔG°_{rxn} for the dissolution of barium sulfate in water at standard conditions:

$$BaSO_4(s) \leftrightarrows Ba^{2+}(aq) + SO_4^{2-}(aq)$$

3. Methane and hexane are both compounds that contain only carbon and hydrogen (hydrocarbons).

   Methane: T_{bp}= 112°C ΔH_{vap}= 8.20 kJ/mol

   Hexane: T_{bp}= 342°C ΔH_{vap}= 28.90 kJ/mol

   a. Calculate the entropy change of the surroundings that occur upon the vaporization of methane and hexane.
   b. Provide an explanation for the difference in entropy changes that you calculated.

4. Consider the reactions shown below:

$2H_2(g) + O_2(g) \rightarrow 2H_2O(l)$

$C(graphite) + O_2(g) \rightarrow CO_2(g)$

$CH_4(g) + 2O_2(g) \rightarrow CO_2(g) + 2H_2O(l)$

a. Using the values given in Appendix II of your textbook, calculate the values of ΔH°_{rxn} and ΔS°_{rxn} for each of the reactions.

b. Using the values that you calculated in part a, calculate ΔG°_{rxn} for each reaction at 25°C.

c. Calculate the value of ΔG°_{rxn} for each reaction at 25°C using the values of ΔG°_f given in Appendix II.

d. Calculate the value of ΔG°_{rxn} for the reaction of graphite and hydrogen gas to produce methane at 25°C.

e. Calculate the equilibrium constant of the reaction using the value of ΔG°_{rxn} that you calculated in part d.

f. Over what temperature range will the reaction in part d be spontaneous?

5. Consider the reactions shown below:

$2O(g) \rightarrow O_2(g)$

$H_2O(l) \rightarrow H_2O(g)$

$2H(g) + O(g) \rightarrow H_2O(g)$

$C(graphite) + 2O(g) \rightarrow CO_2(g)$

$C(graphite) + O_2(g) \rightarrow CO_2(g)$

$C(graphite) + 2H_2(g) \rightarrow CH_4(g)$

$2H(g) \rightarrow H_2(g)$

a. Without doing any calculations, predict the sign of the entropy change for each of the reactions given.

b. Using the values given in Appendix II of your textbook, calculate the values of ΔH°_{rxn} and ΔS°_{rxn} for each of the reactions.

c. Using the values that you calculated in part a, calculate ΔG°_{rxn} for each reaction at 25°C.

d. Calculate the value of ΔG°_{rxn} for each reaction at 25°C using the values of ΔG°_f given in Appendix II.

e. Calculate the value of ΔG°_{rxn} for the reaction combustion of methane at 25°C.

f. Calculate the equilibrium constant for the combustion of methane at 25°C, using the value of ΔG°_{rxn} that you calculated in part e.

g. Over what temperature range will the reaction in part d be spontaneous?

6. Under what conditions will a reaction be spontaneous at all temperatures?

7. Urea is a common fertilizer that is synthesized industrially by reacting carbon dioxide and ammonia:

$$CO_2(g) + 2NH_3(g) \leftrightarrows CO(NH_2)_2(s) + H_2O(l)$$

a. Calculate the standard free-energy change for this reaction using the values given in Appendix II of your textbook.

b. If 50.0 g of urea is placed into a closed 1.0-L container at 25°C, what will the partial pressure of CO_2 be at equilibrium?

c. What percentage of urea has decomposed at equilibrium?

d. At what temperatures is the decomposition of urea spontaneous?

e. Suggest two conditions that would aid in the efficient synthesis of urea using the above chemical reaction and explain how the equilibrium position will change when these conditions are imposed. Note: you should not add or remove material in answering this question.

8. An important reaction in atmospheric chemistry is between nitrogen monoxide and ozone to form nitrogen dioxide and oxygen gas:

$$NO(g) + O_3(g) \leftrightharpoons NO_2(g) + O_2(g)$$

a. Calculate the standard free-energy change for the reaction using the values given in Appendix II in your textbook.

b. Based on your answer in part a, is the reaction product favored, reactant favored, or neither?

c. Calculate the equilibrium constant for the reaction.

d. Calculate the standard entropy change for this reaction using the values given in Appendix II in your textbook.

e. Explain the value of the standard entropy change that you calculated in part d.

f. Do you think that this reaction is entropically or enthalpically driven? Justify your answer.

9. Structural isomers are compounds that have the same molecular formula but different structures. Butane and isobutane (C_4H_{10}) are an example of structural isomers.

a. Calculate the mole percent of butane in a mixture of these isomers at 25°C given that the standard free-energy of formation for isobutane is -18.0 kJ/mol and the standard energy of formation for butane is -15.9 kJ/mol.

b. Based on your answer in part a, which species do you believe is more stable at room temperature?

Concept Questions:

1. The precipitation of AgCl upon the addition of a NaCl solution to $AgNO_3$ seems to be a violation of the second law of thermodynamics because solids are more ordered than ions in solution. Explain why this is not, in fact, a violation of the second law.

2. Another statement of the second law of thermodynamics is that absolute zero can never be reached. Explain this statement using what you have learned in this chapter.

3. Explain why water vapor condenses at low temperatures but not at high temperatures based on the size of ΔS_{surr} for the process.

4. We metabolize sugars in our bodies through a combustion reaction much like the burning of gasoline in a car motor. Our body carries out this reaction in many steps whereas a car carries it out in a single step. In terms of the reversibility of a chemical reaction, explain why more useful work can be obtained by the body than by a combustion engine.

5. The value of ΔG^{o}_{rxn} is proportional to the natural log of the equilibrium constant. This means that when the equilibrium constant is equal to 1, the ΔG^{o}_{rxn} is 0. Explain what this means physically.

6. The standard entropy change for the vaporization of water is +79.2 J/(mol·K), which is smaller than the standard entropy change for many other liquids. Consider the structure of water to explain why this is. Hint: methanol also has a relatively small entropy change associated with vaporization.

7. A substance's heat capacity is related to its entropy. Explain why this makes physical sense using the concept of entropy as a measure of energy dispersion.

Chapter 18: Electrochemistry

Learning Goals:

- Balance redox reactions.
- Identify the anode, cathode, and salt bridge in a galvanic cell.
- Identify spontaneous redox reactions using a table of standard electrode potentials.
- Understand the physical processes behind electrolytic cells.
- Understand the connection between free-energy and cell potential for reactions at standard and nonstandard conditions.
- Construct a concentration cell and calculate its potential.
- Identify the anode and cathode in standard battery types.
- Understand electrolysis and make calculations relating the current, time, and charge necessary to plate out a certain metal.

Chapter Summary:

This chapter is focused on a common and useful reaction type: oxidation-reduction reactions. You will begin by learning how to balance these reactions and then learn how electricity flows in a galvanic cell when half-reactions are separated. You will learn how to construct a galvanic cell and what the driving force is for current flow. A shorthand notation called cell diagrams will be introduced to convey the overall chemistry in a galvanic cell. Next, you will learn how to calculate the cell potential using a table of standard electrode potentials. You will then learn how to relate the free energy concept from last chapter to the cell potential for reactions that are at standard and nonstandard conditions. The concept of nonstandard conditions will be exploited in your exploration of concentration cells. Galvanic cells will be used to explain the operation of batteries. Next, you will learn how external voltage can be used to drive nonspontaneous reactions in electrolytic cells. Finally, you will learn that corrosion is simply an undesirable redox reaction.

Chapter Outline:

I. Pulling the Plug on the Power Grid
    a. Fuel cells use redox reactions to generate electricity.

II. Balancing Redox Reactions
    a. Oxidation is the loss of electrons, which results in an increase in oxidation state.
    b. Reduction is the gain of electrons, which results in a decrease in oxidation state.
    c. In a redox reaction, both mass and charge must be balanced.
    d. Redox reactions can be balanced in a systematic way:
        i. Assign oxidation states to all species.
        ii. Break the reaction up into two half-reactions: one for oxidation and one for reduction.
        iii. Balance the half-reactions for mass.
            1. Oxygen is balanced by adding water.
            2. Hydrogen is balanced by adding hydrogen ions.

3. For a reaction carried out in a basic solution, add enough hydroxide ions to both sides to neutralize all hydrogen ions.

iv. Add electrons to balance charge.

v. Multiply each half-reaction by the smallest whole number that results in the half-reactions having the same number of electrons.

vi. Add half-reactions to get the overall reaction.

► After completing all steps, check that the reaction balances for mass, overall charge, and be sure that all electrons have cancelled out.

EXAMPLE:

Balance the redox reaction between dichromate ions and nitrous acid:

$$Cr_2O_7^{2-}(aq) + HNO_2(aq) \rightarrow Cr^{3+}(aq) + NO_3^-(aq)$$

We will follow the steps given above.

First, we will determine the oxidation state on all atoms:

Reactants:

Cr: +6

O (in $Cr_2O_7^{2-}$): -2

H: +1

N: +3

O (in HNO_2): -2

Products:

Cr: +3

N: +5

O: -2

We see that chromium is being reduced and nitrogen is being oxidized. Note that this is consistent with the fact that the reactant has a nitrogen atom bonded to two oxygen atoms and the product has a nitrogen atom bonded to three oxygen atoms.

Second, we will separate the reaction into two half-reactions:

Oxidation: $HNO_2(aq) \rightarrow NO_3^-(aq)$

Reduction: $Cr_2O_7^{2-}(aq) \rightarrow Cr^{3+}(aq)$

Third, we will balance mass; the oxygen atoms are balanced by adding water and the hydrogen atoms are balanced by adding acid:

Oxidation: $HNO_2(aq) + H_2O(l) \rightarrow NO_3^-(aq) + 3H^+(aq)$

Reduction: $Cr_2O_7^{2-}(aq) + 14H^+(aq) \rightarrow 2Cr^{3+}(aq) + 7H_2O(l)$

Fourth, we will add electrons to balance the charge:

Oxidation: $HNO_2(aq) + H_2O(l) \rightarrow NO_3^-(aq) + 3H^+(aq) + 2e^-$

Reduction: $Cr_2O_7^{2-}(aq) + 14H^+(aq) + 6e^- \rightarrow 2Cr^{3+}(aq) + 7H_2O(l)$

Fifth, we will make the number of electrons equal by multiplying the oxidation half-reaction by 3:

Oxidation: $3HNO_2(aq) + 3H_2O(l) \rightarrow 3NO_3^-(aq) + 9H^+(aq) + 6e^-$

Reduction: $Cr_2O_7^{2-}(aq) + 14H^+(aq) + 6e^- \rightarrow 2Cr^{3+}(aq) + 7H_2O(l)$

Finally, we will add the half-reactions and cancel out species that are on both sides of the reaction:

$$3HNO_2(aq) + 3H_2O(l) + Cr_2O_7^{2-}(aq) + 14H^+(aq) + 6e^- \rightarrow 3NO_3^-(aq) + 9H^+(aq) + 6e^- + 2Cr^{3+}(aq) + 7H_2O(l)$$

$$3HNO_2(aq) + Cr_2O_7^{2-}(aq) + 5H^+(aq) \rightarrow 3NO_3^-(aq) + 2Cr^{3+}(aq) + 4H_2O(l)$$

III. Voltaic (or Galvanic) Cells: Generating Electricity from Spontaneous Chemical Reactions

a. Electrical current is the flow of electrical charge.

   i. An example is the flow of electrons through a wire or ions migrating in a solution.

b. In a redox reaction, electrons can be transferred directly (when both species are combined in a single solution) or can be separated.

   i. When the oxidation and reduction reactions are separated from each other by a wire, electrons can flow and do electrical work.

      1. An electrochemical cell generates electricity in this way.

   ii. A voltaic (or galvanic) cell produces electrical current from a spontaneous chemical reaction.

   iii. Electrolytic cells use electrical current to drive nonspontaneous reactions.

c. A half-cell is a cell in which half of the reaction (oxidation or reduction) takes place.

d. The anode is where oxidation occurs and electrons are released; the anode has a negative charge.

e. The cathode is where reduction occurs and electrons are taken up; the cathode has a positive charge.

   i. Electrons flow from the negative terminal (anode) to the positive terminal (cathode).

f. A salt bridge connects the anode and cathode.

   i. The salt bridge contains a strong electrolyte capped with a semipermeable membrane that allows ions to flow in order to ensure charge neutrality.

      1. Anions travel toward the anode in order to balance the loss of electrons (the build up of positive charge).

      2. Cations travel toward the cathode in order to balance the gain of electrons (the build up of negative charge).

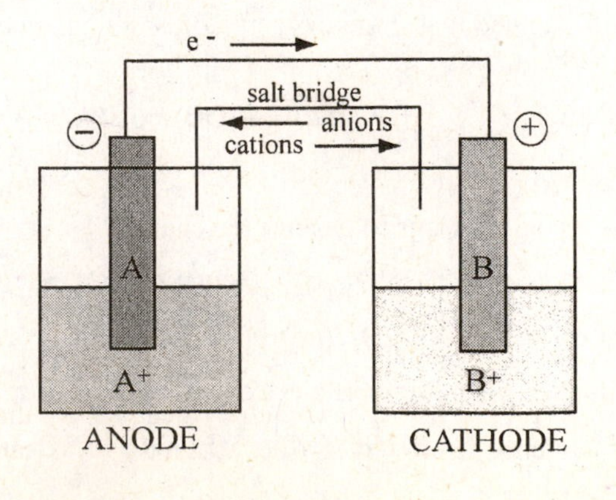

g. The rate of current flow is measured in amperes (A). 1 ampere is equal to one coulomb per second.

h. Electrical current flows due to a potential energy difference and is measured in volts. 1 volt is equal to one joule per coulomb.

i. Electromotive force (emf) is the force that drives electron motion due to a potential energy difference.

j. The difference between the potential energy of each half-reaction is the cell potential (E_{cell}) or the cell emf.

k. The cell potential at standard conditions (1 M aqueous solution and 1 atm pressure for gases) is the potential (E^o_{cell}) or standard condition emf.

l. Cell diagrams or line notation is a shorthand way of writing cells.

   i. In a cell diagram, oxidation is written on the left and reduction is written on the right.

      1. The oxidation and reduction reactions are separated by double vertical lines indicating the salt bridge.

   ii. Phases are separated by a single vertical line.

      1. When there is more than one species in the same phase, the species are separated by a comma.

      2. If all species are aqueous, an inert electrode (such as solid platinum) is used to provide a surface on which electron transfer can occur.

   iii. Phases are written in the same order as seen in the chemical reaction.

EXAMPLE:

Draw the cell diagram for the reaction that was balanced in the last example.

$$3HNO_2(aq) + Cr_2O_7^{2-}(aq) + 5H^+(aq) \rightarrow 3NO_3^-(aq) + 2Cr^{3+}(aq) + 4H_2O(l)$$

The balanced half-cell reactions are:

Oxidation: $3HNO_2(aq) + 3H_2O(l) \rightarrow 3NO_3^-(aq) + 9H^+(aq) + 6e^-$

Reduction: $Cr_2O_7^{2-}(aq) + 14H^+(aq) + 6e^- \rightarrow 2Cr^{3+}(aq) + 7H_2O(l)$

We write oxidation on the left and reduction on the right.

$$HNO_2(aq), NO_3^-(aq), H^+(aq) \mid Pt(s) \mid\mid Cr_2O_7^{2-}(aq), H^+(aq), Cr^{3+}(aq) \mid Pt(s)$$

Notice that we have used platinum electrodes on both sides because all species are aqueous.

IV. Standard Electrode Potentials

a. Half-cell potentials can't be measured directly.

   i. We assign the standard hydrogen electrode (SHE) as the zero of potential energy and measure the potential of all other species against this:

$$2H^+(aq) + 2e^- \rightarrow H_2(g) \qquad E^o_{cell} = 0.00 \text{ V}$$

b. A positive cell potential means that the species under consideration has a greater tendency to undergo reduction than the standard hydrogen electrode does.

i. The species is less likely to undergo oxidation than the SHE.

c. A negative cell potential means that the species under consideration has a greater tendency to undergo oxidation than the standard hydrogen electrode does.

   i. The species is less likely to undergo reduction than the SHE.

d. The cell potentials for half-reactions combined with the SHE are tabulated in a table of standard reduction potentials.

e. The cell potential for an oxidation reaction is found by reversing the reduction reaction and taking the negative value of the reduction potential.

$$E^{\circ}_{oxidation} = -E^{\circ}_{reduction}$$

f. The cell potential is found by adding the cell potentials for the two half-reactions as compared with the SHE.

$$E^{o}_{cell} = E^{o}_{oxidation} + E^{o}_{reduction}$$

g. Cell potentials are intensive properties and are not, therefore, multiplied by factors when changing the stoichiometric coefficients in reactions.

   i. The cell potential is a measure of potential energy differences per unit charge and therefore doesn't depend on the amount.

EXAMPLE:

What is the cell potential for the redox reaction shown below if all species are initially at a concentration of 1 M?

$$Fe^{3+}(aq) + Mg(s) \rightarrow Fe^{2+}(aq) + Mg^{2+}(aq)$$

The half-reactions with the cell potentials from Table 18.1 in your textbook are:

Oxidation: $Mg(s) \rightarrow Mg^{2+}(aq) + 2e^-$ $E^{\circ} = 2.37$ V

Reduction: $Fe^{3+}(aq) + e^- \rightarrow Fe^{2+}(aq)$ $E^{\circ} = 0.77$ V

Notice that the oxidation reaction is the reverse of the reduction reaction listed in the table; the cell potential is, therefore, the reverse of the value given in the table.

The cell potential when these two reactions are combined is the sum of the half-cell potentials, so:

$$E^{o}_{cell} = 2.37 \text{ V} + 0.77 \text{ V} = 3.14 \text{ V}$$

---

h. The table of standard reduction potentials can be used to predict the direction of redox reactions at standard conditions.

   i. Substances that are on the top of the table have a strong tendency to be reduced; they are excellent oxidizing agents.

   ii. Substances that are on the bottom of the table have a strong tendency to be oxidized; they are excellent reducing agents.

   iii. Any half-reaction will be spontaneous when paired with a half-reaction that is below it on the table, provided it is reversed to be an oxidation.

i. Metals can be dissolved in acidic solutions to produce hydrogen gas if the metal is below the SHE on the table of standard reduction potentials.

   i. Nitric acid is a stronger oxidizing agent than other acids because its nitrogen reduces more readily than its acidic hydrogen.

EXAMPLE:

Predict whether or not the following reactions will be spontaneous at standard conditions:

a. $Cu^+(aq) + Ag(s) \rightarrow Ag^+(aq) + Cu(s)$

We look on the standard reduction table and find that Cu^+ is below Ag^+; this means that this reaction will not be spontaneous as written and the reverse reaction will be spontaneous.

b. $2H^+(aq) + Zn(s) \rightarrow Zn^{2+}(aq) + H_2(g)$

On the standard reduction table, H^+ is above Zn^{2+}, meaning that it is more likely to be reduced. This reaction will be spontaneous as written.

c. $2K^+(aq) + OH^-(aq) + H_2(g) \rightarrow 2H_2O(l) + 2K(s)$

On the standard reduction table, K^+ is below H_2O, meaning that it is less likely to be reduced. This reaction will not be spontaneous as written and the reverse reaction will be spontaneous.

V. Cell Potential, Free-energy, and the Equilibrium Constant

a. The cell potential is its potential energy difference per unit charge:

$$E^\circ_{cell} = \frac{\text{Potential Difference (J)}}{\text{Charge (C)}}$$

   i. The potential difference is the maximum work available from the system so it is equal to the free-energy.

   ii. The charge is equal to the number of moles of electrons (n) times the charge of one mole of electrons (F).

$$\Delta G^\circ = -n \cdot F \cdot E^\circ_{cell}$$

b. A process is spontaneous at standard conditions if it has a cell potential that is positive, a free-energy change that is negative, and an equilibrium constant greater than one.

c. A process is nonspontaneous at standard conditions if it has a cell potential that is negative, a free-energy change that is positive, and an equilibrium constant less than one.

EXAMPLE:

Calculate the standard free-energy change for the reaction of copper ions with silver metal at standard conditions:

$$Cu^+(aq) + Ag(s) \rightarrow Ag^+(aq) + Cu(s)$$

Using the table of standard reduction potentials, we can calculate the standard cell potential:

$$E^\circ_{cell} = E^\circ_{oxidation} + E^\circ_{reduction} = +0.52 \text{ V} + -0.80 \text{ V} = -0.28 \text{ V}$$

This reaction is already balanced and there is one electron transferred, so n=1 and the standard free-energy change is:

$$\Delta G^\circ = \text{-n}\cdot F\cdot E^\circ_{cell} = \text{-}1\cdot\left(\frac{96{,}485\ C}{mol\ e^-}\right)\cdot\left(-0.28\frac{J}{C}\right) = +2.7\times10^4\ J$$

Notice that the standard free-energy change is a positive number, in agreement with our prediction that this reaction is not spontaneous as written.

---

d. We can relate the cell potential to the equilibrium constant:

$$E^\circ_{cell} = \frac{RT}{nF}\ln K = \frac{0.0592\ V}{n}\log K$$

EXAMPLE:

Use the cell potential to calculate the equilibrium constant for the reaction in the above example:

$$Cu^+(aq) + Ag(s) \rightarrow Ag^+(aq) + Cu(s) \qquad E^\circ_{cell} = \text{-}0.28\ V$$

We need to rearrange the equation given for the cell potential in terms of the equilibrium constant in order to solve for the equilibrium constant:

$$E^\circ_{cell}\frac{n}{0.0592\ V} = \log K$$

$$K = 10^{E^\circ_{cell}\frac{n}{0.0592\ V}} = 10^{(\text{-}0.28\ V)\frac{1}{0.0592\ V}} = 1.9\times10^{-3}$$

This answer is in agreement with our prediction that this reaction is nonspontaneous at standard conditions since the equilibrium constant is much less than 1 (it favors reactant formation).

VI. Cell Potential and Concentration Cells

a. Using the relationship between the free-energy at standard and nonstandard conditions, we can find an expression for the nonstandard cell potential, called the Nernst equation:

$$E_{cell} = E^\circ_{cell} - \frac{0.0592\ V}{n}\log Q$$

EXAMPLE:

Calculate the potential of a cell that is prepared using a 1.45 M solution of magnesium nitrate and a 2.32 M solution of silver nitrate. In this cell, magnesium is the anode and silver is the cathode.

We will begin by writing the half-reactions along with their cell potentials:

$$Ag^+(aq) + e^- \rightarrow Ag(s) \qquad E^\circ_{cell} = 0.80\ V$$

$$Mg(s) \rightarrow Mg^{2+}(aq) + 2e^- \qquad E^\circ_{cell} = 2.37\ V$$

We will begin by calculating the cell potential at standard conditions:

$$E^\circ_{cell} = E^\circ_{oxidation} + E^\circ_{reduction} = 0.80\ V + 2.37\ V = 3.17\ V$$

The balanced chemical reaction will be:

$$2Ag^{+}(aq) + Mg(s) \rightarrow 2Ag(s) + Mg^{2+}(aq)$$

Now we can use the Nernst equation to calculate the cell potential:

$$E_{cell} = E^{o}_{cell} - \frac{0.0592\ V}{n} \log Q = 3.17\ V - \frac{0.0592\ V}{2} \log \frac{1.45}{2.32^2} = 3.19\ V$$

The cell potential is almost the same as the standard cell potential since the concentrations of both solutions are almost equal initially.

---

b. Concentration cells are those that depend on the concentration of species and not on the standard reduction potentials alone.

   i. A concentration cell can be made by using the same reaction on both sides of a galvanic cell, but with different concentrations.

      1. In this case, the standard cell potential is zero and the cell potential is:

$$E_{cell} = -\frac{0.0592\ V}{n} \log Q$$

   ii. Electrons flow in order to equalize the concentrations of all species on both sides of the cell.

EXAMPLE:

Calculate the cell potential of a cell prepared using a 0.54 M solution of $Cu(NO_3)_2$ on one side of the cell and a 1.43 M solution of $Cu(NO_3)_2$ on the other side. State the direction in which electrons will flow.

We can predict that electrons will flow from the low concentration side of the cell to the high concentration side of the cell in order to equalize the concentrations.

There are two electrons transferred in this reaction, so the cell potential will be:

$$E_{cell} = -\frac{0.0592\ V}{n} \log Q = -\frac{0.0592\ V}{2} \log \frac{0.54}{1.43} = 0.013\ V$$

VII. Batteries: Using Chemistry to Generate Electricity

a. In a battery, redox reactions are used to generate electricity that can be used to do work.

b. An inexpensive battery uses a zinc case as the anode and an ammonium chloride/manganese oxide paste as the cathode:

Anode: $Zn(s) \rightarrow Zn^{2+}(aq) + 2e^{-}$

Cathode: $2MnO_2(s) + 2NH_4^{+}(aq) + 2e^{-} \rightarrow Mn_2O_3(s) + 2NH_3(aq) + H_2O(l)$

c. An alkaline battery is the most common disposable battery:

Anode: $Zn(s) + 2OH^{-}(aq) \rightarrow Zn(OH)_2(s) + 2e^{-}$

Cathode: $MnO_2(s) + 2H_2O(l) + 2e^{-} \rightarrow 2MnO(OH)(s) + 2OH^{-}(aq)$

d. Most car batteries are lead-storage batteries which can be run in the reverse direction with the application of current to reverse the reaction:

Anode: $Pb(s) + HSO_4^{-}(aq) \rightarrow PbSO_4(s) + H^{+}(aq) + 2e^{-}$

Cathode: $PbO_2(s) + HSO_4^-(aq) + 3H^+(aq) + 2e^- \rightarrow PbSO_4(s) + 2H_2O(l)$

e. Nickel-cadmium batteries:

Anode: $Cd(s) + 2OH^-(aq) \rightarrow Cd(OH)_2(s) + 2e^-$

Cathode: $2NiO(OH)(s) + 2H_2O(l) + 2e^- \rightarrow 2Ni(OH)_2(s) + 2OH^-(aq)$

f. Nickel-Metal Hydride (NiMH) batteries:

Anode: $M{\cdot}H(s) + OH^-(aq) \rightarrow M(s) + H_2O(l) + e^-$

Cathode: $NiO(OH)(s) + H_2O(l) + e^- \rightarrow Ni(OH)_2(s) + OH^-(aq)$

g. Fuel cells are a different type of battery because they require a constant supply of fuel in order to continually generate current. The hydrogen-oxygen fuel cell is:

$2H_2(g) + 4OH^-(aq) \rightarrow 4H_2O(l) + 4e^-$

$O_2(g) + 2H_2O(l) + 4e^- \rightarrow 4OH^-(aq)$

VIII. Electrolysis: Driving Nonspontaneous Chemical Reactions with Electricity

a. Electrolysis is the driving of a nonspontaneous reaction forward with the application of an external power source.

i. Electrons are drawn from the anode so that the charge labels are the opposite of what they are in galvanic cells.

b. In the electrolysis of a pure molten salt, the anion is oxidized and the cation is reduced.

i. In a mixture, the cation that is most easily reduced will be reduced first and the anion that is most easily oxidized will be oxidized first.

ii. The electrolysis of aqueous solutions is more complex because water itself can be oxidized or reduced when a voltage is applied.

1. Cations of active metals cannot be reduced in aqueous solutions because water will be reduced first.

EXAMPLE:

Predict the reaction that will occur at the anode and cathode for the electrolysis of a mixture of molten $CuCl_2$ and $FeCl_3$.

The oxidation reaction is:

$2Cl^-(l) \rightarrow Cl_2(g) + 2e^-$

So this reaction takes place at the anode.

The reduction reaction is one of the following:

$Cu^{2+}(l) + 2e^- \rightarrow Cu(s)$

$Fe^{3+}(l) + 3e^- \rightarrow Fe(s)$

Looking at the table of standard reduction potentials, we see that Cu^{2+} is above Fe^{3+} on the table. This means that Cu^{2+} is more likely to be reduced and its reduction reaction will be the one that occurs preferentially at the cathode.

---

c. Over-voltage is the additional voltage that must be applied in order to drive a nonspontaneous reaction forward.

i. The predicted reaction may not occur because of over-voltage.

d. In determining the quantities of metal created, we must consider the stoichiometry of the metal cation and electrons.

e. The current applied for a given time period will give the overall charge transferred:

$$\text{current (C/s)} \times \text{time (s)} = \text{charge (C)}$$

EXAMPLE:

What current is needed in order to plate out 5.0 g of silver in ten minutes?

The half-reaction that is taking place is:

$$Ag^+(aq) + e^- \rightarrow Ag(s)$$

We can convert the mass to time using the current in a conversion factor problem:

$$5.0\text{ g Ag} \times \frac{1\text{ mole Ag}}{107.87\text{ g Ag}} \times \frac{1\text{ mol e}^-}{1\text{ mol Ag}} \times \frac{96{,}485\text{ C}}{1\text{ mol e}^-} = 4.472 \times 10^3\text{ C}$$

This is the total charge that must be transferred in ten minutes. The voltage is the charge per second, so we can continue the problem:

$$\frac{4.472 \times 10^3\text{ C}}{10.0\text{ min}} \times \frac{1\text{ min}}{60\text{ s}} = 7.5\frac{\text{C}}{\text{s}}$$

The units of current are C/s or A, so the current needed is 7.5 A.

IX. Corrosion: Undesirable Redox Reactions

a. Corrosion is the gradual, undesirable oxidation of metals via reaction with oxygen:

$$O_2(g) + 2H_2O(l) + 4e^- \rightarrow 4OH^-(aq) \qquad E^o_{cell} = +0.40\text{ V}$$

b. Oxidation occurs to a more significant extent in an acidic solution:

$$O_2(g) + 4H^+(l) + 4e^- \rightarrow 2H_2O(l) \qquad E^o_{cell} = +1.23\text{ V}$$

c. Some oxides are very strong and form a barrier that resists further oxidation.

d. Iron oxides form a very brittle, porous iron oxide (rust).

i. Rust can be prevented by using a sacrificial electrode that will be more easily oxidized than iron.

Fill in the Blank:

1. The ________________ relates the cell potential at nonstandard conditions to the standard cell potential.
2. A(n) ________________ cell uses electrical current to drive a nonspontaneous reaction.
3. Electromotive force is the ________________.
4. Oxidation corresponds to a(n) ________________ in oxidation state and a ________________ of electrons.
5. In a chemical reaction, both ________________ and ________________ must be balanced.

6. The quantity of charge that flows in an electrochemical cell is given by ________________.
7. The generation of electricity by separating the two half-reactions and connecting them with a wire is done in a(n) ________________ cell.
8. In an electrochemical cell, oxidation occurs at the ________________.
9. The ________________ is a measure of the difference in potential energy per unit charge and has units of ________________.
10. A (n) ________________ battery does not contain large amounts of liquid water.
11. A _______ __________ is needed in a galvanic cell to maintain charge neutrality.
12. ____ ____________ is the additional voltage required to drive a nonspontaneous reaction forward.
13. A half-cell with a greater tendency to undergo reduction than the standard reduction potential will have a ________________ cell potential.
14. Species at the top of the table of standard reduction potentials have a strong tendency to serve as ________________ agents and therefore be ________________.
15. A metal whose reduction half-reaction lies below the reduction of H^+ to form H_2 on the table of ________________ dissolve in acids.
16. A spontaneous reaction will have a negative free-energy change and a ________________ cell potential.
17. A (n) ________________ is different from a standard battery because it needs to have a constant fuel supply.
18. The reduction potential of the standard hydrogen electrode is ________________.
19. The electrode where oxidation occurs is called the ________________ and the electrode where reduction occurs is called the ________________.
20. In a(n) ________________ cell, the standard cell potential is zero while the actual cell potential is nonzero.
21. The standard h ydrogen electrode uses a ________________ electrode.
22. In an aqueous solution undergoing electrolysis, the predicted reduction or oxidation reaction might be incorrect due to the electrolysis of ________________.

Problems:

1. Balance the following redox reactions:
   a. $Al(s) + O_2(g) \rightarrow Al_2O_3(s)$
   b. $NO_3^-(aq) + Al(s) \rightarrow NH_3(aq) + Al(OH)_4^-(aq)$
   c. $Cl_2(g) \rightarrow Cl^-(aq) + ClO_3^-(aq)$
   d. $HNO_3(aq) + H_3AsO_3(aq) \rightarrow NO(g) + H_3AsO_4(aq)$
   e. $NO_2(g) + H_2(g) \rightarrow NH_3(g) + H_2O(l)$
   f. $Cu(s) + HNO_3(aq) \rightarrow Cu(NO_3)_2(aq) + NO(g) + H_2O(l)$
   g. $Cr(OH)_3(s) + ClO_3^-(aq) \rightarrow CrO_4^{2-}(aq) + Cl^-(aq)$ (in a basic solution)
   h. $Au^{3+}(aq) + I^-(aq) \rightarrow Au(s) + I_2(s)$
   i. $FeO(s) + CO(g) \rightarrow Fe(s) + CO_2(g)$
   j. $TiO_2(s) + Cu(s) \rightarrow Cu^{2+}(aq) + Ti(s) + H_2O(l)$

2. A 0.010 M $CrCl_3$ solution is placed in a beaker with a chromium-plated wire and a 0.20 M solution of $CuSO_4$ is placed in a second beaker with a copper wire. The two wires are connected to a voltmeter.

    a. The cell, as described, will not have a voltage reading. What is missing and why does its absence result in zero voltage?

    b. Sketch the complete cell. Label the anode, cathode, direction of electron flow, direction of anion flow, and direction of cation flow.

    c. Write the half-reactions for this cell and label them as oxidation and reduction.

    d. Write the balanced, overall redox reaction and calculate the standard cell potential.

    e. Write the cell diagram for this cell.

    f. What is the standard free-energy change for this cell?

    g. What is the equilibrium constant for this reaction?

    h. Is the potential reading on the voltmeter the same as the standard cell potential that you calculated in part d? Why or why not?

    i. If they are different, calculate the actual cell potential.

3. A cell is prepared using nickel and lead. A 0.10 M solution of nickel(II) nitrate is used with a nickel plated electrode for one half-cell. The other half-cell is prepared using a solution of lead(II) nitrate of unknown concentration and a lead wire. The voltage of this cell is measured to be 0.18 V.

    a. What is the concentration of the lead nitrate solution used?

    b. What will the concentration of the lead nitrate solution be at equilibrium?

    c. Without doing any calculations, how could you increase the cell voltage of this system?

4. Consider electroplating 10.0 g of chromium from a solution of $Cr(NO_3)_3$.

    a. How long (in minutes) will it take for you to carry out this process using a 10 mA current?

    b. What current will be required to electroplate the chromium in 6.4 hours?

    c. What current will be required to electroplate the chromium in one day?

    d. Will it require more or less current to carry out this process in a chromium nitrate solution that is twice as dilute? Explain your answer.

5. An electrochemical cell is prepared using $TiO_2(s)$ in an acidic solution and a titanium electrode in one half-cell and $Cu^{2+}(aq)$ and a copper electrode in the other half-cell. The voltage reading for this cell is measured as a positive number.

    a. Does this reaction occur spontaneously? Explain how you can tell.

    b. Write the half-reactions and label them as oxidation and reduction.

    c. Write the cell diagram for this reaction.

    d. What is the standard cell potential for this reaction? Given that the reduction potential for TiO_2 is -0.870 V and the standard reduction potential for Cu^{2+} is 0.340 V.

    e. What is the cell potential if $[Cu^{2+}]$ = 0.0560 M, the pH of the titanium half-cell is 2.00, and the temperature is 25°C?

    f. What will the cell potential be when the reaction reaches equilibrium?

6. A mixture of NaCl, KCl, and KBr is melted and electrolysis is carried out on the molten salt solution.

    a. What reaction will occur at the anode and the cathode? Explain the reasoning behind your answers.

b. Will the reactions be different if electrolysis is carried out on an aqueous solution of these salts? If not, why not? If so, explain why.

7. Consider the following structures and standard reduction potentials:

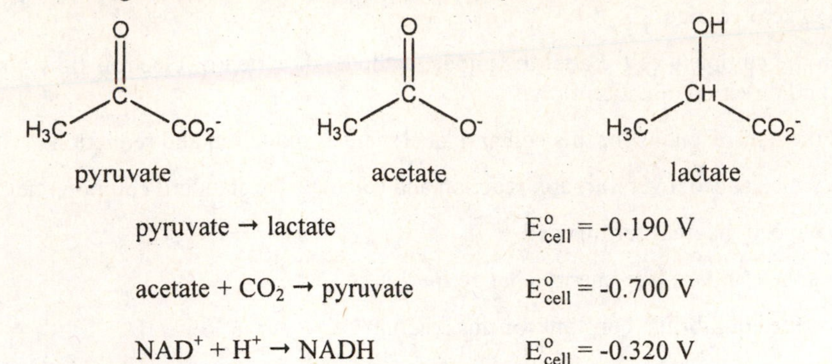

pyruvate → lactate $E^\circ_{cell} = -0.190$ V

acetate + CO_2 → pyruvate $E^\circ_{cell} = -0.700$ V

$NAD^+ + H^+$ → NADH $E^\circ_{cell} = -0.320$ V

NAD^+ is a biologically important nucleotide that is involved in a number of biochemical reactions. It can be produced in metabolism.

a. Write a balanced chemical reaction for the conversion of lactate to acetate and CO_2.
b. Calculate the standard cell potential and the standard free-energy change for the conversion of lactate to pyruvate.
c. Calculate the standard free-energy change for the conversion of NAD^+ to NADH.
d. Which of the reactions given is spontaneous and which is nonspontaneous?
e. The conversion of lactate to acetate and CO_2 is an important step in metabolism. The energy of this conversion can be coupled to the energy associated with the NADH reaction. When these reactions are carried out, will NAD^+ be consumed or produced?
f. How much NAD^+ will be consumed or produced when 1 mole of lactate is converted to acetate and CO_2?

Concept Questions:

1. Consider four metals: A, B, C, and D. Experiments conducted on these metals gave the following results:

- A and C react with hydrochloric acid to give hydrogen gas.
- When C is added to solutions of the other metals, solid A, B, and D are formed.
- Metal D spontaneously reduces metal B.

a. List the metals in order of increasing strength as a reducing agent.

b. A cell composed of which two half-cells will give the largest positive voltage? Be sure to indicate which reaction takes place at the anode and which reaction takes place at the cathode.

2. What are the values of K_c, ΔG°, and E°_{cell} for a product-favored reaction?

3. For certain reactions, the standard cell potential and the measured cell potential are the same.

a. What conditions are necessary in order for this to be true?

b. Is it possible for the standard cell potential to be zero for a reaction other than the standard hydrogen electrode? If not, why not? If so, when?

4. Explain why the terminals of a battery are labeled in the opposite way as the electrodes of an electrolytic cell.

5. In the last chapter, you learned that the free-energy of a spontaneous reaction is the maximum useful work available for the system. Conversely, the free-energy of a nonspontaneous reaction is the minimum amount of work required to drive it forward. Explain how this relates to the concept of over-voltage discussed in this chapter.

Chapter 19: Radioactivity and Nuclear Chemistry

Learning Goals:

- Identify the various types of radioactive decay and write balanced nuclear reactions.
- Predict the relative stability of nuclides using the ratio of neutrons to protons.
- Carry out kinetic calculations for radioactive decay processes.
- Understand radioactive dating methods.
- Understand the principles of nuclear fission and fusion.
- Carry out mass to energy conversions.
- Understand the effects of radiation on life.

Chapter Summary:

In this chapter, you will be introduced to radioactivity and nuclear chemistry. You will begin by learning about the various subatomic particles and how each of them is involved in particular types of nuclear reactions. You will see how the valley of stability is related to the ratio of protons and neutrons in a particular nuclide and how it can be used to predict whether or not a particular nuclide is radioactive or not. Next you will learn how to detect radioactivity and then the kinetics of radioactive decay. All radioactive decay processes obey simple first-order kinetics that can be used in radioactive dating experiments. Next you will learn about nuclear fission processes that can be used in nuclear bombs and for power generation. One of the most famous expressions of all time will then be discussed to relate the mass defect to the energy associated with nuclear reactions. Nuclear fusion will be introduced as a potential energy source that has yet to be utilized. The formation of new elements through transmutation will be described. Finally, the effects of radiation on life and the uses of radiation in medicine will be explored.

Chapter Outline:

I. Diagnosing Appendicitis

   a. Radioactivity is the emission of subatomic particles or high-energy electromagnetic radiation by the nuclei of certain atoms.

      i. Atoms that emit high-energy electromagnetic radiation are radioactive.

II. The Discovery of Radioactivity

   a. Phosphorescence is the long-lived emission of light that sometimes follows the absorption of light.

   b. Radioactivity was discovered by Antoine-Henri Becquerel and Marie Curie.

III. Types of Radioactivity

   a. In radioactivity discussions, a particular isotope of an element is called a nuclide.

   b. Subatomic particles are given symbols in accordance with the atomic symbols for the elements.

      i. Protons have the symbol ${}^{1}_{1}p$ to indicate that they have one proton and contribute to the mass number by a factor of one.

      ii. Neutrons have the symbol ${}^{1}_{0}n$ to indicate that they have no protons and contribute to the mass number by a factor of one.

iii. Electrons have the symbol $^{0}_{-1}e$ to indicate that they remove a proton and do not contribute to the mass number at all.

1. The symbol for the electron will become more obvious in the context of radioactive reactions.

c. Alpha (α) decay occurs when an unstable nucleus emits a particle with two protons and two neutrons.

i. An alpha particle is a helium nucleus: $^{4}_{2}He$

ii. In a nuclear equation, an alpha particle is emitted when a parent nuclide decays into a daughter nuclide:

$$^{238}_{92}U \rightarrow ^{234}_{90}Th + ^{4}_{2}He$$

1. Note that in a nuclear equation, the sum of the atomic numbers must be equal on both sides.

iii. Alpha particles have high ionizing power (ability of radiation to ionize other molecules or atoms) because they have very high energy.

iv. Alpha radiation has very low penetrating ability (ability of radiation to penetrate matter) because alpha particles are so large.

EXAMPLE:

Predict the product of the alpha decay of americium-241.

We look on the periodic table and find that americium is element 95. The atomic symbol is then $^{241}_{95}Am$ and we can determine the identity of the daughter nuclide by subtracting the helium-4 nucleus and identifying the symbol using the atomic number:

$$^{241}_{95}Am \rightarrow ^{4}_{2}He + ^{237}_{93}Np$$

d. Beta (β) decay occurs when an unstable nucleus emits an electron.

i. In radioactivity, a beta particle is emitted along with a proton when a neutron decays:

$$^{1}_{0}n \rightarrow ^{1}_{1}p + ^{0}_{-1}e$$

$$^{228}_{88}Ra \rightarrow ^{228}_{89}Ac + ^{0}_{-1}e$$

ii. Beta radiation has lower ionizing power and higher penetrating ability than alpha decay does because beta particles are smaller.

EXAMPLE:

Predict the product of the beta decay of hydrogen-3.

As in the last example, we will first determine the atomic number of hydrogen which is 1 and write the atomic symbol as $^{3}_{1}H$. We will next determine the identity of the product by considering the loss of a beta particle in the nuclear reaction:

$$^{3}_{1}H \rightarrow ^{3}_{2}He + ^{0}_{-1}e$$

e. Gamma (γ) ray emission is the emission of electromagnetic radiation.

   i. Gamma rays are given the symbol: ${}^{0}_{0}\gamma$

   ii. Gamma rays are usually emitted along with other particles.

   iii. Gamma rays have very high penetrating power which makes exposure to them very dangerous.

f. Positron emission occurs when a positron is emitted.

   i. A positron is the antiparticle of an electron: ${}^{0}_{+1}e$

   ii. When a positron and an electron collide, they annihilate each other and release energy as radiation.

   iii. A positron is formed with the decay of a proton to a neutron:

$${}^{1}_{1}p \rightarrow {}^{1}_{0}n + {}^{0}_{+1}e$$

EXAMPLE:

Predict the product of the reaction when a positron is emitted from a carbon-11 nuclide.

Carbon has an atomic number of 6 so it has an atomic symbol of ${}^{11}_{6}C$. A positron is emitted when a proton splits to give a neutron and an emitted electron:

$${}^{11}_{6}C \rightarrow {}^{11}_{5}B + {}^{0}_{+1}e$$

---

g. Electron capture occurs when a nucleus assimilates an electron from an inner orbital of its electron cloud:

$${}^{1}_{1}p + {}^{0}_{-1}e \rightarrow {}^{1}_{0}n$$

EXAMPLE:

Predict the product of the electron capture reaction of beryllium-7.

Beryllium has an atomic number of 4 so it has an atomic symbol of ${}^{7}_{4}Be$. Electron capture essentially results in the conversion of a proton into a neutron:

$${}^{7}_{4}Be + {}^{0}_{-1}e \rightarrow {}^{7}_{3}Li$$

IV. The Valley of Stability: Predicting the Type of Radioactivity

a. The binding of protons in the nucleus is provided by the strong force, which only occurs at short distances.

   i. Nucleus stability is a balance between the repulsions of the protons and the attraction of all particles via the strong force.

b. A graph of the neutrons versus protons in a nuclide can be used to predict its stability.

i. An N/Z ratio of approximately one is in the valley of stability and is therefore not radioactive. The ratio value increases (deviates from one) as the mass number increases.

ii. When the N/Z ratio is too high, neutrons are at a high-energy state and will be diminished via β decay.

iii. When the N/Z ratio is too low, the repulsions dominate and the number of protons will be diminished via positron emission, α decay, or electron capture.

EXAMPLE:

Predict which of the following is/are likely to decay via positron emission:

a. $^{210}_{84}Po$

Since the atomic number is greater than 83, this is an unstable nuclide. Positron emission can occur:

$$^{210}_{84}Po \rightarrow ^{210}_{83}Bi + ^{0}_{+1}e$$

It is more likely, however, that an alpha particle will be emitted to decrease the atomic number further:

$$^{210}_{84}Po \rightarrow ^{206}_{82}Pb + ^{4}_{2}He$$

b. $^{198}_{79}Au$

The ratio of neutrons to protons is 119/79=1.51. We can see from Figure 19.5 in your textbook that this is above the valley of stability and neutrons must be converted to protons. Positron emission will not accomplish this, so it will not likely occur.

c. $^{37}_{19}K$

The ratio of neutrons to protons is 18/19=0.95. We can see that this is below the valley of stability and can therefore decay via positron emission:

$$^{37}_{19}K \rightarrow ^{37}_{18}Ar + ^{0}_{+1}e$$

---

c. Magic numbers are the numbers of protons and neutrons in nuclides with unique stability.

i. Most stable nuclides have an even number of protons and an even number of neutrons.

ii. The least stable nuclides have an odd number of protons and an odd number of neutrons.

d. Atoms with very high atomic numbers (>83) will decay in multiple steps.

V. Detecting Radioactivity

a. Film-badge dosimeters use photographic film to monitor exposure to radiation.

b. A Geiger-Müller counter uses a chamber filled with argon gas.

i. The argon gas is ionized to give an electrical signal.

ii. The counter will click with each radioactive particle that travels through the chamber.

c. A scintillation counter contains a material that emits ultraviolet or visible light in response to excitation.

VI. The Kinetics of Radioactive Decay and Radiometric Dating

a. All elements above atomic number 83 are radioactive.

b. All radioactive elements decay via first-order kinetics.

c. The rate of radioactive decay is equal to the number of radioactive nuclei (N) multiplied by the rate constant (k):

$$\text{Rate} = k \cdot N$$

d. The half-life of a radioactive element is:

$$t_{1/2} = 0.693/k$$

e. The integrated rate law is given in terms of the number of radioactive nuclei initially, N_0, and time, t:

$$\ln\frac{N_t}{N_0} = -kt$$

f. The integrated rate law can be expressed in terms of the rates:

$$\ln\frac{\text{rate}_t}{\text{rate}_0} = -kt$$

EXAMPLE:

Fluorine-21 decays to neon-21 with a half-life of 4.17 seconds. If you begin with a 14.0 g sample of fluorine-21, what quantity will remain after 1 minute?

We will calculate the rate constant using the half-life formula:

$$t_{1/2} = 0.693/k$$

$$k = 0.693/t_{1/2} = 0.693/(4.17\text{ s}) = 0.166\text{ s}^{-1}$$

We will now rearrange the integrated rate law to solve for the quantity of fluorine-21 after 1 minute (expressed in seconds to match the units of the half-life constant).

$$\ln\frac{N_t}{N_0} = -kt$$

$$N_t = N_0 e^{-kt} = 14.0\text{ g} \cdot e^{-(0.166\text{ s}^{-1})(60\text{ s})} = 6.6 \times 10^{-4}\text{ g}$$

---

g. Radiometric dating is based on the variation of the rate at two different times.

i. Radiocarbon dating uses the decay of carbon-14:

$$^{14}_{6}\text{C} \rightarrow {}^{14}_{7}\text{N} + {}^{0}_{-1}\text{e} \qquad t_{1/2} = 5730\text{ years}$$

1. Since carbon is incorporated into all living organisms and it decays upon death, the age can be determined using the amount of carbon-14 present.

ii. Objects that are very old or were never living can be dated using uranium/lead decay.

EXAMPLE:

Zircon is a mineral that contains large amounts of uranium, contains no lead when formed, and is found throughout the earth. A certain metamorphic rock is found to have 1.52 g of lead-206 for every 1.00 g of uranium-238. How old is this sample? The half-life of U-238 is 4.5 billion years.

We will calculate the rate constant for the decay:

$$t_{1/2} = 0.693/k$$

$$k = 0.693/t_{1/2} = 0.693/(4.5\times10^{9}\text{ years}) = 1.54\times10^{8}\text{ yr}^{-1}$$

To determine the initial concentration of uranium-238, we will carry out a conversion factor problem:

$$1.52\text{ g Pb}\times\frac{1\text{ mol Pb}}{206\text{ g Pb}}\times\frac{1\text{ mol U}}{1\text{ mol Pb}}\times\frac{238\text{ g U}}{1\text{ mol U}}=1.76\text{ g U}$$

This is the quantity of uranium-238 that has decayed, so the original quantity was 2.76 g.

Solving for time in the integrated rate law:

$$\ln\frac{N_t}{N_0} = -kt$$

$$t = -\frac{1}{k}\ln\frac{N_t}{N_0} = -\frac{1}{1.54\times10^{8}\text{yr}^{-1}}\ln\frac{1.76}{2.76} = 2.9\times10^{7}\text{ yr}$$

VII. The Discovery of Fission: The Atomic Bomb and Nuclear Power

a. Nuclear fission is the splitting of a uranium or other heavy atom.

b. When uranium-235 is bombarded with neutrons, it splits into lighter atoms, neutrons, and energy:

$$^{235}_{92}\text{U} + ^{1}_{0}\text{n} \rightarrow ^{140}_{56}\text{Ba} + ^{93}_{36}\text{Kr} + 3\,^{1}_{0}\text{n} + \text{energy}$$

c. A chain reaction can result if there is enough uranium-235 since each nuclear reaction emits three neutrons.

i. The critical mass of uranium is the minimum amount of uranium-235 needed to produce a self-sustaining chain reaction.

d. In a bomb, the energy is released rapidly; the energy can also be released slowly and used in a nuclear power plant.

VIII. Converting Mass to Energy: Mass Defect and Nuclear Binding Energy

a. Matter can be converted to energy in a nuclear reaction:

$$E = mc^2$$

b. The mass defect is the difference in mass between the reactants and products and is related to the nuclear binding energy.

i. One atomic mass unit is equal to 931.5 MeV (mega-electron volts).

ii. The binding energy is usually reported per nucleon, where a nucleon is a particle in the nucleus (protons and neutrons).

EXAMPLE:

What is the binding energy of a Cu-63 nucleus if the actual mass of a copper-63 nucleus is 62.91367 amu?

First, we need to calculate the mass defect. Copper has 29 protons and copper-63 therefore has 34 neutrons. The mass defect is then the difference between the actual mass and the mass of 29 hydrogen atoms and 34 neutrons:

$$\text{mass defect} = 29\times(1.00783\text{ amu}) + 34\times(1.00866\text{ amu}) - 62.91367\text{ amu} = 0.60784\text{ amu}$$

The nuclear binding energy is then calculated by multiplying the mass defect by the energy per amu:

$$\text{nuclear binding energy} = 0.60784\text{ amu} \times \frac{931.5\text{ MeV}}{\text{amu}} = 566.2\text{ MeV}$$

Since there are 63 nucleons in a copper-63 atom:

$$\text{nuclear binding energy per nucleon} = \frac{566.2\text{ MeV}}{63\text{ nucleons}} = 8.987\text{ MeV/nucleon}$$

---

c. The higher the nuclear binding energy per nucleon, the more stable an atom is.

IX. Nuclear Fusion: The Power of the Sun

a. Fusion is the combination of two light nuclei to form a heavier one.

i. Energy is released when the binding energy of the newly formed species is higher.

b. Fusion requires high temperature and materials that can withstand those temperatures.

c. Fusion has not yet been shown to be a viable energy source.

X. Nuclear Transmutation and Transuranium Elements

a. The process of changing one element into another is called transmutation.

i. High energy bombardment is needed in order to carry out transmutation.

ii. Elements that are not found in nature can be made.

b. A linear accelerator accelerates charged particles through an evacuated tube using alternating voltages.

c. A cyclotron accelerates charged particles in a circle using alternating voltages and a strong magnetic field.

XI. The Effects of Radiation on Life

a. Acute radiation damage occurs from exposure to high levels of radiation for a short period of time.

i. Acute radiation kills a large number of cells at one time.

b. An increased cancer risk results from low levels of radiation over an extended period of time.

i. Damage to DNA can result in cancer.

c. Genetic defects result when the DNA of reproductive cells is damaged and then passed on to offspring.

d. Radioactivity is measured in units of curies:

$$1\text{ curie (Ci)} = 3.7\times10^{10}\text{ decay events per second}$$

e. A more useful measure of radioactivity takes the penetrating power into account:

$$1\text{ gray (Gy)} = 1\text{ J per kg of body tissue}$$

1 rad = 0.01 J per kg of body tissue

f. We can also take into account the effect of different types of radiation on biological cells using the biological effectiveness factor:

dose in rads × biological effectiveness factor = dose in rems

XII. Radioactivity in Medicine and Other Applications

a. Radiotracers are radioactive nuclides attached to a compound or included in a mixture to track movement in the body.

b. Positron emission tomography (PET) is a medical technique in which a positron-emitting nuclide such as ^{18}F is attached to a metabolically active substance and given to a patient.

c. Radiation can be used to treat cancer cells, kill microorganisms on food, clean medical devices, and make harmful insects infertile.

Fill in the Blank:

1. Radiocarbon dating measures the rate of the decay of carbon-14 to ________________.
2. ________________ occurs when a particle composed of two protons and two neutrons is emitted.
3. ________________ is the combination of two light nuclei to form a heavier one.
4. In a(n) ________________ a charged particle is accelerated by alternating voltages and kept in circular motion using a magnetic field.
5. In ________________, a heavy atom is split into two or more smaller atoms.
6. In a nuclear reaction, the original atom is called the ________________ and the product atom is called the ________________.
7. The binding energy that holds a nucleus together is called the ________________.
8. Nuclides that lie above the valley of stability have too many ________________ and will tend to undergo ________________.
9. Emission of an electron occurs in ________________.
10. A particu lar isotope of an element is called a(n) ________________ in nuclear chemistry.
11. A(n) ________________ is a radioactivity detector that contains argon gas that is ionized when exposed to a radioactive particle.
12. The difference in mass between the products and reactants in a nuclear reaction is called the ________________. This mass is converted into ________________ when the ________________ of the reactants is larger than that of the products.
13. When an elec tron and a(n) ________________ collide, they annihilate each other.

Problems:

1. Write balanced nuclear equations for the following:

   a. $^{210}_{84}Po$ emits a beta particle

   b. $^{222}_{88}Ra$ emits an alpha particle

   c. $^{234}_{90}Th$ emits a beta particle

   d. $^{138}_{57}La$ emits a positron

e. $^{185}_{79}Au$ emits an alpha particle

f. $^{40}_{19}K$ undergoes electron capture

2. Predict decay pathways for the following radioactive elements. In cases where sequential decay occurs, state this.

 a. technetium-98

 b. potassium-40

 c. einsteinium-252

 d. promethium-145

 e. thallium-201

 f. magnesium-22

 g. molybdenum-99

3. Neon-19 decays to fluorine-19 with a half life of 17.22 s.

 a. Write the nuclear reaction for the decay of neon-19 to fluorine-19.

 b. If 13.2 g of neon-19 is placed in a container, what percentage of the neon-19 will remain after 2 hr? How many fluorine-19 atoms will be in the container?

 c. If 13.2 g of neon-19 is placed in a container, how long will it take for 6.6 g of it to decompose? Note: you should be able to answer this question without doing any calculations.

 d. If 13.2 g of neon-19 is placed in a container, how long will it take for 2/3 of it to decompose?

4. A piece of wood is found buried in the ground. The rate of its decay is found to be 2.34 disintegrations per minute per gram of carbon.

 a. If living organisms have a decay rate of 15.3 disintegrations per minute per gram of carbon, how old is the wood?

 b. A much larger piece of wood is found nearby. If its decay rate is the same, will the age of the wood be older, younger, or the same? Explain your answer.

5. Calculate the mass defect and the binding energy per nucleon for the following:

 a. Chlorine-37 which has a mass of 36.965903 amu

 b. Gallium-71 which has a mass of 70.924701 amu

 c. Palladium-102 which has a mass of 103.904026 amu

 d. Holmium-165 which has a mass of 164.930332 amu

6. A neutron can decay into a proton and an electron. When this happens, energy is released.

 a. Calculate the amount of energy released.

 b. In this case, the energy that is released is turned into kinetic energy of the electron. How fast is the electron moving when it is emitted?

7. Some common household goods used to have considerable amounts of radioactive elements in them. For example, uranium oxides in glazes used in earthenware measured 20 millirad/hour on a Geiger-Müller counter. Calculate radiation absorbed by a 150 lb person from earthenware over one year.

Concept Questions:

1. Which of the various types of radioactive decay is most dangerous and which is the least dangerous? Explain your answers.

2. Explain why it is necessary for neutrons to have energy levels when considering nuclide stability?

3. In this chapter, you learned that all radioactive decay processes follow first-order kinetics. Based on what you learned in Chapter 13 and your knowledge of nuclear processes, why do you think this is?

Chapter 20: Organic Chemistry

Learning Objectives:

- Understand why carbon can uniquely form a variety of compounds.
- Identify hydrocarbons as alkanes, alkenes, alkynes, or aromatics.
- Write hydrocarbons using structural formulas, condensed formulas, and carbon skeletons.
- Identify stereoisomers and their properties.
- Identify functional groups in organic compounds.
- Name simple organic compounds.
- Identify products in basic organic chemical reactions.
- Understand and identify basic polymer structures.

Chapter Summary:

In this chapter, the branch of chemistry that is focused on carbon compounds will be explored. You will first learn why carbon's unique bonding properties lead to such a large variety of compounds. Then you will explore the various organic compounds including hydrocarbons, alcohols, aldehydes, ketones, carboxylic acids, esters, ethers, and amines. Each of these classes of organic compounds will be explored so that you can name them and identify common reactions of each. Finally, you will be introduced to polymers, which will be important in the next chapter.

Chapter Outline:

I. Fragrances and Odors
   a. Most common smells are caused by organic molecules which contain carbon and other elements such as H, N, O, and S.

II. Carbon: Why It Is Unique
   a. Carbon has a strong tendency to form four covalent bonds since it has four valence electrons and an intermediate electronegativity value.
   b. Carbon can form double and triple bonds.
   c. Carbon bonds to itself to form a variety of structures including chains, branched chains, and rings.

III. Hydrocarbons: Compounds Containing Only Carbon and Hydrogen
   a. Alkanes contain only carbon-carbon single bonds.
   b. Alkenes contain one or more carbon-carbon double bonds.
   c. Alkynes contain one or more carbon-carbon triple bonds.
   d. Aromatic compounds contain one or more benzene rings.
   e. Structural isomers have the same formula but different structural connections.
   f. There are five common ways to draw hydrocarbon molecules.
      i. Structural formulas show all atoms and their connections to one another.
      ii. Condensed structures show the connections of carbons only.

iii. Carbon skeletons show carbon-carbon bonds as lines where the end of each line implies a carbon atom; hydrogen atoms are implied based on the octet rule.

iv. Ball-and-stick models show atoms as spheres connected by sticks which represent bonds.

v. The space-filling model shows how electrons fill space.

EXAMPLE:

Draw structural formulas and carbon skeleton formulas for the three structural isomers of pentane, C_5H_{12}.

We begin by sketching the carbon skeletons (without hydrogen atoms) of the isomers:

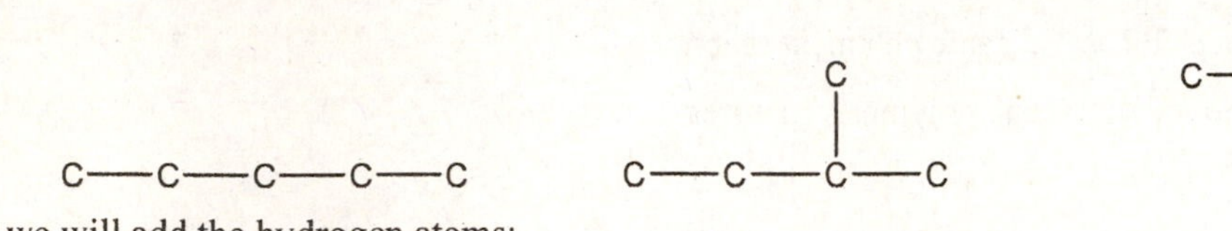

Now we will add the hydrogen atoms:

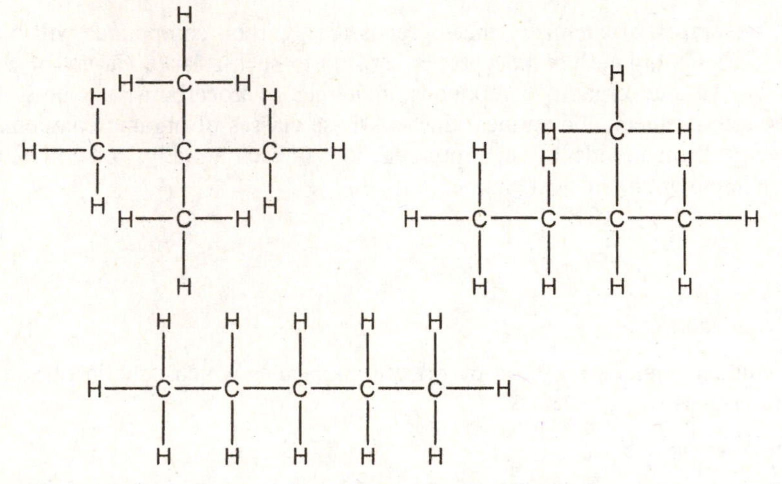

The carbon skeleton formulas are drawn by removing the hydrogen atoms and using lines to represent carbon-carbon bonds:

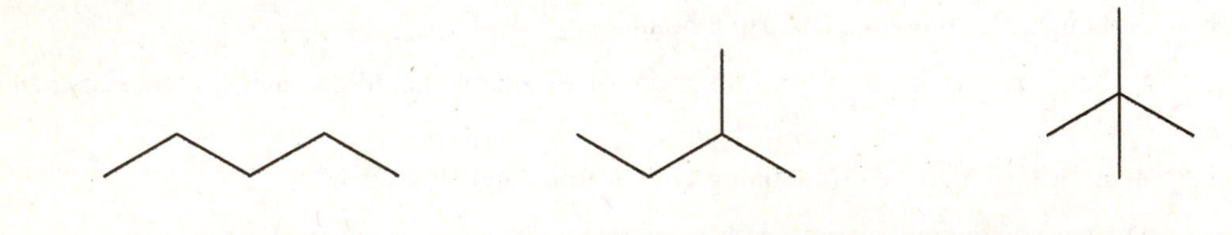

---

g. Single bonds rotate freely at room temperature so the arrangement of atoms does not matter, only the connectivity matters.

h. Double and triple bonds are rigid and therefore the arrangement is important.

i. Stereoisomers have the same connectivity, but have a different spatial arrangement.

1. Geometric isomers are cis/trans isomers and will be discussed later.

2. Optical isomers are non-superimposable mirror images.

a. Optical isomers are also called enantiomers and chiral.

b. Any carbon atom attached to four different substituents is a chiral carbon center.

c. Enantiomers have similar properties to each other except:

  i. They rotate polarized light in different directions.

    1. Right rotation is called dextrorotary (d).

    2. Left rotation is called levorotary (l).

  ii. They display different chemical properties in chiral environments.

    1. This is physiologically very important.

d. An equimolar mix of two enantiomers is called a racemic mixture.

EXAMPLE:

Are any of the isomers in the last example chiral?

None of these molecules has a carbon atom with four different substituents, so none of them is chiral.

IV. Alkanes: Saturated Hydrocarbons

a. Alkanes have only carbon-carbon single bonds; these compounds are "saturated" with hydrogen atoms.

b. The boiling point of alkanes increases with chain length.

c. Name alkanes using prefixes for the number of carbon atoms + "-ane."

  i. Count the number of carbon atoms in the longest continuous chain to find the base name of the alkane.

  ii. The carbon atoms are numbered so that any branches have the lowest possible number.

  iii. The substituent (branch) is named using the same prefixes.

EXAMPLE:

Name the alkanes shown in the first example.

The first alkane is the straight chain structure:

Since there are five carbon atoms, this is pentane.

The next structure is:

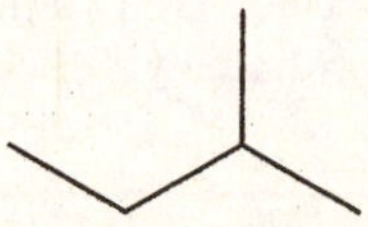

We see that the longest continuous chain is four carbon atoms long, so the base name is butane. There is a single substituent which is a single carbon atom, so it is a methyl group. We number the carbon atoms to give the methyl group the smallest number. So the name of this molecule is 2-methylbutane.

The next isomer is:

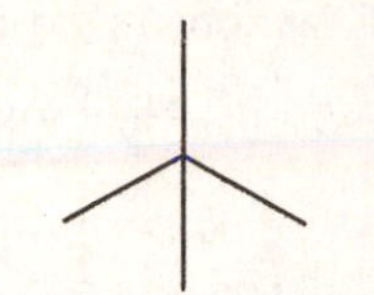

We see that the longest chain has three carbon atoms, so the base name is propane. There are two methyl groups attached to the main carbon chain, both are attached at carbon 2. The name is 2,2-dimethylpropane.

V. Alkenes and Alkynes

a. Alkenes and alkynes are called unsaturated hydrocarbons.

b. Alkenes are named in the same way as alkanes with a prefix + "-ene."

i. The base name is derived from the longest continuous chain containing the double bond.

ii. The position of the double bond is given by a number preceding the name.

iii. Numbering is done to give the alkene the smallest possible number.

EXAMPLE:

Name the following alkenes:

a.

The carbon chain has six carbon atoms and there is a single double bond, so the base name is hexene. We number the carbons so that the double bond starts at the lowest number. The name of this compound is 2-hexene.

b.

The longest carbon chain containing the double bond has seven carbon atoms, so the base name is heptene. The double bond and the methyl group are both on carbon 3, so the name is 3-methyl-3-heptene.

c.

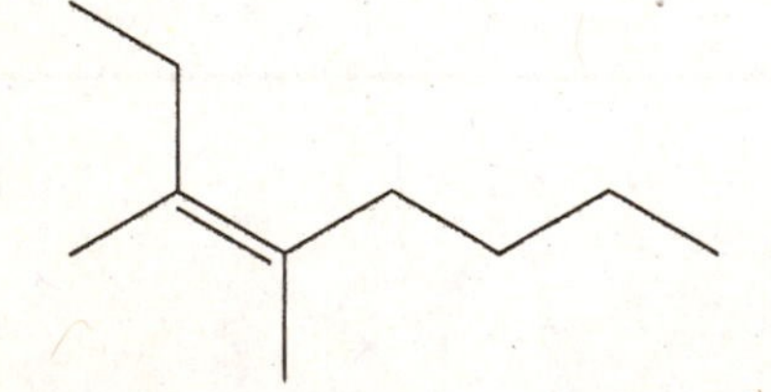

In this molecule, we see that there are eight carbon atoms in the longest carbon chain containing the double bond, so the base name is octene. There are two methyl groups on carbons 3 and 4. The double bond starts on carbon 3, so the name is 3,4-dimethyl-3-octene.

---

c. Alkynes are named exactly as alkenes except use the ending "-yne."

EXAMPLE:

Name the following alkynes:

a.

The longest carbon chain that contains the triple bond is five carbons long, so the base name is pentyne. There is a single ethyl substituent at carbon 2, so the name is 2-ethyl-1-pentyne.

b.

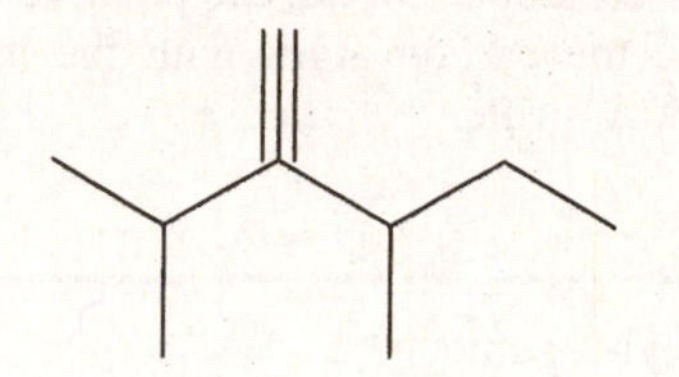

The longest carbon chain containing the triple bond again has five carbon atoms, so the base name is pentyne. There are two groups on this molecule: the isopropyl group at carbon 2 and the methyl group at carbon 3. The name of this compound is 2-isobutyl-3-methyl-1-pentyne.

---

d. In alkenes and alkynes there is both sigma and pi bonding so rotation is hindered.

  i. Geometric (cis/trans) isomerism results.

EXAMPLE:

In the example for naming compounds with a double bond, we named the following compound incorrectly:

Give the full, correct name.

This molecule is a trans isomer because we see that the substituents (carbon chains) are on opposite sides of the double bond. The name should be *trans*-2-hexene.

The cis form of this molecule is:

VI. Hydrocarbon Reactions

a. 85-90% of all energy produced in the United States is produced by the combustion of hydrocarbon compounds.

b. Alkanes commonly undergo halogen substitution:

$$CH_3CH_2CH_3 + Cl_2 \xrightarrow{\text{light or heat}} CH_3CH_2CH_2Cl + HCl$$

  i. This reaction occurs via a reaction mechanism which involves a radical species:

$$Cl_2 \xrightarrow{\text{light or heat}} 2Cl\cdot$$

$$Cl\cdot + R\text{-}H \longrightarrow R\cdot + HCl$$

$$R\cdot + Cl_2 \longrightarrow R\text{-}Cl + Cl\cdot$$

c. Alkenes and alkynes undergo similar chemical reactions.

   i. Halogens can add across a double or triple bond.

   ii. Hydrogen can add across a double or triple bond in order to increase the saturation.

      1. This is called a hydrogenation reaction.

   iii. When a polar reagent is added to an asymmetrical double bond, the positive end of the reagent (the less EN end) will add to the carbon atom with the most hydrogens; this is commonly called Markovnikov's rule.

EXAMPLE:

What is the product of the reaction between hydrochloric acid and 3-methyl-2-pentene?

We can draw the structure of 3-methyl-2-pentene:

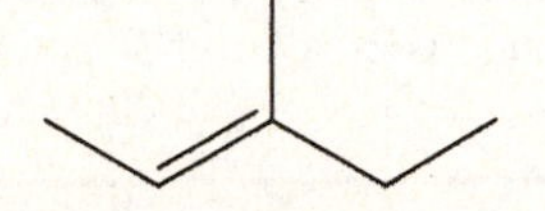

When HCl is added across the double bond, the chlorine atom will add to the more substituted carbon atom and the hydrogen will add to the carbon atom with a hydrogen atom attached to give:

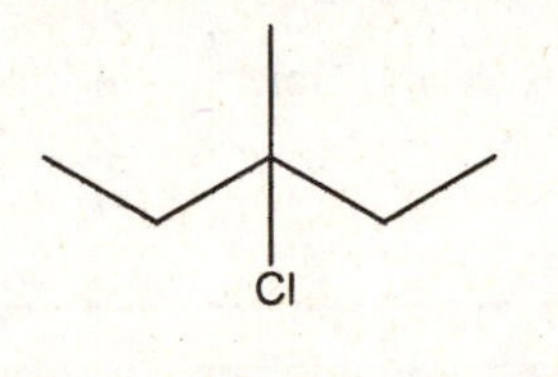

VII. Aromatic Hydrocarbons

a. Benzene is a six-carbon ring with six hydrogen atoms and is represented either by two resonance structures each with three double bonds alternating between carbon atoms or a single ring containing a circle.

   i. Benzene has a delocalized pi orbital system which causes it to be very stable.

b. Compounds containing benzene are named using multiple schemes.

   i. If there is only one substituent, the compound is named using (substituent) benzene.

   ii. Some benzene-containing substances have common names including toluene, aniline, phenol, and styrene.

   iii. When benzene is the substituent it is called phenyl.

   iv. Disubstituted benzene rings are numbered and listed alphabetically; when the substituents are identical, the prefix "di" is used.

      1. Instead of numbering, we can use *o*-, *m*-, or *p*-.

         a. *Ortho* (*o*-) is equivalent to 1,2-.

         b. *Meta* (*m*-) is equivalent to 1,3-.

         c. *Para* (*p*-) is equivalent to 1,4-.

EXAMPLE:

Name the following organic compound:

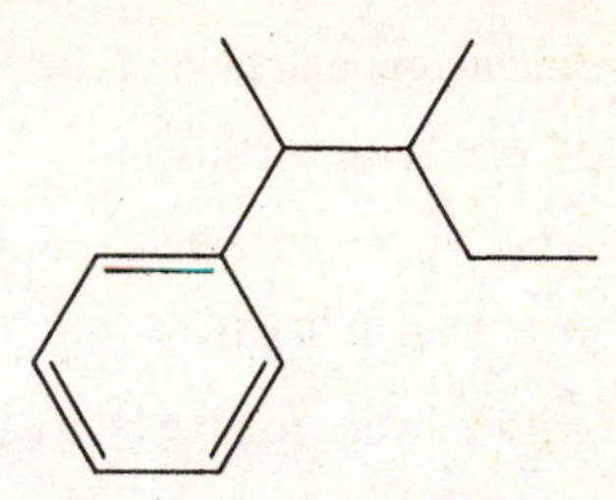

We will treat benzene as a substituent because the alkyl group is large. We see that the carbon chain has five carbon atoms, so the base name is pentane. There is one methyl group and one phenyl group. We number the carbon chain to give the smallest numbers for the substituents. So the name of this molecule is 3-methyl-2-phenylpentane.

---

c. Polycyclic aromatics are compounds that contain fused benzene rings.

d. The reactions of aromatic compounds are different from alkenes because of the pi system.

   i. Substitution of a single hydrogen atom occurs.

VIII. Functional Groups

a. Other organic compounds can be considered hydrocarbons with a functional group attached.

   i. A functional group is a characteristic atom or group of atoms.

b. "R" is used to represent the hydrocarbon.

IX. Alcohols

a. Alcohols contain the -OH (hydroxyl) group and are of the form R-OH.

b. Alcohols are named using the base name of the longest carbon chain containing the OH group. The base name is combined with the ending "-ol" and a number which indicates the location of the OH group on the hydrocarbon chain.

EXAMPLE:

Name the following organic compound:

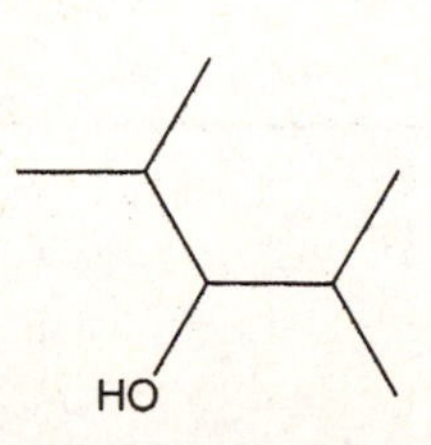

We look for the longest carbon chain containing the hydroxyl group and see that there are five carbon atoms so the base name is pentanol. There are two methyl groups and they are attached to carbons 2 and 4. This molecule is 2,4-dimethyl-3-pentanol.

---

c. Reactions of alcohols fall into four main categories:

i. Substitution reactions where the hydroxyl group is replaced:

$R\text{-}OH + HX \rightarrow R\text{-}X + H_2O$

ii. Elimination or dehydration reactions where an alkene is formed:

$RCH_2CH_2\text{-}OH \rightarrow RCH{=}CH_2 + H_2O$

iii. Oxidation reactions to give aldehydes or carboxylic acids:

$RCH_2OH \rightarrow RCOH \rightarrow RCO_2H$

iv. Reactions of active metals to give hydrogen gas:

$RCH_2OH + M \rightarrow RCOM + \frac{1}{2}H_2$

EXAMPLE:

Predict the product that forms when 2,4-dimethyl-3-pentanol reacts with hydrochloric acid.

The hydroxyl group will be replaced with a chlorine atom to form 3-chloro-2,4-dimethylpentane:

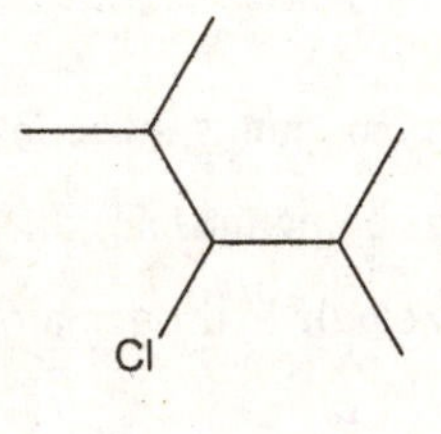

X. Aldehydes and Ketones

a. Both aldehydes and ketones contain the carbonyl (C=O) group in them.

b. Aldehydes have the general form RCOH and the structure:

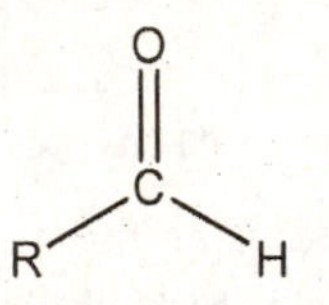

i. Aldehydes are named by dropping the "-e" at the end of the alkane name and replacing it with "-al." The base name is derived from the longest chain that includes the carbonyl group, whose carbon is numbered 1.

ii. Aldehydes are prepared by oxidizing alcohols and can be reduced to form alcohols.

EXAMPLE:

Name the following organic molecule:

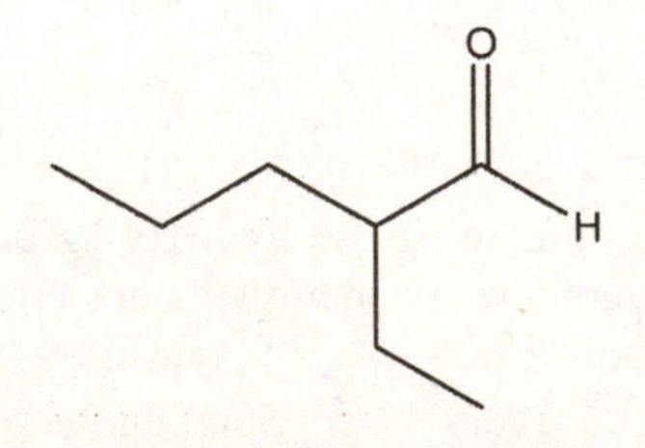

The longest carbon chain that contains the carbon of the aldehyde group has five carbon atoms so the base name is pentanal. There is one substituent: an ethyl group. We number the carbon chain so that the aldehyde is at carbon 1 so the name is 2-ethyl pentanal.

c. Ketones have the general form RCOR′ and the structure:

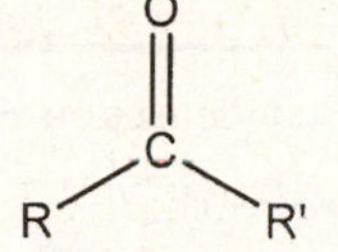

Where R and R′ are the same or different alkyl groups.

i. Ketones are named by dropping the "-e" at the end of the alkane name and replacing it with "-one."

d. Both aldehydes and ketones can react to add across the carbon-oxygen double bond (analogous to the reactions of alkenes).

EXAMPLE:

Name the following organic molecule:

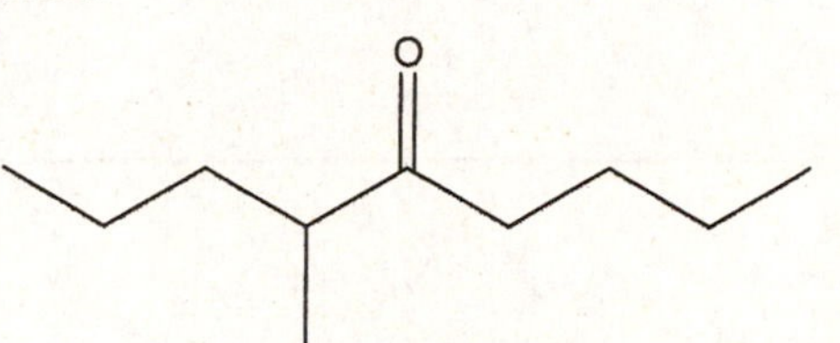

The longest carbon chain containing the carbonyl group has nine carbon atoms and there is a single methyl substituent. We number the carbon chain to give the lowest numbering scheme so this molecule is 4-methyl-5-nonanone.

XI. Carboxylic Acids and Esters

a. Carboxylic acids have the general form RCO_2H and the structure:

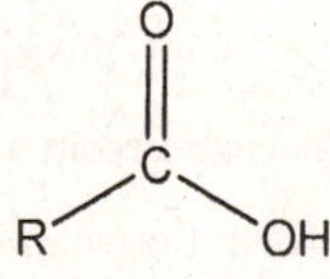

i. Carboxylic acids are named by dropping the "-e" of the alkane name and replacing it with "-oic acid."

ii. Carboxylic acids are weak acids.

b. Esters have the general form RCO_2R' and the structure:

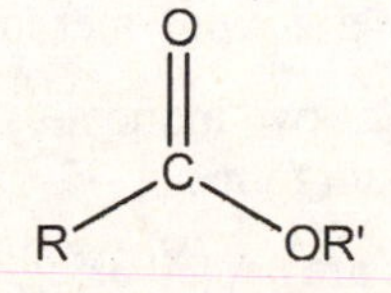

i. Esters are named by listing the two R groups in alphabetical order and replacing the "-e" of the second alkyl group and with "-oate."

c. Carboxylic acids react with alcohols to form esters in condensation reactions (elimination of water).

d. Condensation of two carboxylic acids at high heat results in the formation of an acid anhydride (without water).

EXAMPLE:

Predict the ester that forms from the condensation reaction between butanoic acid and ethanol.

We will first write out the structures of each:

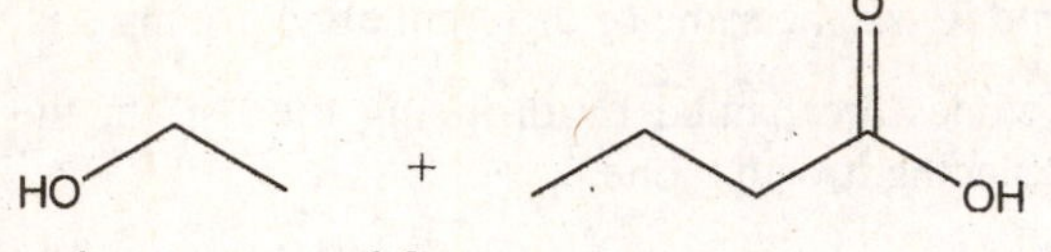

In the condensation reaction, we lose water and form an ester:

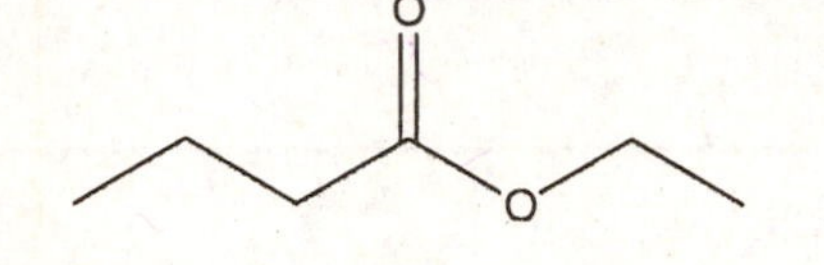

The product is ethyl butanoate.

XII. Ethers

a. Ethers have the general form ROR′ and the structure:

O

R R'

b. Ethers are named by listing the two alkyl groups in alphabetical order followed by "ether" – when the alkyl groups are the same, the prefix "di" is used.

XIII. Amines

a. Amines are derivatives of ammonia with one or more of the hydrogen atoms replaced by alkyl groups.

b. Amines are named by listing the alkyl groups in alphabetical order followed by "amine."

c. Amines are weak bases and react accordingly.

d. Condensation reactions between amines and carboxylic acids are biologically very important and result in protein formation.

XIV. Polymers

a. Polymers are long, chainlike molecules that are composed of repeating units called monomers.

b. An addition polymer is a polymer in which the monomers link together.

i. Reaction of two monomers produces a dimer; as more monomer units are added, a long chain forms.

ii. Polymers are named using the name of the monomer with the prefix "poly-."

c. Condensation polymers eliminate an atom or small group of atoms in polymerization.

d. Copolymers are those that form from more than one type of monomer.

Fill in the Blank Problems:

1. ________________ are organic compounds with the general form of ROR′.
2. When a polar reagent is added to an unsymmetrical alkene, the electropositive end of the reagent adds to the carbon atom with ________________.
3. The ________________ of a molecule shows each atom and how the atoms are bonded to each other.
4. Aldehydes, ketones, carboxylic acids, and esters all contain the ________________ group.
5. Molecules that have the same molecular formula but different structures are called ________________.
6. Nonsuperimposable mirror images are called ________________ and will cause polarized light to rotate in ________________.
7. 1,4-dichlorobenzene is also called ________________.
8. A(n) ________________ is an equimolar mixture of enantiomers.
9. The base chain of an alkene is the longest continuous carbon chain that contains the ________________.
10. Geo metric isomers are also called ________________.
11. ____ ________________ are molecules in which the atoms have the same connectivity but different spatial arrangement.
12. Alco hols contain the ________________ functional group.
13. The most co mmon reactions of aldehydes and ketones are ________________.
14. Al kanes are often referred to as ________________ hydrocarbons.
15. When benze ne is a substituent, it is called a ________________ group.
16. A polymer that forms when monomers simply add together without eliminating any atoms is called ________________.
17. The pr oduct that forms between the reaction of two monomers is called a ________________.

Problems:

1. Draw all of the structural isomers for the alkanes with the molecular formula of C_7H_{16}. Name each of the molecules that you have drawn.
2. Draw all of the structural isomers for the alkenes with the molecular formula of C_5H_{10}. Name each of the molecules that you have drawn.
3. Identify the functional groups in each of the following compounds and provide names for each.

a. O

b. N

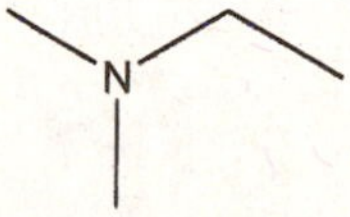

c. O O

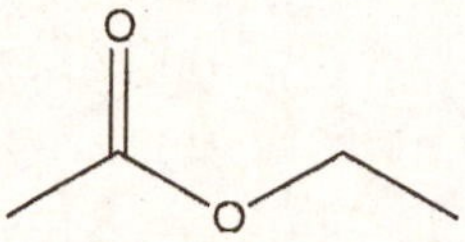

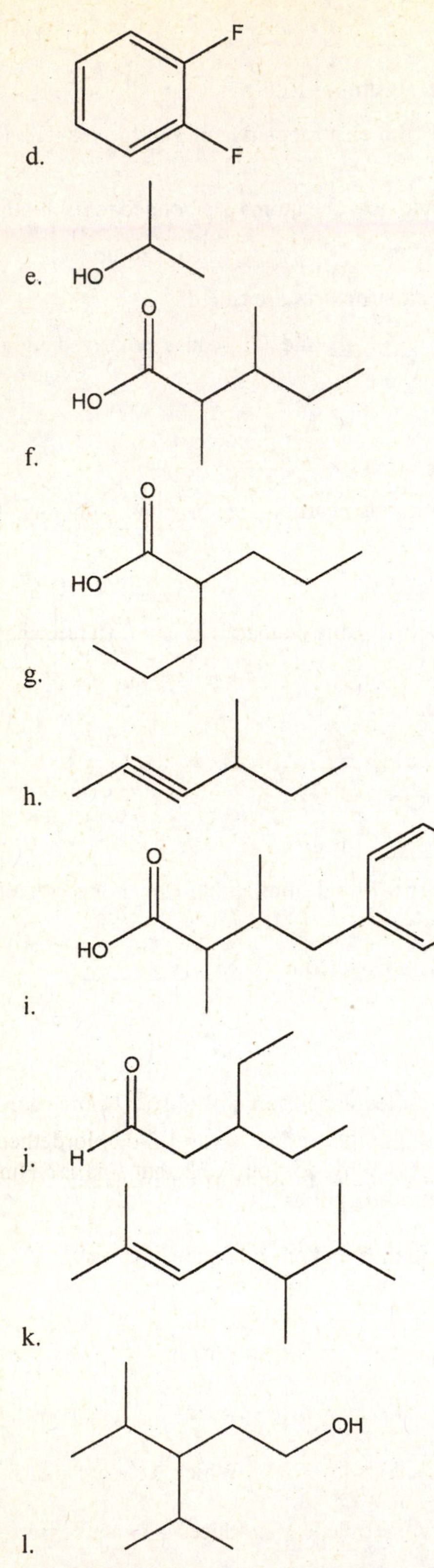

4. Identify which of the compounds in Question 1 have chiral carbons. Identify each of the chiral centers with a star.

5. Draw structures and predict the products that form in the following reactions:

a. F_2 reacts with benzene using iron(III) chloride as a catalyst

b. Combustion of hexane

c. Reaction of 2-chloro-1-propene with hydrochloric acid

d. Condensation of butanoic acid with itself

e. Reaction of ethyl amine with propanoic acid

f. Reaction of hydrogen cyanide with ethanal

g. Oxidation of 2-methyl-4-heptanol (write all oxidation products)

h. Reduction of 2-methyl-pentanal

i. Reaction of benzene with 2-chlorobutane

j. Reaction of diethylamine with nitric acid

6. Polyaramid is a term applied to polyamides that contain aromatic groups. Kevlar is an example of such a polyaramid; it is used in bulletproof vests and many other high-strength composites. The structure of Kevlar is shown below:

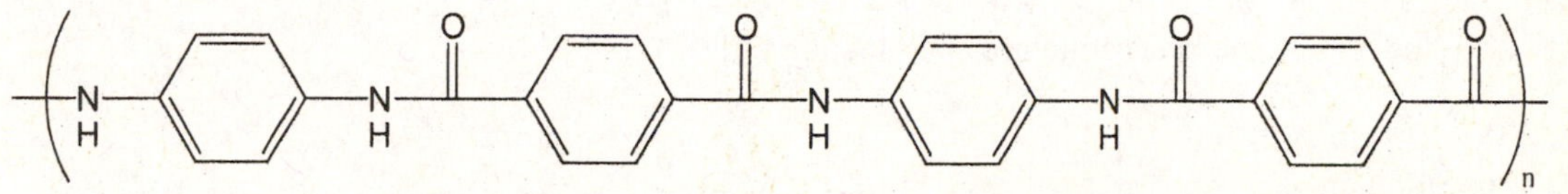

a. Draw the structure of the subunits that are used to make this polymer.
b. Is Kevlar formed in a condensation reaction or in an addition reaction? Explain your answer.
c. Kevlar is very strong because of cross-linking that occurs between the polymer strands. What intermolecular interaction do you expect to hold strands together and provide Kevlar with its unique strength? Explain your answer.
d. The strength of Kevlar is substantially reduced at high temperatures. Explain why you think that this might be.
e. In writing these questions, I wanted to provide as much information as possible and so I searched for the molar mass of Kevlar. I couldn't find it anywhere! Why do you think that this is?

Concept Questions:

1. In the discussion of cis-trans isomerism, you learned that isomers have different physical properties. For example, in Table 20.9 of your textbook you see that the boiling point of *cis*-1,2-dichloroethene has a higher boiling point than *trans*-1,2-dichloroethene does. Why do you think that this is? Hint: consider the intermolecular forces that lead to differences in boiling points.

2. Markovnikov's rule states that the more electronegative atom will add to the more substituted carbon atom of a double bond. Suggest a reason for this.

3. Polyethylene comes in two forms: HDPE and LDPE. LDPE is soft and waxy while HDPE is hard and rigid. The same subunit (ethene – C_2H_4) comprises both polymers, but their structures are different. One form has branches in the chain structure and the other does not.
   a. Explain why these two forms of polyethylene have different properties and identify which polymer is more branched.
   b. One of these polymers melts at 110°C and one of them melts at 130°C. Which polymer melts at the higher temperature? Explain how you arrived at your answer.
   c. Based on the structure of the ethene, do you think that polyethylene is produced in a condensation reaction or an addition reaction? Explain your answer.

Chapter 21: Biochemistry

Learning Objectives:

- Understand the basic structure of fatty acids, fats, phospholipids, glycolipids, and steroids.
- Predict relative melting points for fats.
- Understand the structures of lipid bilayers and why they form.
- Identify the general form of carbohydrates.
- Draw ring structures and chain structures for monosaccharides.
- Understand the linkages between monosaccharides to form complex carbohydrates.
- Understand the basic structure of amino acids and the reasons for their various properties.
- Predict polypeptide structures from the condensation reactions of amino acids.
- Identify the four levels of protein structure.
- Understand the basic structure and function of DNA and RNA.

Chapter Summary:

In this chapter, you will learn about the chemistry of living organisms. You will begin by investigating structural elements of cells such as fatty acids, fats, oils, phospholipids, glycolipids, and steroids. You will learn how fats form from the condensation reaction of glycerol with three fatty acids and then you will learn how slight alterations to the structure of fats results in a large change in the physical properties of these molecules. Next you will learn about an energy storage device of the body, carbohydrates. You will see how sugars exist in both open chain and ring forms. When these sugars link together, large chains result that have very different properties depending on the manner in which they bond to one another. Next you will learn about amino acid polymers that form proteins, the work horses of the body. The four levels of protein structure will then be explored. Another biological polymer, DNA, will then be described and a brief description of its function will be given.

Chapter Outline:

I. Diabetes and the Synthesis of Human Insulin
    a. Biochemistry is the study of the chemistry occurring in living organisms.
    b. Macromolecules are very large molecules.

II. Lipids
    a. Lipids are insoluble in water and soluble in nonpolar solvents.
        i. Lipids serve as structural components and insulators, and they supply long-term energy storage.
    b. Fatty acids, oils, phospholipids, glycolipids, and steroids are examples of lipids.
        i. Fatty acids are carboxylic acids that have long carbon tails.
            1. Fatty acids have the general form of RCO_2H where R is a carbon chain containing 3-19 carbon atoms.
            2. Saturated fats have no double bonds, monounsaturated fats have a single double bond, and polyunsaturated fats have more than one double bond.

3. The melting points of fats increase with increasing chain length.

4. The melting points of fats decrease with an increasing number of double bonds.

ii. Fats and oils are triglycerides which are triesters of glycerol with three fatty acids.

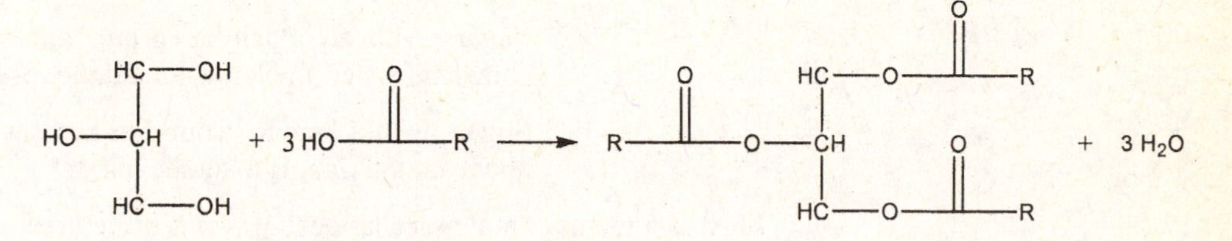

1. The bonds that form between glycerol and the fatty acids are called ester linkages.

2. Saturated fats tend to be solids at room temperature and unsaturated fats tend to be liquids at room temperature.

iii. Phospholipids are molecules that have a phosphate group in place of one of the fatty acids of the fat.

1. Since the phosphate group is polar, the replacement leads to a molecule that has a polar group at one end and a nonpolar group at the other.

2. Phospholipids are represented as polar heads and nonpolar tails:

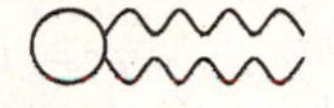

3. Phospholipids form lipid bilayers in which the polar head groups line up in opposing rows with the nonpolar tails aligned on the inside.

a. Lipid bilayers form structures that are soluble in aqueous solutions, but have nonpolar interiors that provide structure.

iv. Glycolipids have the same general structure as phospholipids but have a sugar in place of the phosphate group.

v. Steroids are lipids that have a general four-ring structure with various functional groups.

1. Cholesterol is a steroid that is the starting material for hormones and is a structural element of cell membranes.

III. Carbohydrates

a. Carbohydrates have the general form of $(CH_2O)_n$.

b. Monosaccharides have a single sugar unit such as one of these two structures:

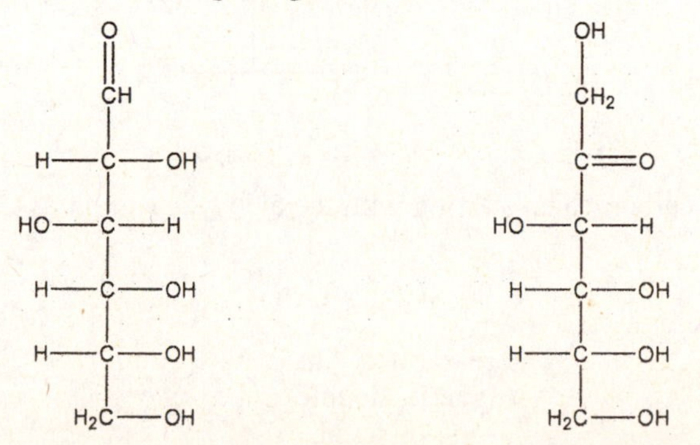

i. Monosaccharides, the simplest carbohydrates, can have 3–8 carbon atoms of which one is the carbonyl carbon of either an aldehyde or ketone functional group.

1. The name of a sugar gives the number of carbons and the type of functional group. All sugars are given the ending "-ose" and a prefix denoting the functional group.

   a. Sugars with an aldehyde group are aldoses. The structure shown above on the left is an aldohexose.

   b. Sugars with a ketone group are ketoses. The structure shown above on the right is a ketohexose.

ii. Most 5–6 member monosaccharides form ring structures.

1. Both open-chain and ring structures exist in equilibrium.

2. The ring structure is formed when the alcohol oxygen of the second-to-last carbon (with carbons numbered from the end with the aldehyde or ketone group, as below) forms a bond with the carbonyl carbon:

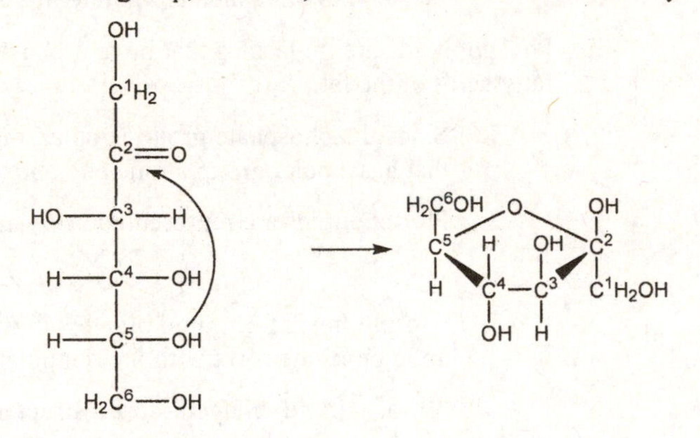

EXAMPLE:

Draw the ring structure of galactose:

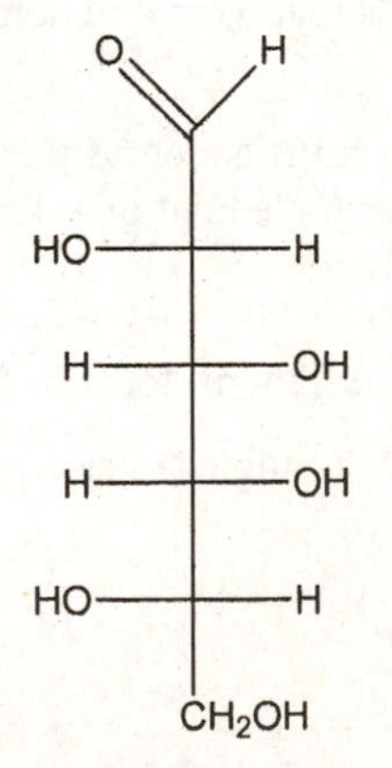

We can number the carbon chain starting with the aldehyde carbon as C^1 and connect C^5 to C^1 through an oxygen atom:

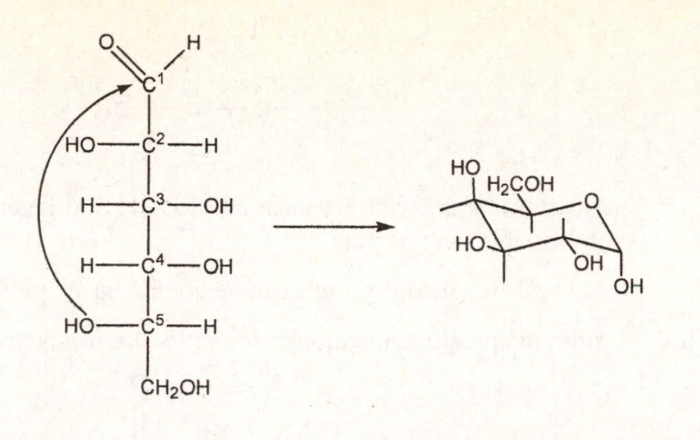

c. Monosaccharides link together to form disaccharides via a glycosidic bond. The reaction between monosaccharides is a dehydration reaction (water is eliminated).

i. Disaccharide linkages are broken through hydrolysis (addition of water across the glycosidic linkage).

▶ Dehydration is the reverse of hydrolysis and vice versa.

d. Polysaccharides have many monosaccharides linked together with glycosidic linkages.

i. There are two types of glycosidic linkages that depend on the orientation of the hydroxyl group on the linking carbon.

1. α-glycosidic linkages form when the oxygen is pointed downward relative to the planes of the rings:

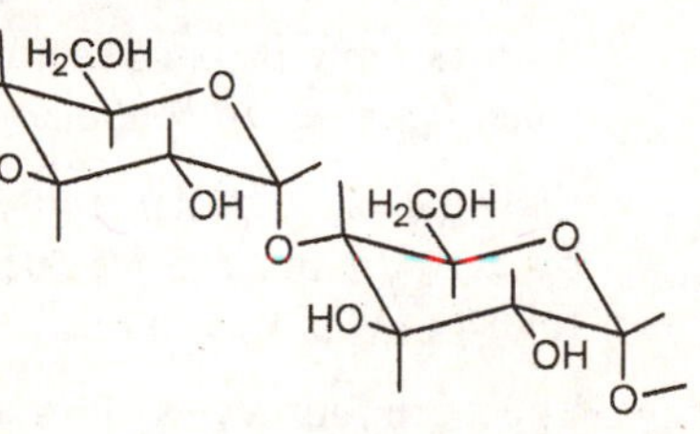

a. Starch and glycogen are polysaccharides with α-glycosidic linkages.

2. β-glycosidic linkages form when the oxygen is pointed roughly parallel with the planes of the rings:

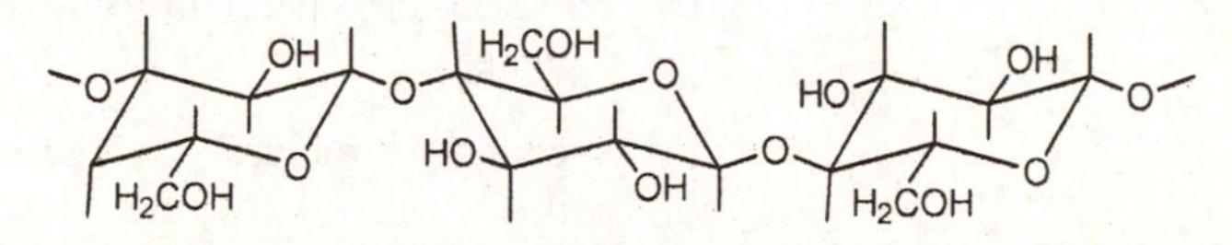

a. Cellulose is a polysaccharide with β-glycosidic linkages.

IV. Proteins and Amino Acids

a. Proteins are polymers of amino acids.

b. Enzymes are proteins that catalyze biochemical reactions.

c. Amino acids all have the same general structure and differ only in their R group:

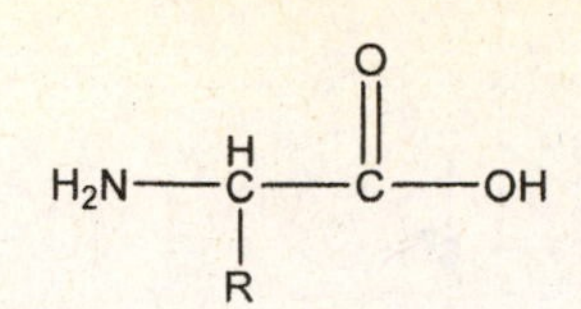

i. The R group is different for each amino acid and determines the properties and reactivity of the amino acid.

1. Amino acid R groups can be acidic, basic, polar, or nonpolar.

ii. At room temperature, the dipolar ion or zwitterion is favored:

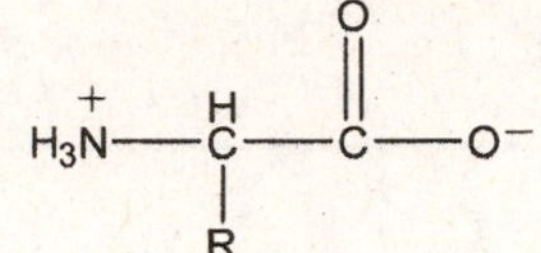

d. Amino acids link together via peptide bonds to form dipeptides.

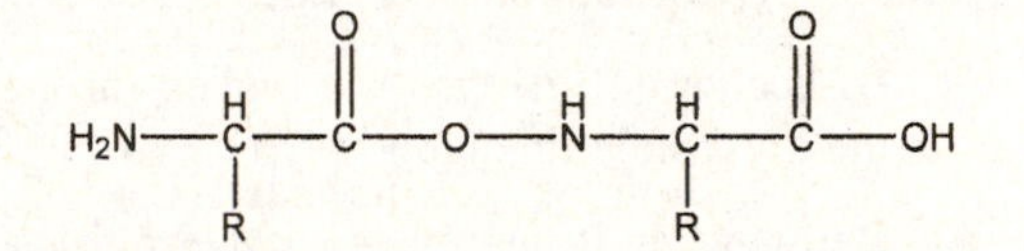

e. The N-terminal end is the end with the amine group and the C-terminal end is the end with the carboxylic acid group.

f. Many amino acids linked together via peptide bonds form polypeptides which make up proteins.

V. Protein Structure

a. The shape (or structure) of a protein dictates its function.

b. Fibrous proteins have relatively simple linear structures and tend to be insoluble in water because the nonpolar groups dictate the intermolecular interactions.

c. Globular proteins have more complex structures and are generally soluble in water because the polar groups point out and the nonpolar groups sequester in the interior folds of the protein.

d. There are four levels of protein structure.

i. Primary structure is the sequence of amino acids in the protein covalently bonded through peptide bonds.

ii. Secondary structure is the periodic or repeating patterns in protein chains.

1. The secondary structure is a result of the amino acids that are close together.

2. Secondary structures include α-helices, β-pleated sheets, and random coils.

iii. Tertiary structure is the large-scale bends and folds that result from the interactions between the R groups of amino acids.

1. Tertiary structure results from the hydrogen bonding, disulfide linkages, hydrophobic interactions, and salt bridges that form.

a. Disulfide bonds are covalent bonds between cysteine amino acids.

b. Salt bridges are the acid-base connections between side chains (R groups).

iv. Quaternary structure describes how individual polypeptide chains fit together.

1. The interactions in quaternary structures are the same as in tertiary structures but occur between different subunits or polypeptide chains.
2. Multimeric proteins have quaternary structure because they have more than one subunit.

VI. Nucleic Acids: Blueprints for Proteins

a. Nucleotides are composed of a sugar, a base, and a phosphate group which serves to link the sugar units.

i. In DNA, the sugar is deoxyribose and the different nucleotides have different base groups that are attached to the sugar on C^1.

1. The nucleotides of DNA are adenine (A), guanine (G), thymine (T), and cytosine (C).

ii. In RNA, the sugar is ribose and again the various nucleotides differ in the base groups that are attached to the C^1 sugar.

1. RNA differs from DNA by the sugar identity and the replacement of thymine (T) with uracil (U).

iii. Bases are complementary, which means that they pair in a precise manner.

1. A pairs with T (or U) and C pairs with G.

b. Codons are sequences of three bases that code for a single amino acid.

c. A gene is a sequence of codons that code for a single protein.

d. Chromosomes are groups of genes.

VII. DNA Replication, the Double Helix, and Protein Synthesis

a. Most cells contain the DNA needed to synthesize all proteins even if only certain proteins are needed in a particular cell.

b. DNA exists as two complementary strands that are wrapped in a coil.

i. When DNA replicates upon cell division, the coil unwinds and DNA strands break apart; a complementary strand is synthesized for each strand.

ii. When a protein is synthesized, part of the DNA strand unravels and copies are made.

1. The copies are messenger RNA (mRNA) molecules, which move to ribosomes where the protein synthesis occurs.

Fill in the Blank Problems:

1. The intramolecular acid-base reaction of amino acids results in the formation of _______________.
2. In DNA, bases are connected to _______________ which are linked together to form polymers through _______________ groups.
3. α-helices are examples of _______________ structures.
4. A monosaccharide with an aldehyde group is called a(n) _______________.
5. Genes are contained in _______________.
6. Monosaccharides link together via _______________ linkages.
7. Complex carbohydrates are also called _______________.

8. Triglycerides are formed in the condensation reaction of _______________ with _______________.

9. Proteins are polymers of _______________ that are linked together via _______________ bonds.

10. ____ ___________ structure results from the amino acid sequence of a protein.

11. Hexose is a s ugar with _______________ carbon atoms.

12. In protein synthesis, complementary strands are synthesized as _______________ which moves out of the cell and to the _______________ where protein synthesis occurs.

13. A(n) _______________ is a carboxylic acid with an R group with _______________ carbon atoms.

14. In DNA the bases are _______________ meaning that A pairs with T and G pairs with C; these pairs are the result of _______________.

15. In a(n) _______________ one of the carboxylic acid groups of the fatty acid is replaced with a(n) _______________ group.

16. A _______________ is a sequence of three bases that codes for a single _______________.

17. A monomeric protein has no _______________ structure.

18. A fatty acid that has one or more double bonds is called a(n) _______________.

Problems:

1. Rank the following fatty acids in order of increasing molar mass:
   a. $CH_3CH_2CH_2CH_2CH_2CH_2CH_2CO_2H$
   b. $CH_3CH_2CH_2CH_2CH_2CO_2H$
   c. $CH_2{=}CHCH_2CH_2CH_2\ CO_2H$
   d. $CH_3CH_2CH_2CH_2CH_2CH_2CH_2CH_2CH_2\ CO_2H$
2. Draw the structure of the fat that results from the condensation of palmitic acid ($CH_3(CH_2)_{14}CO_2H$) and glycerol.
3. Draw the structure of a phospholipid that is related to the fat from question 2.
4. Draw the structure of a glycolipid that has two palmitic acid tails and a glucose sugar.
5. In this chapter, you learned that steroids are lipids.
   a. Draw the structure of cholesterol.
   b. Identify the polar and nonpolar regions of cholesterol.
   c. What reaction (from Chapter 20) would result in the formation of β-estradiol from cholesterol?
6. Draw straight chain structures for the following sugars. Hydrogen atoms are implied in these structures (just like they are in carbon skeleton structures), but they should be included in your chain structures.

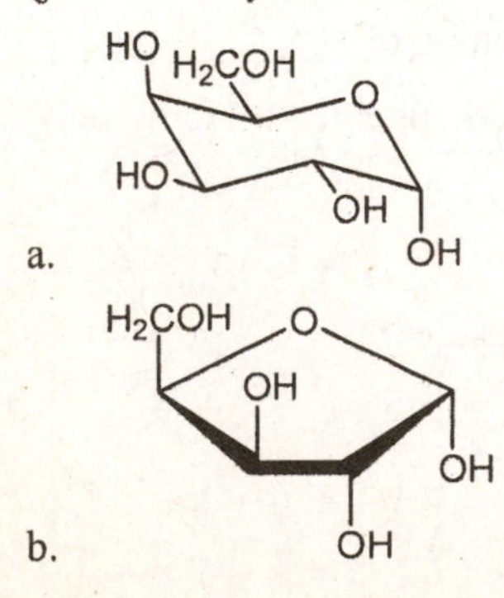

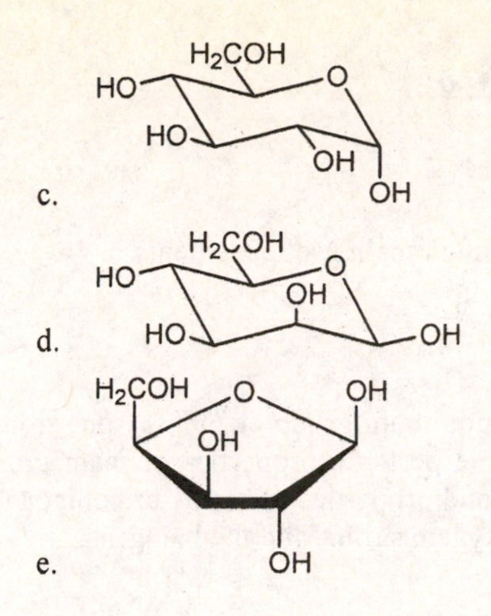

7. Identify each of the straight chain structures that you drew in question 6 as aldoses or ketoses. Name them according to the conventions described in this chapter.
8. Identify any chiral carbon centers in your straight chain figures.
9. Draw the disaccharide that forms when the sugars in parts a and b of question 6 combine. Will the linkage be an α- or β-glycosidic linkage?
10. Dra w the disaccharide that forms when the sugars in parts b and c of question 6 combine. Will the linkage be an α- or β-glycosidic linkage?
11. Ho w many codons do you need in order to synthesize ten proteins? How many amino acids will you need?

Concept Questions:

1. In this chapter, you learned that the melting points of acids increase with increasing carbon chain length and increased saturation. There is also a connection between structure and density as well. How does density change with increasing chain length and increased saturation? Explain the physical reasoning behind your answer.
2. Based on the structures of cellulose, starch, and glycogen that were discussed in this chapter, which do you think has the highest density? the lowest density? Explain the reasoning behind your answers.
3. The structure of a protein is completely determined by the sequence of amino acids in that protein and the function of a protein is determined by the structure. This means that, in principle, you can determine exactly what function a protein will have if you know the amino acid sequence. In this determination, you will need to consider four aspects of the protein's structure.
   a. List and describe each of the four levels of protein structure.
   b. Sometimes, protein structure can be determined with less than four characteristics. Explain why this is true and which of the protein structure considerations are not always important.
4. Sketch the hydrogen bonding interactions that occur between G-C and A-T pairs. Using your sketches, explain why bonding between complementary pairs is specific.
5. The hydrogen bonding network that holds DNA together is the best interaction for the job. Explain why stronger interactions (such as covalent bonds) or weaker interactions (such as dipole-dipole forces) would not be as useful. Hint: stronger and weaker interactions result in different problems for the function of DNA.
6. DNA does not have the structural complexity that proteins do. Explain why you think that this is.

Chapter 22: Chemistry of Nonmetals

Learning Objectives:

- To understand the main sources, uses, and properties of common main group elements.

Chapter Summary:

In this chapter, you will learn about the chemistry of common main group elements from groups 3A to 7A on the periodic table. You will begin by reviewing some periodic properties of main group elements. Next, you will learn about the formation, reactions, uses, and properties of atoms or compounds containing: silicon, boron, carbon, nitrogen, phosphorus, nitrogen, oxygen, sulfur, and the halogens.

Chapter Outline:

I. The Main-Group Elements: Bonding and Properties

a. In this chapter, you will focus on groups 3A to 7A.

b. Atomic radii decrease across a row due to the increase in effective nuclear charge.

c. Ionization energy and electronegativity both increase across a row.

d. The bonding changes across a row; elements on the right (groups 6A and 7A) are more likely to participate in ionic bonding.

II. Silicates: The Most Abundant Matter in Earth's Crust

a. 28% by mass of the earth's crust is silicon, which is the second most abundant element.

b. Silicates are covalent atomic solids made of silicon and oxygen plus a metal.

i. Quartz and glass are examples of silicates.

ii. Silicon forms single bonds with four oxygen atoms in a tetrahedral arrangement.

1. In a regular crystalline form, quartz (SiO_2) has this structure.

2. When quartz is melted and quickly cooled, an amorphous structure, glass, results.

iii. Aluminosilicates have aluminum atoms substituted for silicon in some lattice sites. Aluminum is electron deficient, so an anion forms, requiring a counter-ion to maintain charge neutrality; i.e., aluminosilicates form salts.

iv. Individual silicate units, silicate chains, and silicate sheets can form from SiO_4 units.

1. Orthosilicates have one unit SiO_4^{4-} and form ionic bonds with metals.

2. Disilicates have two SiO_4 units sharing a single oxygen atom to form pyrosilicates with the formula $Si_2O_7^{6-}$.

3. Chains of SiO_4 units connected through shared oxygen atoms are called pyroxenes.

4. Double chains of SiO_4 units are formed when each silicon atom shares two or three oxygen atoms.

a. These are fibrous because the bonding within chains is strong, but bonding between chains is weak. Asbestos is an example.

5. Phyllosilicates are sheet structures in which three oxygen atoms are shared. Mica and talc are examples.

III. Boron and Its Remarkable Structures

a. Boron behaves as a semimetal.

b. There are five different allotropes of elemental boron all based on an icosahedron.

c. Boron is rare and always found in compounds in nature.

d. Boron is added to glass so that the glass can expand upon heating. Pyrex is a glass that contains boron.

e. Boron halides have the general form BX_3 and act as strong Lewis acids.

   i. The empty p orbital of boron allows it to interact with lone pairs.

f. Boron can form three bonds to oxygen atoms to give B_2O_3.

g. Boranes are boron-hydrogen compounds.

   i. *closo*-boranes have the general form $B_nH_n^{2-}$.

   ii. *nido*-boranes have the general form B_nH_{n+4}.

   iii. *arachno*-boranes have the general form B_nH_{n+6}.

IV. Carbon, Carbides, and Carbonates

a. Carbon has the most versatile bonding of all of the elements on the periodic table.

b. Carbon forms two common allotropes: diamond and graphite.

   i. Graphite has sp^2 hybridized carbon atoms that form sheets.

      1. The extended pi network of graphite makes it a good conductor.

      2. The density of graphite is 2.2 g/mL.

   ii. Diamond has sp^3 hybridized carbon atoms.

      1. The density of diamond is 3.5 g/mL.

c. Coal is formed from the decomposition of plants and contains carbon, hydrogen, oxygen, and sulfur. Coal is classified by the percentage of each element present.

d. Coke is formed by heating coal in the absence of air and is used to reduce iron to iron ore (carbon is oxidized to form CO).

e. Burning wood in the absence of air produces charcoal.

   i. Activated charcoal is finely ground charcoal and is used as a filter.

f. Soot is amorphous carbon that is formed during incomplete combustion.

g. Fullerenes are carbon clusters that contain between 36 and 100 carbon atoms.

   i. Carbon nanotubes are graphite sheets that are rolled up into cylindrical tubes.

      1. Single-walled nanotubes (SWNT) and multi-walled nanotubes (MWNT) have both been produced.

      2. Nanotubes are 100 times stronger than steel and 1/6 as dense, making them important structural elements.

h. Carbides are compounds that contain carbon and one other (less electronegative) element.

   i. Ionic carbides usually contain the dicarbide ion, C_2^{2-}, which is called the acetylide ion and is used as a source for acetylene.

ii. Covalent carbides include silicon carbide, SiC.

1. Often used in cutting and polishing metals.

iii. Metallic carbides are metallic lattice structures with holes that fit carbon atoms in them.

1. Metallic carbides have metallic properties but are stronger, harder, and less malleable.

i. Two carbon oxides, CO and CO_2, commonly form.

i. CO is a common reducing agent and produces CO_2 upon oxidation.

j. Carbonates are weak bases that form when carbon dioxide is dissolved in water.

V. Nitrogen and Phosphorus: Essential Elements for Life

a. Nitrogen and phosphorus are both group 5A nonmetals.

b. Diatomic nitrogen, N_2, is 78% of the earth's atmosphere by volume.

i. Nitrogen is relatively inert because of the triple bond.

ii. Nitrogen is often utilized to provide an unreactive environment.

c. Phosphorus is found in three main forms.

i. White phosphorus is a tetrahedron of phosphorus atoms, P_4.

1. White phosphorus is highly reactive and unstable.

ii. Red phosphorus has P_4 units connected to each other and is less reactive than white phosphorus.

iii. Black phosphorus has a layered structure similar to graphite and is the least reactive of the three phosphorus allotropes.

d. The most common nitrogen compound is the hydride ammonia, NH_3.

i. Ammonia is produced commonly using the Haber-Bosch process, which utilizes high pressure and a catalyst to form ammonia from nitrogen and hydrogen gas.

e. Hydrazine, N_2H_4, is similar to hydrogen peroxide, H_2O_2, but is a reducing agent.

f. Hydrogen azide, HN_3, is a weak acid but is highly reactive.

g. Most nitrogen oxides are thermodynamically unstable.

i. Nitrogen monoxide, NO, has diverse biological functions.

h. Nitric acid is an important commercial producer of N_2.

i. Nitric acid is a strong acid.

ii. Nitric acid is used to produce fertilizers and is used in explosives.

i. Metal nitrates are used in fireworks where the nitrate is responsible for the explosive nature and the metal results in color.

j. Nitrites are used as preservatives in food.

k. Phosphorus is in the same column as nitrogen and has similar oxidation states.

l. Phosphine, PH_3, is a poisonous gas. It is less polar than NH_3 and is not a base.

m. Phosphorus halides, PX_3 and PX_5, react with water to produce phosphorous acid and halogen acids.

i. Phosphorus halides are starting materials used to form many important phosphorus-containing compounds.

n. Phosphorous oxides form cage structures.

o. Phosphoric acid is important in fertilizer production and as a food additive.

p. Phosphates are used in detergents and as food additives.

VI. Oxygen

a. Oxygen has a high natural abundancy and reactivity.

b. Oxygen is found as a diatomic, nonpolar gas.

c. Oxygen is produced by fractionation of air, electrolysis of water, and decomposition of oxygen-containing compounds.

d. Oxygen is used to enrich air for combustion and in medical applications.

e. Oxygen is a strong oxidizing agent and will form metal and nonmetal oxides.

f. Ozone, O_3, is a toxic diamagnetic gas.

i. Ozone is a strong oxidizing agent; it is used to kill bacteria and disinfect water.

ii. Ozone is classified as good or bad for human health depending on where in the atmosphere it is.

VII. Sulfur: A Dangerous but Useful Element

a. Sulfur is larger and less reactive than oxygen is.

b. Sulfur's most common allotrope is a ring structure, S_8.

i. Sulfur is generally found underground.

1. Frasch process uses superheated water to melt sulfur deposits, which are then pushed to the surface with the water.

ii. Sulfur is also a common byproduct in chemical reactions.

iii. Sulfur can also be obtained from sulfide minerals and the Claus process (reaction with oxygen).

c. Hydrogen sulfide is a toxic gas that is formed in the reactions of anaerobic bacteria on organic substances.

d. Metal sulfides are sparingly soluble and are useful because they are toxic to bacteria.

e. Sulfur dioxide produces acid rain when it reacts with water in the atmosphere and is used as a preservative for food.

f. Sulfuric acid is the most abundantly produced commercial chemical in the world.

i. Sulfuric acid is a strong acid, an oxidizing agent, and a dehydrating agent.

VIII. Halogens: Reactive Elements with High Electronegativity

a. Halogens are the most electronegative of the elements in their period and are very reactive.

b. The primary source for halogens (other than fluorine) is dissolved salts in sea water.

c. Fluorine is the most electronegative of the halogens and has an oxidation state of 0 or -1.

i. Fluorine is the most reactive halogen due to its weak F-F bond and the high lattice energy that results when it forms an ionic compound due to its small size.

ii. Fluorine forms hydrofluoric acid, which is a weak acid but is a strong oxidizing agent.

d. All halogens other than fluorine can have oxidation states from -1 to +7.

i. All halogens other than fluorine form strong acids when bonded to hydrogen.

e. Interhalogen compounds have the general form AB_n where A is the larger of the two halogens and n = 1, 3, 5, or 7.

f. Halogen oxides are extremely unstable; they are generally explosive.

i. Oxides of chlorine are powerful oxidizing agents.

Fill in the Blank Problems:

1. The most reactive of the halogens is ________________ due to its ________________.
2. The most abundantly produced chemical in the world is ________________.
3. Very fine carbon particles that have a high surface area are called ________________.
4. Silicates that are comprised of a single SiO_4^{4-} unit are called ________________.
5. ________________ are silicates that can form sheets of material as a result of the silicate structure.
6. *closo*-Boranes have the formula ________________.
7. ________________ are covalent atomic solids that contain silicon, oxygen, and various metal atoms.
8. The two most common allotropes of carbon are ________________ and ________________.
9. The most abundant element in the earth's atmosphere is ________________.
10. The form ula unit for quartz is ________________.
11. The weakest halogen-halogen bond is between ________________ atoms.
12. The most co mmon allotrope of sulfur is ________________.

Problems:

1. In the Haber-Bosch process, ammonia is formed at high pressures and with the use of a catalyst.
   a. Write the balanced chemical reaction for the synthesis of ammonia from hydrogen and nitrogen.
   b. Explain why high pressures are used to facilitate ammonia formation.
   c. Calculate the enthalpy change of this reaction using standard heats of formation.
   d. Why is a catalyst used instead of simply increasing the temperature?
2. Table 22.4 in your textbook states that hydrogen azide has an oxidation state of -1/3. How is this possible and what does it imply about the structure of this compound?
3. Carbon reacts with elemental sulfur to form carbon disulfide.
   a. Write the balanced chemical reaction.
   b. Assign oxidation states to all species.
   c. Determine which species is being oxidized and which species is reduced.
   d. Draw Lewis structures for carbon dioxide.
   e. Can you suggest any other compounds that form between carbon and sulfur? If you can, draw them. If you can't, explain why there are no other possibilities.
4. The tarnishing of silver is the formation of silver sulfide. Is silver oxidized or reduced in this reaction?
5. Oxygen will react with fluorine to form OF_2.
   a. Write the balanced chemical reaction.

b. Assign oxidation states to all species.

c. Determine which species is being oxidized and which species is being reduced.

d. Do you expect that Br_2 will react in a similar way with O_2? Why or why not?

Concept Questions:

1. Explain the differences between carbides, covalent carbides, and metallic carbides with an emphasis on the bonding types and physical properties.
2. Using the principles of bonding present in each form, explain why red phosphorous is more stable than white phosphorous is.
3. Explain why boron-halogen bonds are shorter than expected based on the length of boron-hydrogen bonds.
4. Why are metal atoms generally required in silicates? Would the silicate form a solid without the metal?
5. Main group nonmetals are usually reduced when they react with metals. Explain why this is based on the general properties of nonmetals.
6. The enthalpies of formation for nitrogen oxides are all positive. Explain why this is.
7. Nitrogen and phosphorus have the same electron configuration, but form very different compounds. For example, nitric acid is a strong acid and phosphoric acid is a weak acid. Draw the Lewis structures for these acids and explain why they are so different.
8. The chemistry of the halides is dominated by oxidation-reduction reactions. Explain why this is.

Chapter 23: Metals and Metallurgy

Learning Objectives:

- Understand and identify the various metallurgy processes.
- Understand the structures of the two types of alloys.
- Utilize phase diagrams for alloys and use them to determine melting points and primary structures.
- Identify the sources and uses of some common metallic elements.

Chapter Summary:

This chapter focuses on the part of the periodic table that was not discussed in the last chapter, the metals. The metals, you will see, are obtained from minerals and ores; the processes of mining, separating, and refining these minerals and ores, as well as the subsequent production of pure metals and alloys, will be discussed in detail. Various metallurgical processes will then be discussed with examples of metals that are obtained in each manner. Next you will learn about substitutional and interstitial alloys. Phase diagrams for substitutional alloys will be discussed and you will learn how to use them in determining the alloy composition. Finally, you will learn about the sources, uses, and properties of some common commercial metals.

Chapter Outline:

I. Vanadium: A Problem and an Opportunity
   a. Metallurgy includes all of the processes associated with mining, separating, and refining metals as well as the production of pure metals and mixtures of metals called alloys.

II. The General Properties and Natural Distribution of Metals
   a. Metals are opaque, good conductors of heat and electricity, malleable, and ductile.
   b. Most metals are found in nature as minerals, which are homogeneous, naturally occurring crystalline solids.
      i. Ores are rocks that contain a large quantity of a given metal.

III. Metallurgical Processes
   a. Extractive metallurgy extracts elemental metals from their compounds.
   b. Refining is the purification of raw material.
   c. The steps involved in extraction and refining are:
      i. Crush the ore and separate minerals from gangue (remaining material).
      ii. Separate the metal from the mineral.
         1. Pyrometallurgy uses heat to carry out the separation.
            a. Calcination uses heat to drive off a volatile substance.
            b. Roasting causes a chemical reaction to occur which allows separation of sulfides.
            c. Smelting is roasting that results in a liquid.
         2. Hydrometallurgy uses an aqueous solution to extract metals from their ores.
            a. Leaching is the process of selective dissolution.

3. Electrometallurgy uses electrolysis in separation.

    a. The Hall process and the Bayer process are both used to separate aluminum from bauxite, $Al_2O_3 \cdot nH_2O$.

4. Powder metallurgy uses powdered metal in the production of metal components.

IV. Metal Structures and Alloys

a. Metals commonly form face-centered cubic, body-centered cubic, and hexagonal closest-packed structures.

    i. The structure of a metal can change with temperature and pressure.

b. Alloys are metallic materials with more than one element present.

    i. Substitutional alloys occur when one metal atom substitutes for another in the crystal structure.

        1. The metal atoms must have similar atomic radii.

        2. The phase diagram for a binary mixture depends on the amount of each substance in the mixture.

            a. There are some mixtures that have a melting point that is lower than the melting points of each of the individual components.

EXAMPLE:

What is the melting point of a chromium-vanadium mixture that has a 0.10 mole fraction of vanadium?

Using Figure 23.6 from your textbook, we can see that the melting point of the solution will be approximately 1815°C.

---

3. When two metals have different structures, their alloys can change structure at a given temperature.

4. Some alloys have a two-phase region in their phase diagram; the predominant phase will depend on the composition.

    a. The lever rule states that the phase that is closest to the composition of the alloy is the more abundant phase.

EXAMPLE:

Determine the phases present and the relative amounts of each at 1100°C and a 60% mole percent of chromium in a chromium-nickel alloy.

Using Figure 23.7 from your textbook, we can see that 1100°C and a 60% mole percent of chromium corresponds to the two-phase region so there will be the nickel-rich alloy phase of face-centered cubic and the chromium-rich alloy phase of body-centered cubic. According to the lever rule, the mixture will be predominantly nickel alloy (face-centered cubic) since this point in the two-phase region is closer to the nickel alloy.

---

ii. Interstitial alloys contain atoms that fit into the lattice holes of the other.

    1. If the radius of the second metal is up to 41.4% of the first metal, then it can fit into an octahedral hole, which will cause it to be surrounded

by six atoms. The number of octahedral holes is equal to the number of metal atoms in the main structure.

2. If the radius of the second metal is up to 23% of the first metal, then it can fit into a tetrahedral hole. The number of tetrahedral holes is twice the number of metal atoms in the main structure.

V. Sources, Properties, and Products of Some of the 3d Transition Metals

a. Titanium is the ninth most abundant element in the earth's crust.
    i. Titanium is very reactive – it is oxidized by O_2 and N_2.
    ii. The sources of titanium are rutile, TiO_2, and ilemite, $FeTiO_3$.
    iii. Titanium is collected by arc-melting; an electric voltage is used in a controlled environment and the molten metal is collected in a water-cooled copper pot.
    iv. Titanium quickly oxidizes to form an oxide that coats the surface and inhibits further oxidation.
        1. Titanium is very strong and light so it is commonly used in the airline industry for jet engine parts.
    v. Titanium dioxide is the most common use of titanium. It is a white powder used in paints and other common household goods.

b. Chromium is found in a variety of compounds that are commonly brightly colored.
    i. The main source of chromium is chromite, $FeCr_2O_4$.
    ii. Chromium is commonly used in steel alloys called stainless steel.
        1. Chromium forms an oxide layer that protects iron from rusting.
    iii. Chromium can have oxidation states of +1 to +6.
        1. Chromium that serves as a cation has a low oxidation state.
        2. When chromium is covalently bound to oxygen as part of a polyatomic ion, it has a high oxidation state.
            a. Chromate, CrO_4^{2-}, and dichromate, $Cr_2O_7^{2-}$, are both strong oxidizing agents.

c. Manganese can have oxidation states from +1 to +7.
    i. Manganese is found in pyrolusite (MnO_2), hausmannite (Mn_3O_4), and rhodochrosite ($MnCO_3$) minerals.
    ii. Manganese is added to steel to make it more malleable at high temperatures.
    iii. Manganese is added to copper, aluminum, and magnesium in order to make these metals stronger.
    iv. Manganese is reactive and can serve as an oxidizing agent in oxides (such as MnO_4^-), in which it has a high oxidation state.
    v. Manganese is added to glass to "decolorize" it.

d. Cobalt is found in ores that contain other metals.
    i. Cobalt is most commonly found in sulfide-containing minerals such as cobaltite (CoAsS).
    ii. Cobalt is ferromagnetic and is therefore commonly used in magnet production.
    iii. Cobalt is a steel additive because it provides additional strength.

iv. Cobalt is part of the vitamin B_{12}.

e. Copper is found in its elemental form and in chalcopyrite ($CuFeS_2$) and malachite ($Cu_2(OH)_2CO_3$).

i. Bronze is a copper-tin alloy that resists corrosion.

ii. Copper is easily recycled.

iii. Copper is a good conductor of electricity (only Ag is better) and is therefore used in wire production.

iv. Copper is a good conductor of heat and is used in components such as car radiators.

v. The rust of copper is unique and desirable for its beauty – a green patina.

vi. Copper is weak, so many alloys of it are used, such as bronze, which is made from copper and zinc.

f. Nickel is found in deposits in Canada and with sulfur in iron sulfides.

i. Nickel is unreactive and resists corrosion, which makes it useful as an alloying metal in stainless steels.

g. Zinc is found in sphalerite (ZnS), smithsonite ($ZnCO_3$), and franklinite (an oxide of zinc, manganese, and iron).

i. Zinc is commonly used as an anticorrosion coating for steel.

1. Steel is dipped in molten zinc to form galvanized steel.

ii. Zinc is used in paints that are intended for steel to inhibit corrosion.

iii. Zinc has replaced chromium and lead in steel alloys because it is considered non-toxic.

<u>Fill in the Blank Problems:</u>

1. Selective dissolution of a metal is done in a process called ________________.
2. Because titanium is highly reactive with atmospheric gases, it must be ________________ in order to purify it.
3. All of the processes associated with mining, separating, and refining metals and the subsequent production of pure metals and mixtures of metals are collectively called ________________.
4. The two different types of holes between atoms in a closest-packed crystal structure are ________________ and ________________ holes.
5. ________________ is the process by which a crude metal material is purified.
6. Homogeneous, naturally occurring, crystalline inorganic solids are called ________________.
7. When one atom substitutes for another in the crystal structure, a(n) ________________ is formed.
8. ________________ is a process by which metallic components are made from powdered metal.
9. A(n) ________________ is a metallic material that contains more than one type of element.
10. Two of the most important alloys of copper are ________________ and ________________; both of these alloys resist corrosion better than copper does.
11. When two metals have different crystal structures, the phase diagram of their mixture will have a ________________.
12. ____ ______________ is the heating of an ore in order to decompose it and drive off volatile product.
13. Cobalt is a(n) ______ ___________ metal which results in its importance in magnet production.

14. The undesired material of a metal ore is called the ________________.

Problems:

1. Gold and silver both have cubic closest-packed structures. The melting point of gold is 1340°C and the melting point of silver is 1230°C.
   a. Sketch the phase diagram for the mixture of gold and silver assuming that there is no decrease in the temperature as the mole fraction of gold is increased.
   b. Estimate the melting point of a solution that is exactly 50% by mass silver.
   c. Estimate the melting point of a solution that has a weight percent of silver 24%.
   d. Estimate the composition of gold in a solution that has a melting point of 1280°C.
2. The phase diagram of a platinum-ruthenium mixture is shown below:

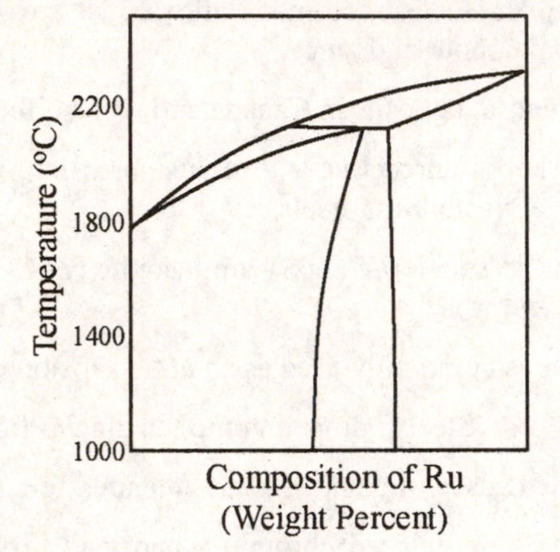

   a. There are six distinct phase regions on the phase diagram: a homogeneous liquid, a liquid with solid ruthenium, liquid with solid platinum, a two-phase solid region, the platinum alloy, and the ruthenium alloy. Label each on the phase diagram.
   b. Estimate the melting points of ruthenium and platinum.
   c. Platinum has a cubic closest-packed structure and ruthenium has a hexagonal closest-packed structure. What is the structure of the ruthenium alloy?
   d. What is the structure of the platinum alloy?
   e. What will the structure most be like at a 50% weight percent of Ru?
   f. What will the structure most be like at a 0.50 mole fraction of Ru?
3. Steel is an interstitial alloy of iron and carbon (with other trace elements).
   a. Look up the atomic radius of carbon and iron.
   b. Based on the radii, will carbon occupy tetrahedral or octahedral holes?
   c. One phase of the iron-carbon mixture is called cementite and has the formula Fe_3C. Approximately what percentage of the holes are occupied in cementite?

Concept Questions:

1. Discuss four common traits of metals and explain why each one exists.
2. Main group metals are generally oxidized in the reactions of their elements.
   a. This is another way of saying what common trait of metals?
   b. Show that this is true by writing reactions for five metals from this chapter and giving the oxidation states of all species.

Chapter 24: Transition Metals and Coordination Compounds

Learning Objectives:

- Identify and explain periodic trends of transition metals including electron configuration, atomic radius, electronegativity, ionization energy, and oxidation state(s).
- Understand the bonding principles of coordination compounds.
- Name coordination compounds from formulae and write formulae from names.
- Identify and draw isomers of coordination compounds.
- Understand crystal field theory and use it to explain the colors and magnetic properties of coordination compounds.
- Identify common uses of coordination compounds and explain how they relate to their properties.

Chapter Summary:

This chapter focuses on the properties and chemistry of transition metals. You will begin this exploration by looking at the electron configurations of transition metals and then move on to a discussion of periodic properties including atomic size, ionization energy, electronegativity, and oxidation state. Next you will learn basic definitions of transition metal bonding so that you can name these compounds in a systematic fashion. As you draw coordination compounds, you will see that they can have both structural isomers and stereoisomers, which you will learn how to identify. Your understanding of coordination compound bonding will be expanded with a discussion of crystal field theory, which explains the varied colors associated with transition metals as well as their magnetic properties. Finally, you will finish this chapter by learning about some common applications of coordination compounds.

Chapter Outline:

I. Properties of Transition Metals

   a. Electrons are added to the (n-1)d orbitals as you move from left to right across the periodic table.

      i. The ns and (n-1)d orbitals are very close in energy, resulting in some exceptions to the normal order for obital filling.

      ii. Transition metals form ions by losing the ns electrons before losing the (n-1)d electrons.

   b. There is little variation in atomic radius across a row because the effect of added electrons (in the n-1 orbital) is minimal.

      i. There is a small increase in the atomic radius down the periodic table from the first to the second row of transition metals, but there is no meaningful change between the second and third row.

         1. The third transition row has additional electrons in the (n-2)f subshell. This is called the lanthanide contraction.

   c. Ionization energies increase slowly across a row, but the increase is small compared to the main group elements.

      i. There is an increase in the ionization energy for the third row of the periodic table due to the lanthanide contraction.

   d. Electronegativity values slowly increase across a row on the periodic table.

i. The electronegativity generally increases down from the first to second row, but there is no change from the second to third row.

e. Transition metals often exhibit multiple oxidation states.

II. Coordination Compounds

a. A complex ion contains a central metal ion bound to one or more ligands.

i. Ligands are Lewis bases that form bonds with metals.

ii. The primary valence is the oxidation state of the central metal atom.

iii. The secondary valence is the number of ions or molecules that are directly bound to the central metal atom; the secondary valence is called the coordination number.

b. A coordination compound is one in which a complex ion combines with one or more counterions, which do not serve as ligands.

EXAMPLE:

What is the primary valence and the coordination number of cobalt in $[Co(NH_3)_5Cl]Cl_2$?

With three negative ions present in this compound, the primary valence must be +3 to give an overall neutral compound. There are six ligands directly bound to the metal, so the coordination number is 6.

---

c. Coordination compounds form when an empty orbital on the metal accepts an electron pair from a ligand.

i. This type of bond is often referred to as a coordinate covalent bond.

ii. Ligands that donate a single electron pair are called monodentate.

iii. Some ligands can donate two or more electron pairs to the metal; these are called chelating agents and the complex ion is called a chelate.

1. Bidentate ligands bond to a metal with two electron pairs; they must be large enough to attach to the metal using two different atoms each with a lone pair.

2. Ethylenediaminetetraacetate (EDTA) can donate up to six electron pairs to a central metal atom. EDTA is a polydentate ligand.

d. Linear, square planar, tetrahedral, and octahedral geometries are the most common for transition metals.

e. The rules for naming a complex ion or coordination compound are different than those for other ionic compounds (discussed in Chapter 3 of your book).

i. First the ligands are named.

1. Neutral ligands are named as molecules, except for water, ammonia, and carbon monoxide.

a. Water, ammonia, and carbon monoxide ligands are named as aqua, ammine, and carbonyl respectively.

2. Anionic ligands are named using the ion name with a change of ending:

a. "–ide" becomes "–o"

b. "–ate" becomes "–ato"

c. "–ite" becomes "–ito"

ii. The ligands are listed in alphabetical order.

iii. A prefix is used to denote the number of each ligand type as in molecular compounds.

1. If a prefix is already used in the ligand name, the prefixes are altered to be "bis-" (instead of bi-), "tris-" (instead of tri-), or "tetrakis-" (instead of tetra-). The ligand name follows in parenthesis.

iv. The metal is named.

1. If the complex ion is a cation, the metal is named with the oxidation state written as a Roman neutral.

2. If the complex ion is an anion, the ending of the metal is dropped and "-ate" is added followed by the oxidation state of the metal as a roman numeral.

v. The entire name is written as a list of the ligands followed by the metal name in a single word.

EXAMPLE:

What is the name of the cobalt compound from the last example, $[Co(NH_3)_5Cl]Cl_2$?

First we will name the ligands and determine the number of each. There are ammine ligands and there is a chloro ligand.

Next we will determine the listing of the ligands and the prefix for each. Ammine is listed first with the penta- prefix since there are five of them. Chloro comes after ammine, and since there is only one chloro ligand, no prefix is needed.

Next we will name the ligand as cobalt(III) since we determined that the oxidation state is +3.

The name of this compound is: pentaamminechlorocobalt(III) chloride.

---

f. When a formula is written for a complex ion, the symbol of the metal is written first, then the neutral molecules, and then the anions; when more than one anion or neutral molecule is present, they are listed in alphabetical order based on the chemical symbol.

EXAMPLE:

What is the formula of dichlorobis(ethylenediamine)platinum(IV) chloride?

The symbol for platinum is Pt and the ligands are Cl and $H_2NCH_2CH_2NH_2$; there are two of each ligand present. Since ethylenediamine is neutral, it is listed first. There are two chlorine counterions necessary to maintain an overall neutral charge. The formula is therefore: $[Pt(H_2NCH_2CH_2NH_2)_2Cl_2]Cl_2$.

III. Structure and Isomerization

a. There are two main types of isomerism for coordination compounds: structural isomers and stereoisomers.

i. Structural isomers differ in the ways that atoms are connected to one another.

1. Coordination isomers are those in which ligands and uncoordinated counterions exchange places.

2. Linkage isomers occur when ligands can connect to metals in different ways.

ii. Stereoisomers differ in their spatial arrangement.

1. Geometric isomers include cis-trans isomers in square planar of the form (MA_2B_2) and octahedral complexes of the form (MA_4B_2) and also fac-mer isomers which occur in octahedral complexes of the form (MA_3B_3).

2. Optical isomers are nonsuperimposable mirror images of one another.

a. These molecules are chiral and are called enantiomers; they rotate polarized light in different directions.

EXAMPLE:

Does the following square planar molecule have a stereoisomer? If so, what kind?

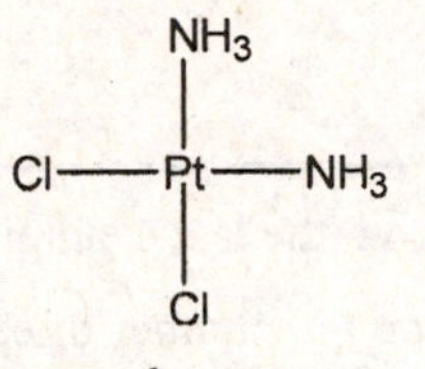

We can see that there is a trans isomer of this complex:

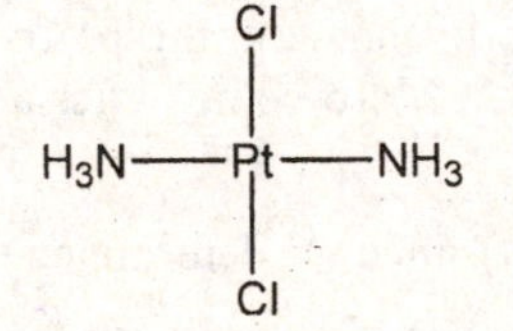

We can also check to see if the mirror image of the original complex is superimposable:

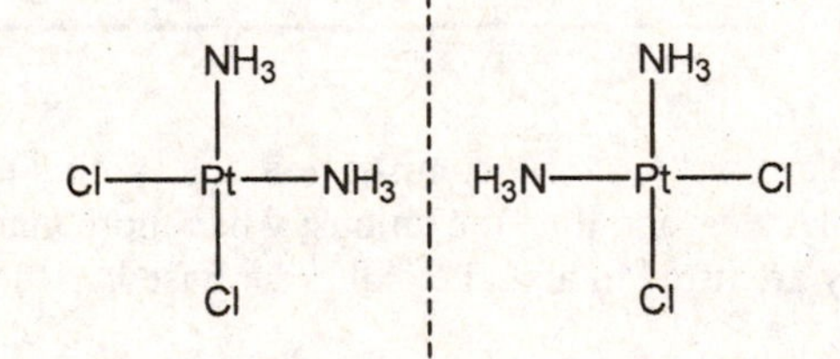

We can see that the mirror image can be superimposed on the original molecule by rotating by 90° so these are not enantiomers. Therefore, the original molecule has a geometric isomer but no optical isomer.

IV. Bonding in Coordination Compounds

a. Bonding in coordination compounds can be described in terms of valence bond theory in which the atomic orbitals of the metal are hybridized.

b. Crystal field theory is a bonding model for transition metals that accounts for the colors and magnetism of coordination compounds.

i. Crystal field theory focuses on the repulsions between the electron pairs of the ligands and the unhybridized metal d orbitals.

1. The degenerate d orbitals of the metal are split into higher and lower energy levels because of the ligand arrangement.
2. In strong-field complexes, the splitting is large and in weak field complexes, the splitting is small.

ii. The color of metal complexes is a result of the metal complex's absorption of some wavelengths of light and the reflection of others.

1. The color of metal complexes results when all colors but one are absorbed; in this case the metal will look like the reflected color.
2. The color of metal complexes can also result when only one color is absorbed and all others are reflected; the metal will look like the color that is complementary to the absorbed color.

iii. The energy of an absorbed photon is related to the splitting of the d orbitals.

EXAMPLE:

An unknown metal complex is green in aqueous solution. What is the splitting of the d orbitals in this complex?

We first calculate the energy of the absorbed light. Since red is complementary to green on the color wheel, the wavelength of absorbed light is approximately 700 nm. Converting this to energy:

$$E = h\nu = \frac{hc}{\lambda} = \frac{(6.626 \times 10^{-34}\,\mathrm{J \cdot s})(3.00 \times 10^{8}\,\mathrm{m/s})}{700 \times 10^{-9}\,\mathrm{m}} = 2.84 \times 10^{-19}\,\mathrm{J}$$

This is the energy per ion; the energy per mole of ions is:

$$2.84 \times 10^{-19}\,\mathrm{J/ion} \times \frac{6.022 \times 10^{23}\,\mathrm{ions}}{1\,\mathrm{mol}} \times \frac{1\,\mathrm{kJ}}{1000\,\mathrm{J}} = 170\,\mathrm{kJ/mol}$$

---

iv. The magnitude of the crystal field splitting in a complex ion is dependent on the ligands that are attached to the central metal ion and the identity of the metal.

1. The ability of an ion to split the d orbitals is given in a spectrochemical series from highest splitting ability to lowest:

$$CN^- > NO_2^- > en > NH_3 > H_2O > OH^- > F^- > Cl^- > Br^- > I^-$$

2. The splitting increases as the charge on the metal increases.

v. The magnetic properties of transition metal complexes are affected by d orbital splitting because once the low energy d orbitals are half-filled, electrons can either be paired or occupy higher energy orbitals.

1. If the splitting of the orbitals is weak, the electrons will occupy the higher energy orbital and form high-spin complexes.
2. If the splitting of the orbitals is large, the electrons will pair up prior to occupying the higher energy orbitals and will form low-spin complexes.

vi. The crystal field splitting is dependent on the geometry of the coordination compound.

1. For octahedral complexes, the z^2 and x^2-y^2 orbitals are higher in energy than the xy, yz, and xz orbitals are.
2. For tetrahedral complexes, the xy, xz, and yz orbitals are higher in energy than the z^2 and x^2-y^2 orbitals are.
3. For square planar complexes, the x^2-y^2 orbital is highest in energy, the xy orbital is next highest in energy, the z^2 orbital is next, and the xz and yz orbitals are lowest in energy.

V. Applications of Coordination Compounds

a. Ligands selectively bind to metals, so they can be used to identify the presence of a given metal (similar to selective precipitation).

b. Coordination compounds are used as coloring agents.

c. Many metals are involved in biological systems.

i. In hemoglobin and in cytochrome c, an iron complex called a heme is connected to a protein.

1. A heme is an iron ion coordinated to a flat, polydentate ligand called a porphyrin.
2. In the blood vessels of the lung, oxygen molecules coordinate with the iron ion in hemoglobin. Hemoglobin carries oxygen to body tissues, where the coordinated oxygen is replaced by water molecules.

ii. In chlorophyll, a different porphyrin ring is coordinated to magnesium; chlorophyll is used by plants in photosynthesis.

iii. Carbonic anhydrase has a zinc ion bound to three nitrogen atoms from surrounding nucleic acids. The fourth site of the tetrahedral zinc complex can bind a water molecule.

1. Carbonic anhydrase catalyzes the reaction between carbon dioxide and water in respiration by making the water molecule more acidic than it normally is.

iv. Coordination compounds are also used in commercial drugs.

Fill in the Blank Problems:

1. In ________________ complexes, the splitting of d orbitals is very large and in ________________ complexes, the splitting of d orbitals is small.
2. The splitting of a ligand field ________________ with increasing metal charge.
3. Complex ions contain a central metal atom bound to one or more ________________.
4. ________________ isomers occur when a coordinated ligand exchanges places with an uncoordinated ligand.
5. The list of ligands in order of their ability to split a ligand field is called the ________________.
6. The ________________ is the charge of the central atom and the ________________ is the number of ligands bound directly to the central atom.
7. Elements in the third row of the transition metals are approximately the same size as their counterparts in the second row due to ________________.
8. ________________ isomers result when the ligands bound to the metal have a different spatial arrangement. There are two types: ________________ and ________________.

9. ________________ complexes exhibit magnetic properties due to their weak field ligand.

10. Li gands that have the ability to donate more than one electron pair to the central atom are called ________________ or ________________.

11. In an octahedral field, the ________________ orbitals are higher in energy than the ________________ orbitals.

12. In coordination compounds, the metal acts like a(n) ________________ and the bound species act like ________________.

Problems:

1. Provide names for the following compounds:
   a. $K_4[Fe(CN)_6]$
   b. $Pt(NH_3)_2Cl_4$
   c. $Cr(CO)_6$
   d. $[Cr(NH_3)_3(H_2O)_3]Cl_3$
   e. $[Co(NH_3)_4Cl_2]Cl$
   f. $[PtNH_3Cl_3]^-$
   g. $[Ni(en)_3]Cl_2$
   h. $[Co(NH_3)_4ClBr]$
2. Write formulas for the following compounds.
   a. Pentaamminebromocobalt(III) sulfate
   b. Pentaamminesulfatocobalt(III) ion
   c. Hexafluorocobalt(III) ion
   d. Diaquatetracyanocopper(VI) chloride
   e. Tris(ethylenediamine)cobalt(III) chloride
   f. Pentaamminechlorochromium(III) sulfate
   g. Diamminedichloroplatinum(II)
3. Determine the type of isomers each of the following are examples of:
   a. $[Co(NH_3)_4ClBr]Cl$ and $[Co(NH_3)_4Cl_2]Br$
   b. $[Cr(H_2O)_5CN]^{2+}$ and $[Cr(H_2O)_5NC]^{2+}$
   c. $[Cr(NH_3)_5(OSO_3)]Br$ and $[Cr(NH_3)_5Br]SO_4$
4. Draw all isomers and identify each type for the following complexes. If no isomers exist, state that.
   a. $[Co(NH_3)_4Cl_2]^+$
   b. $[Co(en)]^{2+}$
   c. $[PtBr_4]^{2-}$
   d. $[Co(en)_2Cl_2]^+$
   e. $Co(NH_3)_3Cl_3$
   f. $[Cr(H_2O)_5Cl]^{2+}$

5. Determine the number of unpaired electrons that you would expect to see for the following compounds:
   a. $[PtBr_4]^{2-}$
   b. $[Co(en)_3]^{3+}$
   c. $Cr(CO)_6$
   d. $[Co(H_2O)_6]^{2+}$
   e. $[Co(NH_3)_3(CN)_3]$
   f. $[CoCl_6]^{3+}$
6. Rank the cobalt compounds from question 5 in order of increasing magnetic properties.
7. Which of the complex cobalt ions from question 5 is/are most likely to be green in aqueous solution?

Concept Questions:

1. Isomers can have different colors. For example, $[Cr(NH_3)_5(OSO_3)]Br$ is a red complex while $[Cr(NH_3)_5Br]SO_4$ is violet.
   a. Explain the difference in their colors.
   b. Do you think that all isomer types will have different colors? If so, explain. If not, which isomers will and which will not?
   c. Which compound do you expect to be more strongly attracted to a magnetic field? Explain.
   d. Which exerts a stronger ligand field, SO_4^{2-} or Br^-? Explain your answer.
   e. Calculate the ligand field splitting for each of the compounds in order to verify your answer from part d.
2. Draw ligand field splitting diagrams for octahedral, square planar, and tetrahedral complexes and explain the magnetic property differences that you expect to see as a result of these diagrams. In other words, do you expect one geometry to promote magnetism more than others? Explain your answer.
3. An experiment is carried out on two different transition metal complexes: MX_6 and MY_6. Both molecules are excited with electrical energy and MX_6 appears red while MY_6 appears green.
   a. Which ligand, X or Y, exerts a stronger crystal field on the metal? Explain the reasoning behind your choice.
   b. Another experiment is done and it is determined that only one of these complexes is magnetic. Which compound is it? Explain the reasoning behind your choice.
   c. How does the color of emitted light for each of these relate to the color of light that they will be when illuminated with white light? Hint: think about the relationship between the emission in the original problem and absorption discussed in this chapter.
   d. Carbonyl groups are often used as tags to indicate the electron density of the metal. The metal donates electrons to the antibonding orbital of the carbonyl group, weakening the C-O bond. When one of the ligands (X or Y) is replaced with a carbonyl group, the energy of the carbonyl bond changes more in MX_5CO than in MY_5CO. Which of the two ligands, X or Y, donates more electron density to the metal? Explain the reasoning behind your choice.

Chapter 1:

Fill in the Blank Problems:

1. experiments
2. potential
3. Scientific laws
4. gases, liquids, and solids.
5. homogeneous
6. physical
7. element
8. chemical
9. law of conservation of energy
10. compou nd
11. kilogra m
12. inte nsive
13. te mperature; Kelvin
14. cry stalline
15. accurac y; precision

Problems:

1. a. mixture
   b. mixture
   c. mixture
   d. pure substance
2. a. heterogeneous mixture
   b. heterogeneous mixture
   c. homogeneous mixture
   d. homogeneous mixture
3. a. physical change
   b. physical change
   c. chemical change
   d. chemical change
4. a. 9.8 Mm
   b. 8.95 ns
   c. 5.71mm
5. a. 6.548×10^7
   b. 9.800×10^{-4}
   c. 5.804×10^{-3}
   d. 1.529×10^{10}
6. a. 5 sig. figs.
   b. 7 sig. figs.
   c. 4 sig. figs.
   d. 4 sig. figs.
7. a. 5.05×10^2
   b. 5.6119×10^1
   c. 3.48×10^3
   d. 8.883×10^8
8. 4 g
9. a. 2.0×10^{-4}
   b. 3.460×10^5
   c. 2.96×10^5
   d. 6.7580×10^{-2}
10. The average is 5.51 mL. This result is very accurate since it agrees exactly with the actual value when two significant figures are considered.
11. $ 6,000
12. 5.9×10^{12} miles
13. 76 days

Concept Questions:

1. As discussed in this chapter, theories are the backbone of science. Theories are used to explain the world around us and provide the best answer for the available experimental evidence. Throwing out all ideas that are only theories would result in, essentially, throwing out all of science.
2. a. No, the students did not use the same measuring device since the number of significant figures varies from two to three.
   b. We cannot comment on the accuracy since the actual value is not provided. The results are fairly precise however since they all agree up to two significant figures (which is the number of significant figures for the average based on the least precise value).
3. The density of the metal is 0.9303 g/mL. Since the slug is less dense than water, it will float when placed in a glass of water.
4. A simple example is gas mileage where we can find the distance traveled and divide it by the amount of gas in a full tank to determine the miles per gallon of a vehicle.
5. On a microscopic scale, the energy is released to the environment as heat. The temperature of the air around the fire increases and the kinetic energy (as well as the internal energy) of the air molecules increases.

Chapter 2:

Fill in the Blank Problems:

1. mole
2. law of definite proportions
3. electrons
4. electrons
5. atomic mass
6. created or destroyed
7. nucleus
8. group
9. cations
10. ato mic mass unit
11. protons
12. natural ab undance
13. ratio of s mall whole numbers
14. Metals
15. ato ms
16. m ass spectrometry
17. Cation s
18. carbon -12

Problems:

1. The first set of masses gives 2.91 mol C, 5.82 mol H, and 2.91 mol O. The second set of masses gives 0.0372 mol C, 0.0745 mol H, and 0.0373 mol O. Both of these give a ratio of moles C:H:O of 1:2:1.
2.

Symbol	Ion	Number of Electrons	Number of Protons	Number of Neutrons
^{40}Ca	Ca^{2+}	18	20	20
^{9}Be	Be^{2+}	2	4	5
^{79}Se	Se^{2-}	36	34	45
^{37}Cl	Cl^{-}	18	17	20

3. Metals: All elements other than those listed below.
   Metalloids: B, Si, Ge, As, Sb, Te, At
   Non-metals: H, He, C, N, O, F, Ne, P, S, Cl, Ar, Se, Br, Kr, I, Xe, Rn
4. Halogens are group 7A on the periodic table. Halogens are reactive non-metals.
5. a. Na^{+}
   b. O^{2-}
   c. Mg^{2+}
   d. Al^{3+}
   e. S^{2-}
   f. Cl^{-}

6. 7 amu
7. The mass of one mole of Na-23 atoms is 23.19005459 g and one mole of Na-23 ions is 23.18950601 g which differ in the ten-thousandths place. It is appropriate, therefore, to use the same mass for both as long as the mass doesn't determine the significant figures in the final result.
8. 5.86×10^{22}
9. 2.64×10^{8} g

Concept Questions:

1. All are metals, all three tend to lose a single electron to form a +1 ion, and all three are highly reactive.
2. 88.9 g of oxygen must be produced. The mass of products (hydrogen and oxygen) must equal the mass of reactants (water) according to the law of conservation of mass. The mass of the oxygen, therefore, is simply the mass of the water minus the mass of the hydrogen produced.
3. Carbon is arbitrarily defined to be exactly 12 amu but protons and neutrons do not have a mass of exactly 1 amu and electrons do not have zero mass. These small deviations result in non-integer masses.
4. Silicon should have a molar mass of approximately 28 g/mol since most of the silicon atoms (92.3% of them) in a sample will have a mass of 28 amu.
5. Macroscopic samples are huge collections of atoms (on the order of Avogadro's number) so we can assume that we have an average sampling of isotopes.
6. If both isotopes have the same charge upon ionization, then U-238 will be heavier and therefore will be deflected less (or move slower) by the magnet in the experiment.

Chapter 3:

Fill in the Blank Problems:

1. formula mass
2. combustion
3. formula unit
4. ionic
5. molecular formula
6. electrostatic
7. Binary compounds
8. functional
9. mass percent composition
10. covalent
11. diatom ic
12. h ydrocarbons

Problems:

1. H_2, O_2, N_2, F_2, Cl_2, Br_2, I_2
2. a. NaBr
   b. MgS
   c. S_2O_4
   d. H_2CO_3
   e. HI
   f. $HClO_4$
   g. SeN_2
   h. Fe_2O_3
   i. AgCl
   j. $SiCl_4$
   k. $Ca(HCO_3)_2$
   l. $CoPO_4 \bullet H_2O$
3. a. diphosphorus pentoxide
   b. rubidium oxide
   c. titanium(IV) oxide
   d. beryllium iodide
   e. dinitrogen tetraoxide
   f. iron(III) sulfate
   g. xenon tetrafluoride

h. phosphorous acid
i. lithium hydroxide
j. copper(II) fluoride
k. copper(II) sulfate pentahydrate
l. chlorous acid

4. a. 141.94 amu
   b. 186.94 amu
   c. 79.87 amu
   d. 262.81 amu
   e. 92.02 amu
   f. 399.91 amu
   g. 207.29 amu
   h. 81.99 amu
   i. 23.95 amu
   j. 101.55 amu
   k. 249.70 amu
   l. 68.46 amu
5. 68.13%
6. 2.6×10^{23} sodium atoms
7. a. $N_2O_4 \rightarrow 2NO_2$
   b. $2NaOH + H_2SO_4 \rightarrow 2\ H_2O + Na_2SO_4$
   c. $2NaCl + BeF_2 \rightarrow 2NaF + BeCl_2$
   d. $4Mg + Mn_2O_4 \rightarrow 4MgO + 2Mn$
   e. $2HCl + Na_2CO_3 \rightarrow 2NaCl + H_2O + CO_2$
   f. $CH_3CH_2OH + 3O_2 \rightarrow 2CO_2 + 3H_2O$
   g. $2C_2H_6 + 7O_2 \rightarrow 4CO_2 + 6H_2O$
   h. $Fe_2(SO_4)_3 + 12KSCN \rightarrow 2K_3Fe(SCN)_6 + 3K_2SO_4$
   i. $CuO(s) + C(s) \rightarrow Cu(s) + CO(g)$
   j. $2Co(NO_3)_3(aq) + 3(NH_4)_2S(aq) \rightarrow Co_2S_3(s) + 6NH_4NO_3(aq)$
8. 1.0×10^{23} C atoms
9. $CuCl_2 \cdot 6H_2O$
10. 22.1% $CaCO_3$
11. 27.37% Na, 1.202% H, 14.30% C, and 57.14% O. 2.9 g of oxygen in a 5.0 g sample of $NaHCO_3$.
12. CH_3 is the empirical formula of the hydrocarbon and 107.97 g of oxygen were used.
13. You will need to consume 3.5 mg of NaF.

Concept Questions:

1. The number of molecules is not conserved – only the number of each type of atom is conserved on each side of the equation.
2. The chemical formula indicates the number of each type of atom, not the grams of each type of atom.
3. In ionic compounds, electrons are transferred resulting in positively charged cations and negatively charged anions. The charged species interact with each other in all directions. Conversely, in molecular compounds the electrons are shared between atoms and their localization results in discrete units.
4. Molecular compounds combine in various ways whereas ionic compounds form specific formula units based on the charges of the ions.
5. When a chemical reaction is balanced, the numbers of atoms of each type are equal on the reactant and product sides of the reaction, and no atoms are gained or lost. If the numbers of atoms are equal on both sides, then the masses are equal since each type of atom has a specific mass.
6. Functional groups: alcohol, ether, and amine.
   One of the OH groups in structure a is replaced with an ether group in structure b; since this is the only difference, it must be this difference that results in the different reactivity.

Chapter 4:

Fill in the Blank Problems:

1. limiting reagent
2. Oxidation-reduction
3. nonelectrolyte

4. polyprotic
5. percent yield
6. concentrated or dilute
7. precipitation reaction
8. stoichiometry
9. decreases
10. acid -base indicator
11. the ion c harge
12. Spectator ions
13. strong electrolyte
14. stock sol ution

Problems:

1. a. $2NH_3 + CO_2 \rightarrow CH_4N_2O + H_2O$
   b. 240.5 kg CH_4N_2O
   c. The ammonia is completely consumed and 35.19 kg of CO_2 remain.
   d. 70.03%
2. a. $Na_2CO_3(aq) + 2HCl(aq) \rightarrow H_2O(l) + CO_2(g) + 2NaCl(aq)$
   b. $CO_3^{2-}(aq) + 2H^+(aq) \rightarrow H_2O(l) + CO_2(g)$
   c. 741 mL
   d. 3.10 g
3. a. 1.67 M
   b. 0.167 M
   c. $2NaOH(aq) + H_2SO_4(aq) \rightarrow 2H_2O(l) + Na_2SO_4(aq)$
   $2Na^+(aq) + 2OH^-(aq) + 2H^+(aq) + SO_4^{2-}(aq) \rightarrow 2H_2O(l) + 2Na^+(aq) + SO_4^{2-}(aq)$
   $2OH^-(aq) + 2H^+(aq) \rightarrow 2H_2O(l)$
   d. 0.0510 M
4. a. The concentration of Na_2CO_3 will be 0.0 M since sodium salts are completely soluble.
   b. $Mg(C_2H_3O_2)_2\ (aq) + Na_2CO_3\ (aq) \rightarrow 2NaC_2H_3O_2\ (aq) + MgCO_3\ (s)$
   c. 3.2 M Na^+
   d. 3.2 M Na^+
   e. 0.84 g $MgCO_3$
   f. The light bulb would be brighter before the reaction took place since there will be more aqueous ions in the solution before the formation of the solid (which removes ions from solution).
5. a. $2H_2\ (g) + O_2\ (g) \rightarrow 2H_2O\ (l)$
   b. 342 g of air
   c. 11.3 g of H_2O, 0.0 g of O_2 and 8.7 g H_2
   d. Hydrogen in H_2 is zero; oxygen in O_2 is zero; hydrogen in H_2O is +1; oxygen in H_2O is -2.
   e. Oxygen is the oxidizing agent in this reaction.

Concept Questions:

1. The number of molecules on each side is not necessarily conserved, only the number of atoms of each type is conserved.
2. A homogeneous mixture is independent on the size of the sample, meaning that an expression which gives the amount of material in a liter will be a scaled-up quantity from the milliliter sample.
3. If one is looking at qualitative results such as whether a precipitate forms or not in a reaction, then simply knowing an approximate solution concentration would be sufficient.
4. In the solution, the two species (Ag^+ and Cl^-) are able to move freely and can physically come into contact and react to form a solid precipitate (AgCl).
5. a. A side reaction occurs, resulting in the formation of another product. Some of the product is lost when it is transferred to a weighing vessel.
   b. It is unlikely that you would make more product than you calculate theoretically since the theoretical yield assumes that all of one reactant is used up.
   c. If a precipitate is wet when it is weighed, then water is being included in the product mass.
6. The actual nature of the reactants is the formula of the substance when it is prepared; for example, sodium chloride is always found as a solid, neutral salt even though it is completely soluble in water

and sodium ions are often spectator ions. The net ionic equation, however, gives information about species that change in the reaction.

7. Sulfates might not interact with water as strongly as they will interact with the cation of the salt.

Chapter 5:

Fill in the Blank Problems:

1. Dalton's law of partial pressures
2. mean-free-path
3. temperature; Kelvin.
4. Pressure
5. indirectly
6. Charles's
7. repulsions
8. faster
9. mmHg
10. pascal (Pa)
11. 22.4 L
12. vapor pressure
13. 8.314 5 L·atm/J·K

Problems:

1. 1 mmHg = 13.22 mm salt water
2. a. 0.29 atm, 4.4 psi, 230 torr, and 230 mmHg
   b. 6.3 kPa
   c. 9.5 kPa
   d. 34 kPa
3. 2.1 atm
4. 2.6 atm
5. a. 1.3 g He
   b. 0.17 g/L
   c. 1.2 g/L
   d. 420 days
6. a. 0.85 atm
   b. 33 L
7. a. 3.90 atm
   b. 4.13 atm
8. 140 atm
9. 2.8 L
10. a. $Zn + 2HCl \rightarrow ZnCl_2 + H_2$
    b. 767 mmHg
    c. 97.7% Zn
11. Particle size is negligible. Average kinetic energy is proportional to temperature in kelvins. Collisions are perfectly elastic.
12. 576.2 m/s
13. Hydrogen gas will effuse 3.7 times as fast and therefore nitrogen will take 3.7 times as long.
14. a. 2.9 g
    b. 2.9 g
    c. Methane is appropriately described as an ideal gas under these conditions as demonstrated by the fact that both calculations gave the same numerical result.

Concept Questions:

1. a. The pressure inside your tires is greater than atmospheric pressure. You can tell because when a tire pressure gauge is put on the valve stem, air releases spontaneously.
   b. The tires don't collapse or explode because the rubber of the wheels exerts a force against the air pressure.
   c. Underinflated tires are dangerous because there will be pressure against the side of the tires which weakens them. Underinflated tires also get hotter which weakens them.
2. a. The two gases have the same kinetic energy since they are both at the same temperature.

b. The two gases exert the same partial pressure since there are the same numbers of moles of each.
c. Gas A is lighter than gas B which means that if the two gases have the same kinetic energy, gas A moves faster.
d. Gas B contributes more to the average density since it is heaver.

3. As the temperature of the gas increases, it moves faster. The faster the gas moves, the more often it hits the walls of the container. Since pressure is directly related to the number of times the gas hits the walls of the container, the pressure increases as the temperature increases.
4. Carbon tetrachloride has a much larger value of a than carbon dioxide does. This indicates that carbon tetrachloride has stronger intermolecular forces than carbon dioxide does.
Carbon tetrachloride has a b value that is larger than the value for carbon dioxide meaning that carbon tetrachloride has a larger volume correction term than carbon dioxide does.

Chapter 6:

Fill in the Blank Problems:

1. Kinetic energy
2. joule (J)
3. calorie (cal)
4. constant
5. state
6. positive
7. on
8. thermal
9. heat capacity
10. negat ive
11. heat
12. endother mic
13. Hess 's law
14. a pressure of 1 at m
15. p ure substances in their standard state

Problems:

1. a. Heat is negative. Work is negative. The energy change is negative.
b. Heat is zero (unless you count friction). Work is negative. The energy change is negative.
c. Heat is zero. Work is negative. The energy change is negative.
d. Heat is positive. Work is zero. The energy change is positive.
2. a. The reaction is exothermic since the enthalpy change is negative.
b. The temperature in the calorimeter will increase since heat is released in the reaction.
c. The internal energy is higher for the reactants. Energy is released in the reaction so the products will be at lower energy than the reactants are.
3. a. The enthalpy change is -2658 kJ.
b. The internal energy change is -2661 kJ.
c. The work that is done is the butane pushing against the atmosphere.
d. Yes, it is reasonable since the energy and the enthalpy are almost the same and, to three significant figures, the two are equal.
4. a. For C(s): -393.5 kJ/mol
For $CH_4(g)$: -802.5 kJ/mol
For $C_8H_{18}(g)$: -5074 kJ/mol
b. For C(s): -393.5 kJ/mol CO_2
For $CH_4(g)$: -802.5 kJ/mol CO_2
For $C_8H_{18}(g)$: -634.3 kJ/mol CO_2
c. Methane is the most environmentally friendly because it releases the most heat with the smallest amount of carbon dioxide.
5. a. 24.2 J/mol°C
b. 24.5 J/mol°C
c. 25.4 J/mol°C
d. 24.8 J/mol°C
e. 25.3 J/mol°C

f. 25.1 J/mol°C

6. a. 0.897 J/g°C
   b. Aluminum
   c. If the metal was placed in the boiling water when it was transferred, hot water would also be transferred. The mass of the hot substance will be higher than measured and the temperature change will be different due to the high heat capacity of the water.
7. a. The enthalpy change for this reaction should be negative.
   b. ΔH_{rxn}=-285.8 kJ/mol
   c. 4 mol H_2 and 2 mol O_2
   d. 4 mol H_2 and 2 mol O_2
   e. 5 mol H_2 and 2 mol O_2
   f. Driving habits will have a larger effect since the distance traveled in the calculation of part d is three times larger. Even though the heavier car requires more fuel for the same conditions, accelerating more gradually will save more gas.
8. a. $H_2(g) + I_2(g) \rightarrow 2HI(g)$ $\Delta H° = -9.4$ kJ/mol
   b. -9.4 kJ/mol since there is no PV work done in the bomb calorimeter.
   c. 0.57 °C
9. $CH_4(g) + 2O_2(g) \rightarrow CO_2(g) + 2H_2O(g)$ $\Delta H = -803$ kJ/mol
10. $CH_4(g) + 2O_2(g) \rightarrow CO_2(g) + 2H_2O(l)$ $\Delta H = -847$ kJ/mol
11.

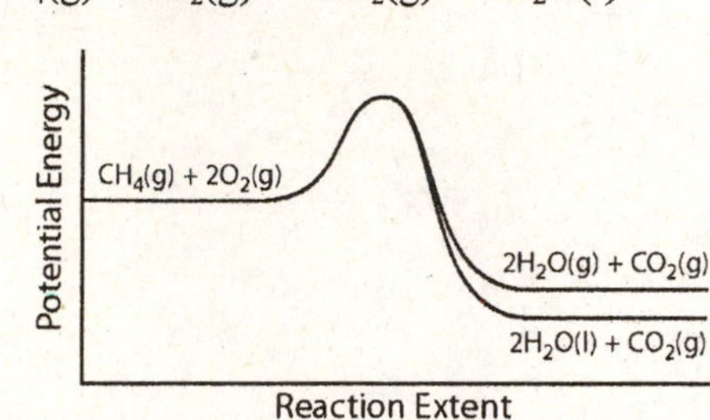

Concept Questions:

1. Gas stove tops are the least efficient: first, the burning of gas heats the air around the pot significantly. Second, some of the energy is lost as work. Electric stoves are more efficient because there is direct contact between the heating element and the pan; there is also not as much energy lost as work. Finally, the induction cooktops are the most efficient because the energy is transferred directly from the stove-top to the pan with no heat being generated except through the pan.
2. A gold pot would be best because gold has the smallest specific heat capacity, meaning that it requires the least amount of heat in order to change the temperature.
3. a. The molecular complexity of all of the metals is approximately the same since they all have approximately the same molar heat capacity value.
   b. This is a result of differences in the molar mass which are specific to a specific metal.
   c. 18.0 cal/mol°C is equal to 75 J/mol°C which, compared to the metals, shows that liquid water is much more complex than the metal samples.
4. a. Liquid water has a higher heat capacity since the temperature of the ice changes more.
   b. The water molecules in the solid will have a larger change in molecular motion assuming that the temperature of the solid water changes more than the liquid water temperature changes.
   c. Heat will be transferred from the ice to the liquid and heat will also be absorbed from the surroundings. As heat is transferred, the molecular motion of the ice increases as expected for all processes.
5. Most processes in the laboratory are carried out open to the environment, which is a constant pressure condition.

Chapter 7:

Fill in the Blank Problems:

1. diffraction
2. magnetic quantum number
3. Electromagnetic radiation
4. frequency

5. node
6. more
7. threshold frequency
8. gamma rays.
9. Constructive interference
10. mo ves to a lower energy level
11. w avelength
12. quant um-mechanical model
13. The Heisenberg u ncertainty principle
14. e mission spectrum
15. comple mentary
16. three ; -1, 0, and +1.
17. vacu um
18. princip al quantum number
19. indeter minate
20. w ave properties
21. proba bility of finding an electron at a given location.
22. quant um numbers
23. ang ular momentum quantum number
24. photoelectric effect
25. decreases
26. energ y difference

<u>Problems:</u>

1. a.

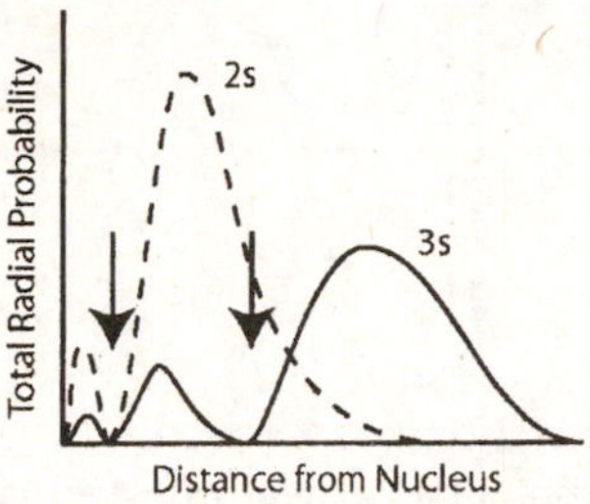

   b. Nodes are indicated with arrows on the figure.

   c. An electron in the 3s orbital can be closer to the nucleus, but it is less likely to be closer.

2. a. 284 nm

   b. 1.06×10^{15} Hz

   c. Since 500 nm light is longer wavelength (lower energy) than the wavelength of light that will eject an electron, no electrons will be ejected from the metal.

   d. When the intensity of the light is increased tenfold, there will be ten times as many electrons ejected with zero kinetic energy.

   e. 7.00×10^{-19} J

   f. The difference in energy between the n=1 and n=4 levels is greater than the energy required to ionize an electron from the n=4 level.

3. 32 electrons total

   $n = 4$; $l = 3, 2, 1, 0$; $m_l = -3, -2, -1, 0, +1, +2, +3$; and $m_s = +1/2, -1/2$

4. a. 3.63×10^{-35} m.

   b. The wavelength is so small that it could never be detected.

   c. Since the transition really comes down to what can be measured, measuring the smallest possible wavelength is the most quantum system that can be detected.

   d. No interference pattern would be observed since the wavelength is too small and the pattern will be too compressed to see anything.

   e. The zero-point energy of a baseball will actually be zero because the ball is too large to measure a smaller amount – we can think of the baseball as a non-quantum mechanical system because of its size.

5. a. 6.45×10^{14} Hz

   b. 257 kJ/mol

c. The rate of photon emission has changed.

6. The wavelength of the photons will be 400 nm and the energy will be 4.97×10^{-19} J.

Concept Questions:

1. The zero of energy for an electron is the energy of the ionized electron.
2. a. Satellites are very large and the uncertainty is very small so it will not have an effect on the measurement of its position.
   b. If an electron were in an orbit, it would follow a trajectory which would allow for precise knowledge of the position and velocity of the electron.
3.

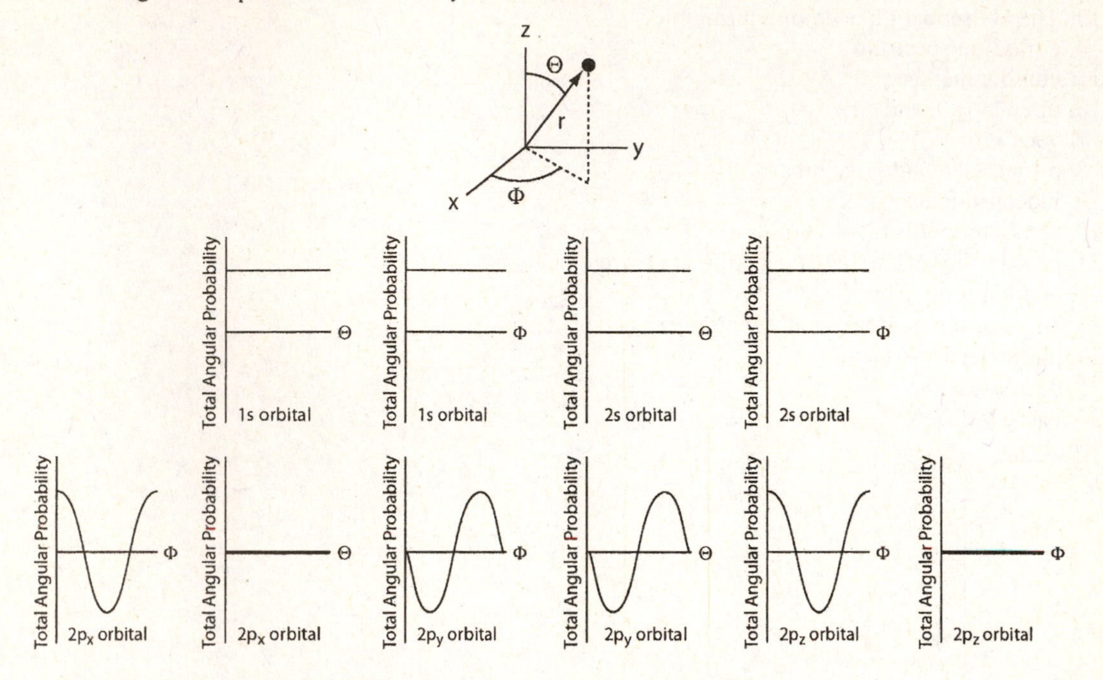

4. a. If I were to turn on a radio, I would be able to pick up a radio station.
   b. Radio waves have very long wavelengths and low energy, which makes them safe for general use.
5. a. When the slits move apart, the bright spots become more closely spaced.
   b. A longer wavelength will result in a pattern with the bright spots spaced further apart.
   c. Making the slits smaller increases the certainty in the position so there will be an increased uncertainty in the momentum in the y direction. The smaller slits will result in a pattern with the bright spots spaced further apart.
6. Once the spacing becomes very small, it will not be detectable. Once the levels are no longer detectable, they become essentially meaningless.
7. The clicks correspond to a single photon since it is emitted from a single decay process. The single photons relate directly to single particles.
8. Electrons must have a smaller wavelength than visible light if using electrons allows one to image smaller things.

Chapter 8:

Fill in the Blank Problems:

1. ground state
2. Cations
3. paramagnetic
4. decreases; increases
5. n-1
6. always larger
7. adding an electron
8. spin up; spin down.
9. quantum numbers
10. shieldi ng

11. spins
12. [X]ns ^{1}
13. periodic
14. xeno n; krypton
15. average bondin g radii
16. ionization e nergy
17. Valence electron s
18. not bonded to another ato m
19. decreases ; increases
20. energ y

<u>Problems:</u>

1. a. B: $1s^2 2s^2 2p^1$
   O: $1s^2 2s^2 2p^4$
   F: $1s^2 2s^2 2p^5$
   He: $1s^2$
   Na: $1s^2 2s^2 2p^6 3s^1$
   K: $1s^2 2s^2 2p^6 3s^2 3p^6 4s^1$
   Rb: $1s^2 2s^2 2p^6 3s^2 3p^6 4s^2 3d^{10} 4p^6 5s^1$

   b. B: 3
   O: 6
   F: 7
   He: 2
   Na: 1
   K: 1
   Rb: 1

   c. B, O, F, Na, K, Rb
   d. Rb, K, Na, He, B, O, F
   e. He, F, O, B, Na, K, Rb
   f. Rb, K, Na, B, O, F, He
   g. F, O, B, Na, K, Rb

2. a. P: [Ne]$3s^2 3p^3$
   [↑↓] [↑↓] [↑↓|↑↓|↑↓] [↑↓] [↓|↓|↓]
   1s 2s 2p 3s 3p

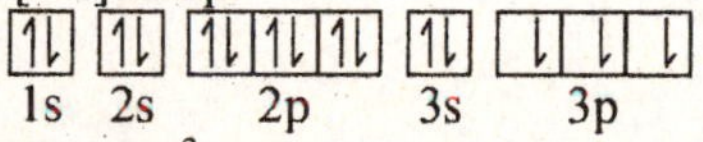

   Ca: [Ar]$4s^2$
   [↑↓] [↑↓] [↑↓|↑↓|↑↓] [↑↓] [↑↓|↑↓|↑↓] [↑↓]
   1s 2s 2p 3s 3p 4s

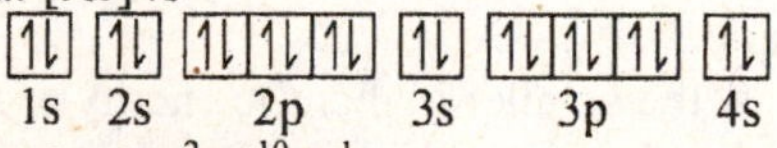

   Ga: [Ar]$4s^2 3d^{10} 4s^1$
   [↑↓] [↑↓] [↑↓|↑↓|↑↓] [↑↓] [↑↓|↑↓|↑↓] [↑↓] [↑↓|↑↓|↑↓|↑↓|↑↓] [↑| |]
   1s 2s 2p 3s 3p 4s 3d 4p

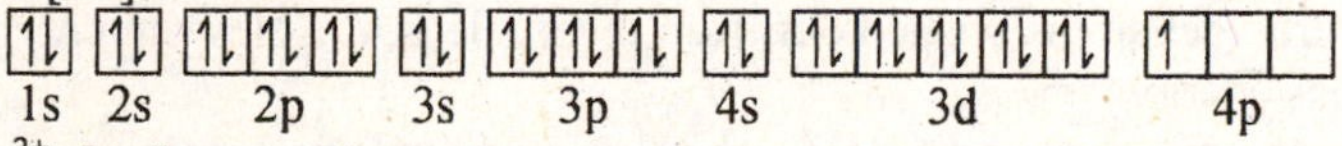

   Ca^{2+}: [Ar]
   [↑↓] [↑↓] [↑↓|↑↓|↑↓] [↑↓] [↑↓|↑↓|↑↓]
   1s 2s 2p 3s 3p

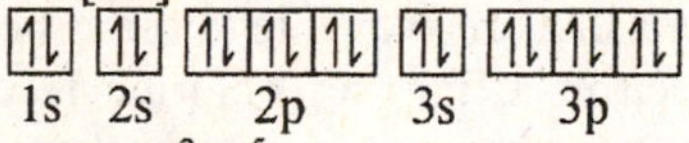

   Cl: [Ne]$3s^2 3p^5$
   [↑↓] [↑↓] [↑↓|↑↓|↑↓] [↑↓] [↑↓|↑↓|↑]
   1s 2s 2p 3s 3p

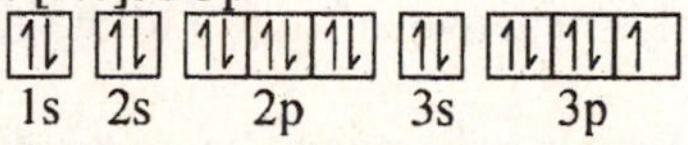

   b. P: 5
   Ca: 2
   Ga: 3
   Ca^{2+}: 0 or 8
   Cl: 7

   c. P, Ga, Cl

   d. P: 5
   Ca: 2
   Ga: 3
   Ca^{2+}: N/A
   Cl: 7

e. Ca^{2+}, Cl, P, Ga, Ca
f. Ca^{2+}, Cl, P, Ga, Ca
g. Ca^{2+}, Cl, P, Ga, Ca
h. The smaller the atom/ion is, the closer the electrons are to the positively charged nucleus. According to Coulomb's law, the closer the electrons are to the nucleus, the more energy will be required to remove the electron.
i. Ca^{2+} will have the highest ionization energy since there are an excess of protons in the nucleus, which pulls the electron density closer to the nucleus.

3. a. S^{2-}: $1s^2 2s^2 2p^6 3s^2 3p^6$
   Cl^-: $1s^2 2s^2 2p^6 3s^2 3p^6$
   Ar: $1s^2 2s^2 2p^6 3s^2 3p^6$
   K^+: $1s^2 2s^2 2p^6 3s^2 3p^6$
   Ca^{2+}: $1s^2 2s^2 2p^6 3s^2 3p^6$
   b. Ca^{2+}, K^+, Ar, Cl^-, S^{2-}
   c. Ca^{2+}, K^+, Ar, Cl^-, S^{2-}
   d. S^{2-}, Cl^-, Ar, K^+, Ca^{2+}
   e. These species are expected to have very similar chemical reactivity since they have the same number of valence electrons (the same electron configuration).
4. a. $1s^2 2s^2 2p^6 3s^2 3p^6 4s^2 3d^{10}$
   b. Zn will be diamagnetic since all of its electrons are paired in the ground state.
   c. Zinc will form a +2 ion since it will lose electrons from the 4s orbital.
   d. $n=3$, $l=2$, $m_l=+2$, $m_s=+½$ (m_l and m_s can have multiple values)
   e. $1s^2 2s^2 2p^6 3s^2 3p^6 4s^2 3d^9 4p^1$
   f. The ionization energy of the ground state will be higher than the ionization energy from the excited state. In the excited state configuration, the highest energy electron is further away from the nucleus (lower force is felt by the electron) and the energy of the electron is closer to zero (the ionized electron).
5. The electron affinity of Li^+ is -520 kJ/mol.
6. The larger the atom is, the further the electrons are from the nucleus and the more they are able to interact with other species. Kr and Xe are large noble gases and therefore can interact with other atoms to form compounds.

Concept Questions:

1. See Figure 7.25 for plots
   a. The higher value of n, the further away from the nucleus the electron density is, meaning that the atom is larger when higher energy orbitals are occupied.
   b. The further away from the nucleus that an electron is, the smaller/higher the energy of the electron will be (Coulomb's law). The higher the energy of the electron, the closer it is to the ionized state (zero energy). So, as electron density is pushed out from the nucleus, the electron energy increases and ionization requires less energy.
2. Transition metals have electrons in the ns and (n-1)d orbitals which are very close in energy. When electrons are removed, there are multiple stable configurations that are available to the system.
3. a. Ionization energy for sodium is positive, but the electron affinity of chlorine is negative. In addition, the positive and negative ions form a crystal structure in which there are strong attractions between the ions which contributes favorably (negative enthalpy) to the formation of sodium chloride.
   b. The magnesium ion will have to lose two electrons which will require more energy, but two chlorine ions will form and the interaction energy between the +2 ions and the -1 ions will be stronger (Coulomb's law). The formation of $MgCl_2$ is, therefore, more exothermic.
4. a. The distance between the valence electrons and the nucleus changes more dramatically from n=1 to n=2.
   b. The d orbital is half full for Mn so there is a jump.
   c. As you get further from the nucleus, the stability associated with a half-full shell becomes less important.
5. a. They will be the same in that there will be discrete energies associated with transitions from one level to another, but they will have many more lines and those lines will be much closer together.

b. The difference in energy becomes smaller because the additional electron/proton causes a smaller relative energy change. For example, going from +1 to +2 is a doubling of the Coulombic force while going from +3 to +4 is a smaller change.

6. a. If the spins were randomly oriented, then there would be an increase in energy associated with the opposing magnetic fields (associated with the opposing spins).

   b. Two like charges, which repel each other, are in close proximity and electrons in the same orbital more effectively shield one another.

   c. Since the s and d orbitals are close in energy, it is possible that the pairing energy is larger than the difference between the orbital energies (chromium has half-filled s and d orbitals). Also, as the electrons fill the orbitals, their energies shift and can actually switch order (copper has a filled d orbital and half-filled s orbital).

Chapter 9:

Fill in the Blank Problems:

1. positive; decreases
2. electronegativity
3. octet (or duet in the case of He)
4. ionic
5. ionic
6. bond
7. Lewis structures
8. bonding pair
9. lattice energy
10. for mal charge
11. Born -Haber cycle
12. bond energ y
13. bond lengt h
14. polar covalent
15. polar
16. Fluorine
17. Radicals
18. t hird
19. decreases
20. triple bond

Problems:

1. a. $BaCO_3$

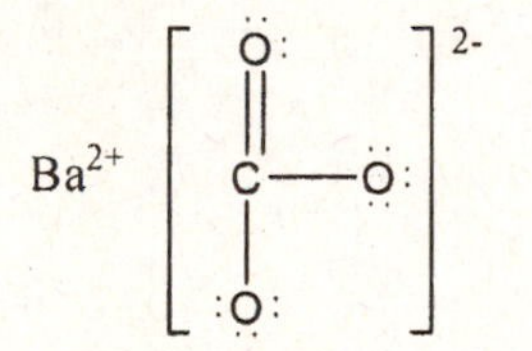

The bond between the barium ion and the carbonate ion is ionic. The bonds between carbon and oxygen atoms are covalent.

Formal Charges: Ba: 2+, C: 0, double bonded O: 0, and single bonded O: -1

b. KNO_3

K^+ [O=N(–O)–O]$^-$

The bond between the potassium ion and the nitrate ion is ionic. The bonds between nitrogen and oxygen atoms are covalent.

Formal Charges: K: +1, N: +1, double bonded O: 0, and single bonded O: -1

c. Na_2SO_4

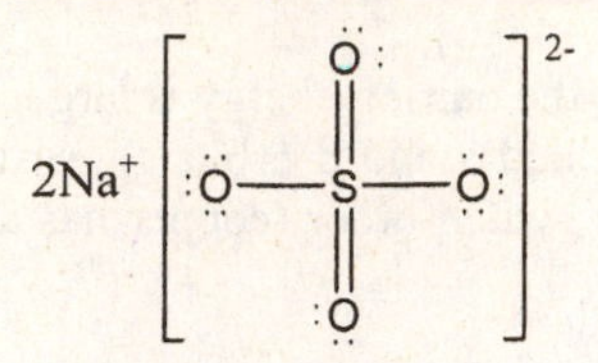

The bonds between the sodium ions and the sulfate ion are ionic. The bonds between sulfur and oxygen atoms are covalent.

Formal Charges: Na: +1, S: 0, double bonded O: 0, and single bonded O: -1

d. LiIO

Li^+ [: I—O:]$^-$

The bond between the lithium ion and the hypoiodite ion is ionic. The bond between iodine and oxygen is covalent.

Formal Charges: Li: +1, I: 0, O: -1

e. $Mg(ClO_3)_2$

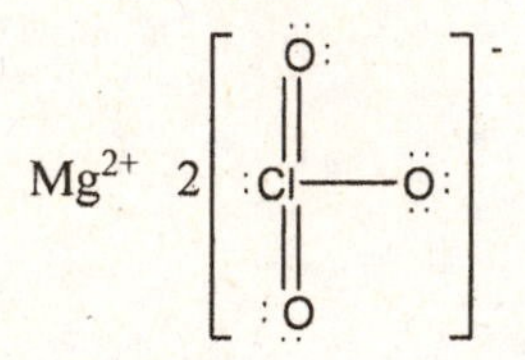

The bonds between the magnesium ion and the chlorate ion are ionic. The bonds between chlorine and oxygen atoms are covalent.

Formal Charges: Mg: +2, Cl: 0, double bonded O: 0, single bonded O: -1

2. a. ΔEN C-O: 1.0, dipole moment ~0.88, ~25%
   b. ΔEN N-O: 0.5, dipole moment ~0.44, ~12%
   c. ΔEN S-O: 1.0, dipole moment ~0.88, ~25%
   d. ΔEN I-O: 1.0, dipole moment ~0.88, ~25%
   e. ΔEN Cl-O: 0.5, dipole moment ~0.44, ~12%

3. a. $[C(=O)(O)(O)]^{2-} \longleftrightarrow [C(O)(=O)(O)]^{2-} \longleftrightarrow [C(O)(O)(=O)]^{2-}$

   b. $[N(=O)(O)(O)]^{-} \longleftrightarrow [N(O)(=O)(O)]^{-} \longleftrightarrow [N(O)(O)(=O)]^{-}$

c.

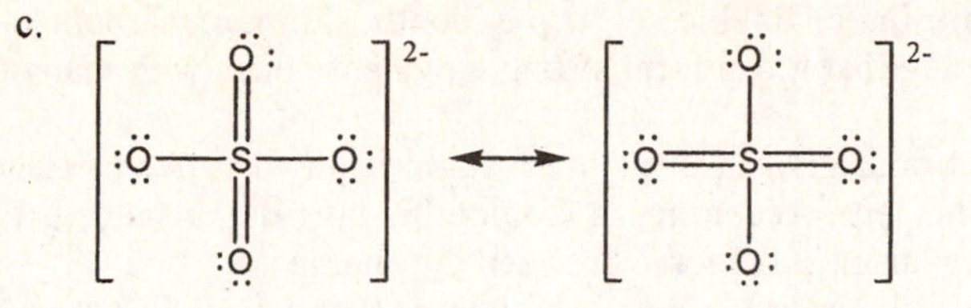

d. No resonance structures

e.

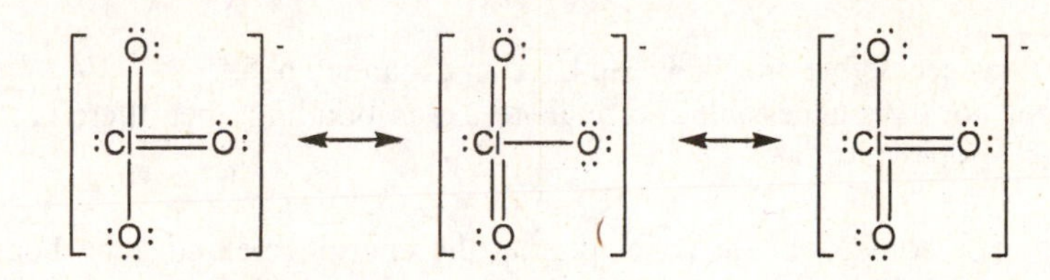

4. a. $\Delta H = -142$ kJ/mol; this reaction is exothermic
   b. $\Delta H = -170$ kJ/mol; this reaction is exothermic
   c. $\Delta H = 0$ kJ/mol; this reaction is neither endothermic nor exothermic
5. a. $\Delta H = +64.4$ kJ/mol; the difference in values is a result of the fact that the phase of carbonic acid in the book is aqueous whereas the phase in the reaction from problem 4 is gas.
   b. $\Delta H = -164.7$ kJ/mol; the values are quite close, the small difference is negligible
   c. $\Delta H = 39.3$ kJ/mol; the values are quite close, the small difference is negligible
6. a. C = 0 and H = 0
   b.

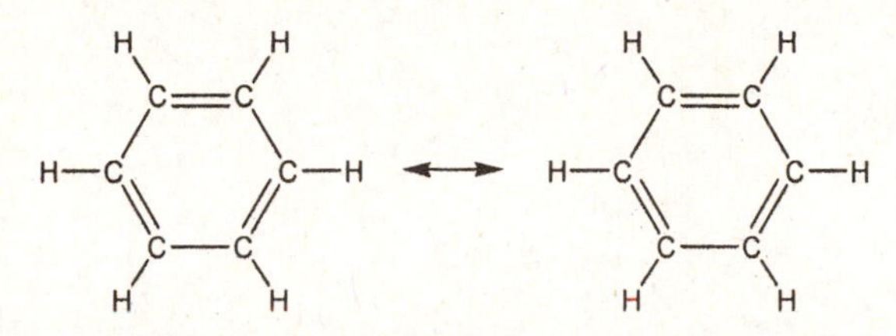

   c. The bond length should be between 154 pm and 134 pm: ~144 pm since the bonds are not double bonds or single bonds, but an average of the two.
7. a. -3836 kJ/mol
   b. 692 kJ/mol
   d. -606 kJ/mol

Concept Questions:

1. a.

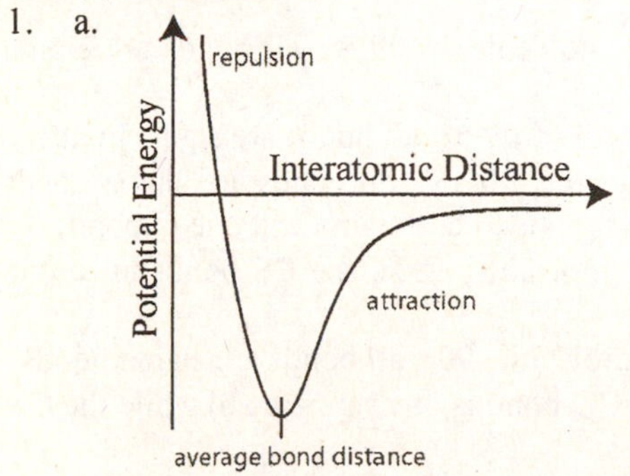

   d. As the distance between the atoms gets very large, they have a smaller effect on one another and the energy between them goes to zero.
2. Looking at the values in the book we see that the distance is not much different across a row, so the large change in charge has a much larger impact.

3. When we look at metals and nonmetals we see that they have large electronegativity differences, so both definitions apply.
4. Carbon forms four bonds and can form single, double, or triple bonds. Further, carbon has an intermediate electronegativity value, meaning that it can form strong, covalent bonds with many other elements.
5. The smaller the atom is, the larger the electronegativity value since a small atom will have its electrons relatively close to the positively charged nucleus. According to Coulomb's law, the distance between oppositely charged species is a determining factor in the force between the charges.
6. Central atoms tend to share their electrons with multiple atoms. An element that is very electronegative will not be as likely to share electrons with other atoms as an element that has a low electronegativity.
7. Boron only has three valence electrons, so it only has three electrons that can be shared to form three bonds.
8. a. Any element that has accessible d orbitals can have an expanded octet.
   b. If an element does not have accessible d orbitals to use in bonding, then there is not any "place" for the extra electrons to be.
9. $2H_2 + O_2 \rightarrow 2H_2O$
   Since the reaction is exothermic, we can conclude that the energy released upon bond formation is greater than the energy needed to break the reactant bonds. We see that we must break two H-H bonds and one O-O double bond while making four H-O bonds. Since the oxygen molecule has a strong oxygen-oxygen double bond, we can conclude that the H-O bond is much stronger than the H-H bond.

Chapter 10:

Fill in the Blank Problems:

1. sp^2; three; one
2. a radical
3. five
4. trigonal planar
5. electrostatic
6. uneven
7. head-to-head; side-to-side
8. destructive
9. equals
10. spread
11. <109.5 °

Problems:

1. a. tetrahedral, trigonal pyramidal, sp^3, polar NH bonds, polar molecule, <109.5°, all bonds are sigma bonds.
   b. Linear, linear, sp, polar bonds, polar molecule, 180°, the C-H bond is a sigma bond and the CN bond is one sigma and two pi bonds.
   c. Tetrahedral, trigonal pyramidal, sp^3, polar bonds, polar molecule, <109.5°, all bonds are sigma bonds.
   d. Tetrahedral, tetrahedral, sp^3, polar bonds, non-polar molecule, 109.5°, all bonds are sigma bonds.
   e. Tetrahedral, tetrahedral, sp^3, polar bonds, non-polar molecule, 109.5°, all bonds are sigma bonds (the actual structure has one PO double bond which corresponds to one sigma and one pi bond).
   f. Linear, linear, C: sp and S: sp^2, polar bonds, non-polar molecule, 180°, the CS bonds have one sigma and one pi bond each.
   g. Octahedral, square planar, sp^3d^2, polar bonds, non-polar molecule, 90°, all bonds are sigma bonds.
   h. Linear, linear, sp, polar bonds, polar molecule, 180°, the CO bond is one sigma bond while the CN bond is one sigma bond and two pi bonds.
   i. Tetrahedral, tetrahedral, sp^3, polar bonds, polar molecule, 109.5°, the SCl bonds are one sigma bond and the SO bonds are one sigma and one pi bond.
   j. Linear, linear, sp, polar bonds, polar molecule, 180°, the SC bond and the CN bond each have one sigma and one pi bond.
   k. C: tetrahedral and O: tetrahedral, C: tetrahedral and O: bent, polar bonds, polar molecule, HCH: 109.5°, COH: <109.5°, all bonds are sigma bonds.

l. C: tetrahedral and O: tetrahedral, C: tetrahedral and O: bent, polar bonds, polar molecule, HCH: 109.5°, COC: <109.5°, all bonds are sigma bonds.

m. Trigonal bipyramidal, t-shaped, polar bonds, polar molecule, 90°, all bonds are sigma bonds.

2. a.

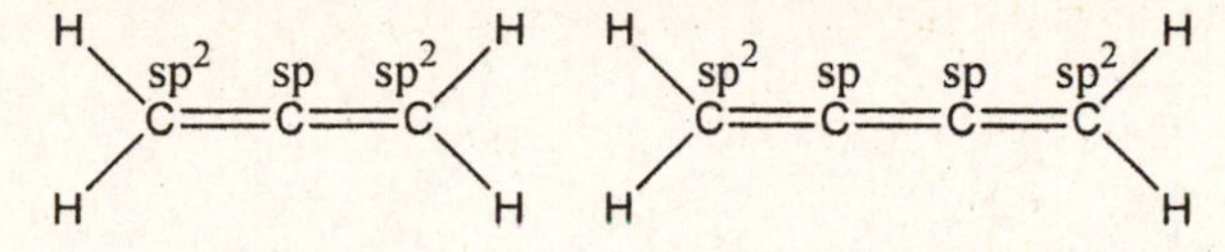

b. CH_2CCH_2 has six sigma bonds and two pi bonds. CH_2CCCH_2 has seven sigma bonds and three pi bonds.

c.

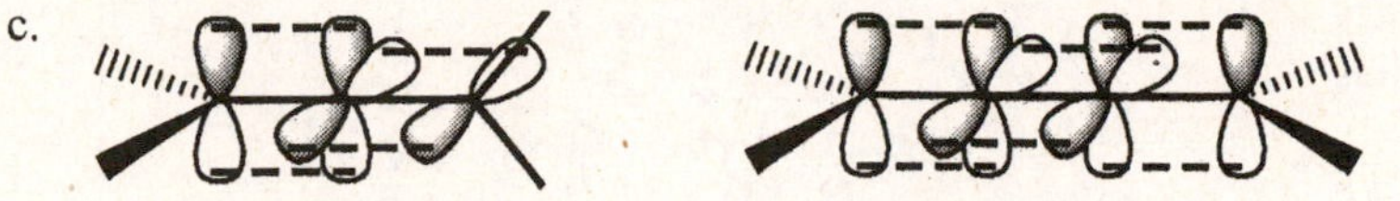

d. As can be seen in the drawings, the CH_2CCH_2 molecule will have CH_2 groups that are perpendicular to one another while the CH_2CCCH_2 molecule will have CH_2 groups that are parallel.

3.

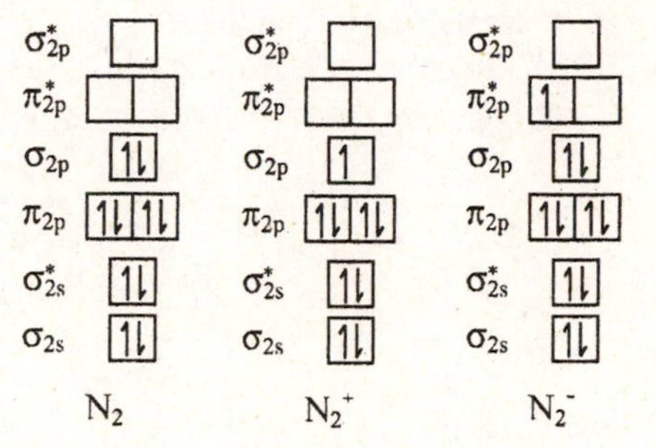

a. N_2 is the most stable.

b. N_2^+ and N_2^- are both equally stable and are less stable than N_2.

c. N_2 - 3, N_2^+ - 2½, N_2^- - 2½

d. N_2^+ and N_2^- are both paramagnetic.

4. a. CO has the same electron configuration (same arrangement of electrons in the molecular orbitals) as N_2 does.

b. I would expect N_2 to have a stronger bond since the orbitals will overlap more fully.

c. The CO bond energy is 1072 kJ/mol and the N_2 bond energy is 942 kJ/mol. The CO bond is polar, which results in a stronger bond.

5.

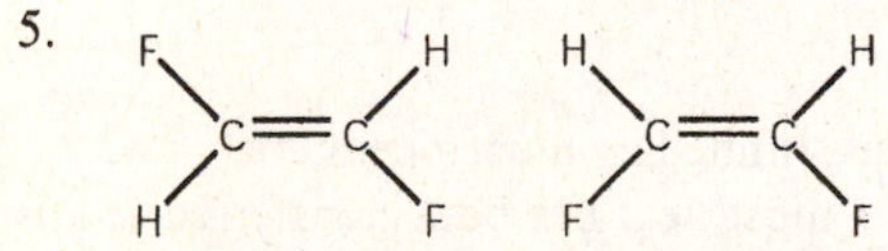

a. The geometry of both carbon atoms is trigonal planar.

b. The hybridization of each carbon atom is sp^2.

c. The first molecule shown will not have a net dipole while the second molecule shown will have a dipole moment. The difference is due to the difference in symmetry.

Concept Questions:

1. Delocalized electrons have more space over which they can be, which leads to a lower energy configuration. In pi bonding, the electrons will be delocalized and will therefore be lower in energy.

2.

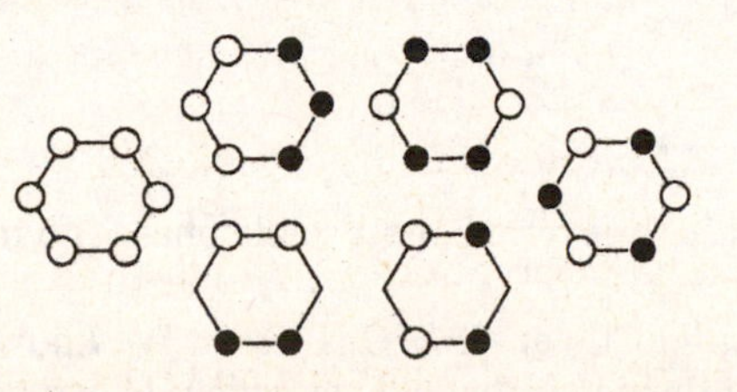

Chapter 11:

Fill in the Blank Problems:

1. increased
2. critical point
3. heat of fusion
4. decreases N, F, and O.
5. band gap
6. instantaneous dipole
7. surface tension
8. capillary action
9. network covalent atomic solids
10. perma nent dipoles
11. packin g efficiency
12. positive ; negative
13. bo iling temperature
14. subli mation
15. greater
16. cry stalline
17. increase
18. ionic solids; high

Problems:

1. a. Dispersion
   b. Dispersion
   c. Ion-Dipole
   d. Dipole-Dipole
   e. Dipole-Dipole & H-bonding
   f. Dispersion
   g. Dipole
   h. Ion-Dipole
2. a. $NaCl< NH_3<NF_3< He$
   b. $NH_3<NF_3<NCl_3<NBr_3$
   c. $HF<ClF<F_2$
   d. $CH_3CH_2CH_2CH_2CH_2CH_2CH_2CH_3< CH_3CH_2CH_2CH_2CH_3<CH_3CH_2CH_3$
3. a. $AlCl_3<NaBr<NaCl< MgS$
   b. $CO<CO_2<CCl_4$
   c. $O_2<CO_2<SO_2<O_2$
   d. $SO_2< P_2O_3< SiO_2< Al_2O_3$
4. a. The gas at 123°C has the largest kinetic energy, which is determined by the temperature.
   b. The gas at 123°C has the largest internal energy since the most heat has been transferred at this point without any input or output of work.
   c.

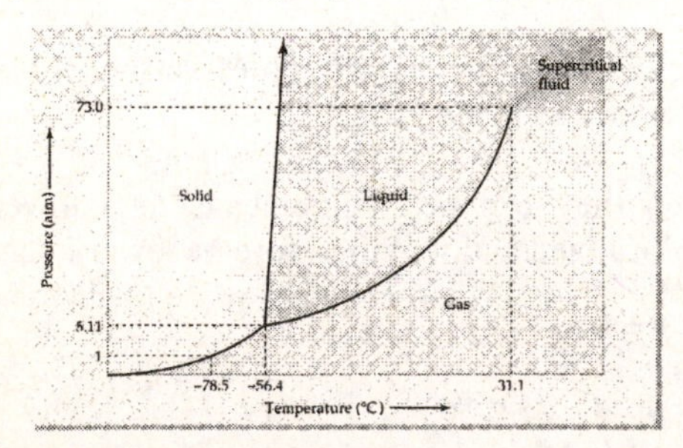

   d. At -60°C, the liquid phase is never the most stable phase (no matter the pressure) as can be seen in the phase diagram.
   e. The vapor pressure of solid CO_2 at -78.5°C is 1 atm. We know this because this temperature is the normal sublimation point which is the point at which the vapor pressure of the solid is equal to the normal pressure (1 atm).

f. No, the sample has changed to a gas.

g. In this case, the sample will go from solid to liquid instead of going from a solid to a gas.

5. a. NH_3 is polar and therefore has dipole-dipole interactions which cause the pressure to be less than the ideal gas law predicts.

   b. The gas with the highest boiling point is the gas with the strongest intermolecular forces which, in this case, is ammonia.

   c. H_2 will have the lowest critical temperature since it is an ideal gas at the temperature used in the tabulated data.

6. a.

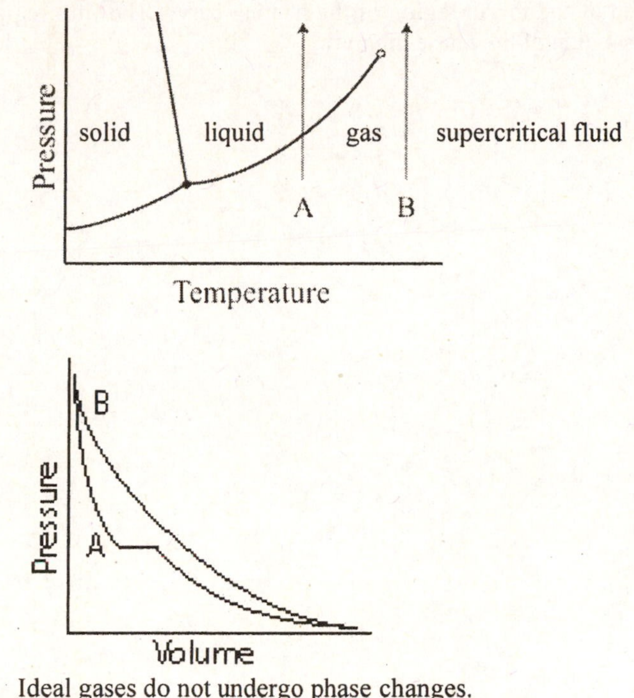

   c. Ideal gases do not undergo phase changes.

7. a. 24.2 kJ/mol

   b.

   c. 21.1 kJ

8. Simple cubic: 48% empty — Example: Polonium — Density: 9.32 g/mL
   Body-centered cubic: 32% empty — Example: Sodium — Density: 0.968 g/mL
   Face-centered cubic: 26% empty — Example: Silver — Density: 10.5 g/mL
   The more empty space there is, the lower the density will be if the mass of the atom is kept constant. Since polonium is the most massive (209 g/mol), it is more dense than sodium (with the lowest molar mass).

9. a. This substance is expected to be a conductor since the higher energy band has electrons in it.

   b. If all of the electrons are removed from the conduction band, the material will become a semiconductor since a small amount of energy will be required in order for electrons to move from the valence band to the conduction band.

   c. The conductivity would increase with increasing temperature since energy will be supplied to surmount the band gap.

Concept Questions:

1. a. The pentane beaker will empty out first since the intermolecular forces between pentane molecules are weaker than the intermolecular forces between pentanol molecules.

   b. Since the pentane has a higher vapor pressure due to the weaker intermolecular forces, the concentration of vapor over pentane will be higher.

   c. Due to the weaker intermolecular forces, pentane will boil at a lower temperature.

2. Covalent bonds are too strong and DNA must be able to separate in replication processes. Dispersion forces are too weak and DNA could come apart and be susceptible to damage.

3. A higher heat of vaporization corresponds to stronger intermolecular forces. When vaporization occurs, the intermolecular interactions are effectively destroyed, so the stronger the interaction, the more energy is required to break that interaction.
4. A sodium chloride solution should have a higher boiling point since the interactions between the ions of sodium chloride and the polar water molecules are very strong.
5. The first region of the heating curve corresponds to the solid region of the phase diagram. The line between the solid and liquid phases is represented by the second region of the heating curve. The third region of the heating curve corresponds to the liquid region of the phase diagram. The line between the solid and liquid phases is represented by the fourth region of the heating curve. The fifth region of the heating curve corresponds to the gas region of the phase diagram.

Chapter 12: Solutions

Fill in the Blank Problems:

1. miscible
2. Henry's law constant
3. aqueous
4. Osmotic pressure
5. Tyndall effect
6. higher
7. Entropy
8. less
9. saturated
10. ho mogeneous
11. ideal solution
12. negat ive manner
13. heat of sol ution
14. van 't Hoff factor
15. decrease ; increase
16. mo lality

Problems:

1. a. Since this is ionic, it will dissolve in water.
   b. Since CO_2 is non-polar, it does not dissolve significantly in water. A non-polar solvent would be more suitable.
   c. Since hexane is non-polar, it does not dissolve significantly in water. A non-polar solvent would be more suitable.
   d. Lithium hydroxide is ionic, so it will dissolve in water.
   e. The nitrate ion is charged and should therefore dissolve in water.
2. a. The original solution was unsaturated since more ammonium chloride dissolved. The solution after the addition is saturated since there is solid present.
   b. The heat of hydration must have a smaller absolute value than the lattice energy.
   c. Ammonium ions are larger than sodium ions are so the charges are farther apart.
3. 0.33 M
4. a. When the ethanol is added to the water, the hydrogen bonding network is disrupted and so the packing of the water molecules is less efficient.
   b. I do not expect an ethanol/water solution to behave ideally since the intermolecular interactions will be different. It will deviate in a positive fashion since the interactions between the ethanol and water will be weaker than the interactions between water molecules.
   c. 3.4 M
   d. 4.3 *m*
   e. 16%
   f. 0.072
   g. 0.93
5. a. 0.22 atm
   b. 0.0306 atm

c. I expect that the actual value will be less than the value calculated since ammonia and water should have strong interactions.

6. a. P = 58.8 torr
   Freezing point = -114.5°C
   Boiling point = 78.5°C
   Osmotic pressure = 3.60 atm
   b. P = 55.6 torr
   Freezing point = -118.7°C
   Boiling point = 81.1°C
   Osmotic pressure = 43.00 atm
   c. P = 59.0 torr
   Freezing point = -114.1°C
   Boiling point = 78.3°C
   Osmotic pressure = 0.103 atm
   d. P = 56.7 torr
   Freezing point = -115.8°C
   Boiling point = 79.4°C
   Osmotic pressure = 17.0 atm
   e. P = 25.8 torr
   Freezing point = -169.8°C
   Boiling point = 112.4°C
   Osmotic pressure = 540.5 atm
   NOTE: The values calculated for this problem are unrealistic since the methanol solution has a mole fraction that is greater than 50%.
7. 52 g of NaCl per kg of water, which is a large number. Based on the size of this number, we can assume that salt is added to water in order to flavor food, not to change the temperature of the water.

Concept Questions:

1. Solids do not have particles that move freely and therefore they will not mix completely on a molecular level with gas particles.
2. Once there are more than six carbon atoms, the dispersion forces become more important than the potential hydrogen bonding that can occur with the OH group.
3. As the temperature increases, the gas particles move faster. These fast-moving particles access the surface of the liquid more often, where they can escape into the surroundings. The fast-moving particles also have more kinetic energy which allows them to overcome the intermolecular forces holding them in the solution phase.
4. a. Carbon tetrachloride will have a larger freezing point change.
   b. The intermolecular forces will affect the values. The stronger the intermolecular forces of the solvent, the more the freezing point will be altered upon the addition of a solute.
   c. The heat of vaporization is much larger than the heat of fusion so there is a smaller impact on the boiling point than there is on the freezing point.
5. a. If the solvent-solvent interactions are about the same as the solvent-solute interactions, then when the mixing occurs, there will be no real contraction (stronger interactions) or expansion (repulsions dominate).
   b. No, this is not always a valid assumption. A solution that has a negative deviation of vapor pressure is one that has strong solute-solvent interactions and we could expect that the solution volume will deviate from the additive volume.
   c. The solvent-solute interactions will be strong.
6. a. I would expect more NaCl to dissolve in the pentanol because the OH group of pentanol can interact with the sodium and chloride ions.
   b. The boiling point changes because the vapor pressure changes. A lower vapor pressure results in a higher boiling point.
   c. The boiling point changes by a different amount for two reasons – first, the two solvents have different boiling point elevation constants and second, the NaCl will dissociate to a different extent. Based on the dissolution of the NaCl, we can expect that the pentanol solution will have a higher boiling point because there will be more dissolved particles.

7. This process is not favorable since the sodium ions are going from low to high concentration or from a more disordered state (with mixing of solvent and sodium) to a more ordered state (with less mixing).
8. a. A water-ethanol solution will deviate in a positive fashion because the hydrogen bonding network of the water will be disrupted with the introduction of the ethanol.
   b. It will be less than the ideal solution because the vapor pressure will be higher.
   c. The nitric acid will deviate in a negative fashion because it will dissociate into ions.
   d. It will be higher than the ideal solution because the vapor pressure will be lower.
   e. There will be a different change in the boiling point elevation because of the dissociation of the acid and because of the differences between the interactions between solvent and solute.
9. Soap molecules align themselves in a sphere with the non-polar side oriented inward where they can trap the dirt. The nonpolar sides of the molecules are on the outside of the sphere where they allow for the soap sphere to be dissolved in water.

Chapter 13:

Fill in the Blank Problems:

8. rate law
9. first-
10. slo w step
11. doubles
12. tange nt
13. proven true
14. catal yst
15. ti me
16. inter mediate
17. Che mical
18. decreases
19. increases
20. reaction order
21. ele mentary step
22. activatio n energy
23. Enz ymes
24. p ; z,
25. inver se concentration
26. mo larity
27. mo lecularity
28. prop ortional to
29. m echanism
30. kine tic energy of the sample
31. activatio n energy divided by R
32. decrease

Problems:

1. Rate = (0.41 M-2s-1)[H2][NO]2
2. Mechanism 3 is the best choice. Mechanism 1 has a different rate law and mechanism 2 includes a termolecular step which is not as plausible as bimolecular steps such as those given in mechanism 3.
3. rate = k$[H_2][I_2]$

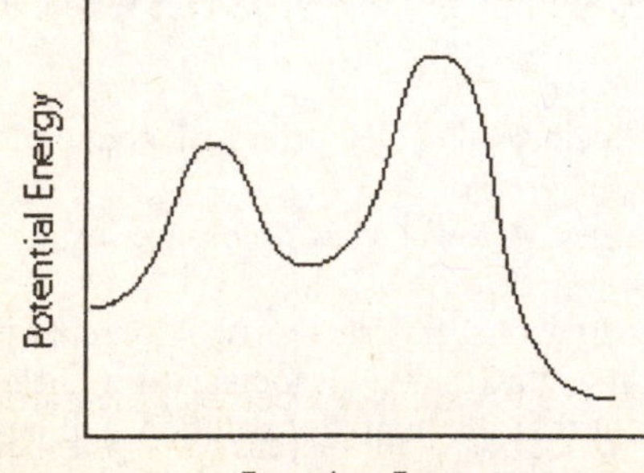

4.

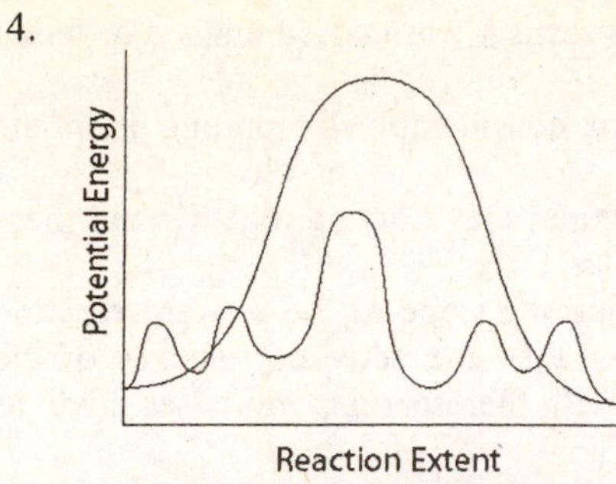

5. a. [phenol acetate] vs. time

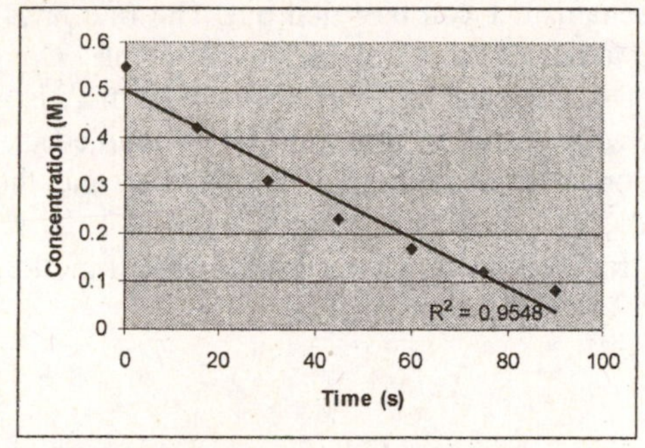

b. 1/[phenol acetate] vs. time

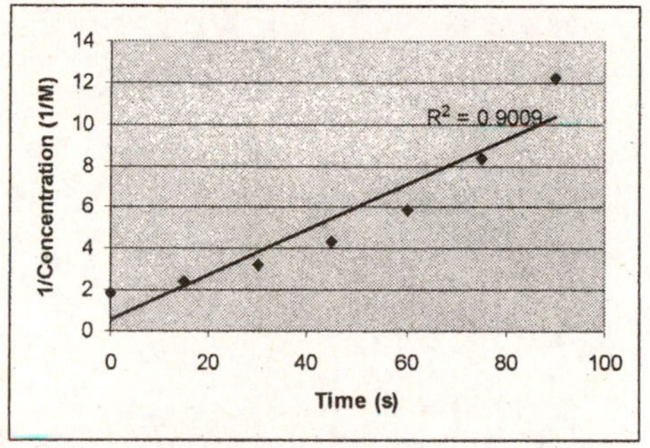

c. ln[phenol acetate] vs. time

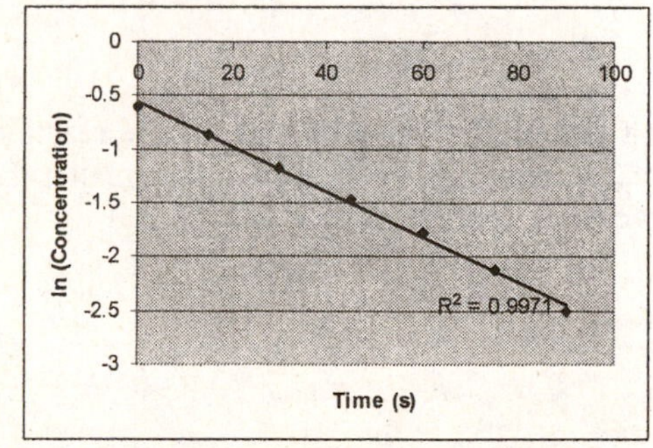

The plot of the natural log of concentration versus time is most linear. This corresponds to a first order rate law.

6. a. 4.9×10^{-9}
   b. 2.86
   c. 497°C
7. 1.1×10^{11} atoms
8. 2.6 atm

Concept Questions:

1. An increase in temperature will result in an increase in the number of collisions because the particles will move faster; it will have no impact on the geometry criteria; it will increase the number of particles with sufficient energy to overcome the reaction barrier.

2. Experiments 2 and 3 have a different initial concentration. Experiments 1 was carried out at a different temperature from experiments 2 and 3.
3. The glow stick is saved because the reaction is slowed down considerably by putting it in the refrigerator.
4. This reaction must occur in multiple steps because there are nine particles reacting together and there are no known reactions that are more complicated than termolecular.
5. The slope of the reverse reaction has a larger absolute value than the slope of the forward reaction because the activation energy of the reverse reaction is greater than the activation energy of the forward reaction. The rate constant of the reverse reaction will therefore be more sensitive to temperature than the rate of the forward reaction.
6. The particle that removes excess energy does not have any orientational restrictions.
7. If NO and/or NO_3 are observed, then mechanism 1 can be ruled out. The rate laws of mechanisms 2 and 3 are different and therefore can be tested using an initial rates experiment.
8. We can write $[A]^2$ as [A][A] and think of an increase in the concentration of A is an increase in two reactant species so the rate will quadruple when the concentration of A is doubled.
9. Energy is not consumed in the generation of a large supply of enzymes and there is no need to maintain a large supply of them in the body.
10. Substa nce X will have the highest concentration if it decays via second order kinetics.

Chapter 14: Chemical Equilibrium

Fill in the Blank:

1. do not change
2. number of gas particles
3. reactants
4. reaction quotient
5. equal
6. products
7. inversed
8. equilibrium constant
9. minimizes
10. equilibriu m

Problems:

1. $K_c = 1.2\times10^{-25}$
2. The reaction is not at equilibrium, it will proceed in the forward direction in order to reach equilibrium.
3. a. 0.109
   b. 0.95
   c. 0.91
   d. An increase in the pressure will result in a shift toward the side of the reaction with fewer gas particles, toward the reactants in this case. This shift is reflected in the smaller fraction of reactant particles that have decomposed in part c as compared with part b.
4. a. $P_{NH3} = P_{H2S} = 0.332$ atm
   b. 1.73%
   c. $P_{NH3} = 2.490$ atm
   $P_{H2S} = 0.044$ atm
5. a. The reaction will shift toward reactants.
   b. No shift occurs.
   c. The reaction will shift towards products.
   d. No shift occurs.
   e. No shift occurs.
   f. No shift occurs.
   g. The reaction will shift towards reactants.
6. a. $AgHCO_3$, H_2O, Ag^+, HCO_3^-, H_3O^+, and CO_3^{2-}
   b. The addition of hydrochloric acid would increase the quantity of the silver bicarbonate present.
   c. Adding NaOH would result in a decrease in the concentration of H_3O^+.

Concept Questions:

1. We could use isotopes and prepare an equilibrium mixture. If the isotopic distribution equalized itself on both sides of the reaction, equilibrium must be dynamic.
2. The concentrations are not always equal when K=1. The concentrations will only be equal when they are equal to 1; otherwise the concentration of C is equal to the product of the concentrations of A and B.
3. Definitionof equilibrium : $\text{rate}_1 = \text{rate}_{-1}$

$$\text{rate}_1 = k_1[A]^a[B]^b$$

$$\text{rate}_{-1} = k_{-1}[C]^c$$

Combiningthese: $k_1[A]^a[B]^b = k_{-1}[C]^c$

Rearranging give: $\frac{k_1}{k_{-1}} = \frac{[C]^c}{[A]^a[B]^b}$

$K_{eq} = \frac{[C]^c}{[A]^a[B]^b}$ therefore $K_{eq} = \frac{k_1}{k_{-1}}$

4. When the concentration of a reactant or product is changed, the rate of the forward or reverse reaction is changed. The rate constants, however, do not change. So, the rates will change until equilibrium is reestablished and that equilibrium will have a new set of concentrations associated with it, but the ratio of the product concentrations raised to their stoichiometric coefficients to reactant concentrations raised to their stoichiometric coefficients will remain unchanged.
5. Using the Arrhenius plot, we can see that a given temperature change will result in a change of the rate constant that is larger with a larger activation energy (slope of the graph). For an endothermic reaction, the activation energy of the forward reaction is larger than the activation energy of the reverse reaction. Since the equilibrium constant is a ratio of the rate constants, we see that the numerator changes more than the denominator and the equilibrium constant value increases with an increase in the temperature.
6. Let's call the number of particles with sufficient energy to overcome the barrier in the forward direction N_1 and the number of particles with sufficient energy to overcome the barrier in the reverse direction N_2. For an endothermic reaction, the activation energy of the forward reaction is greater than the activation energy of the reverse reaction. As can be seen in the distribution, the value of N_2 will be less affected than the value of N_1 by a given temperature change.

Chapter 15:

Fill in the Blank:

1. polyprotic
2. percent ionization
3. Lewis
4. proton acceptor
5. 1.0×10^{-7} M; 7.00
6. ionizes
7. weaker
8. acid ionization constant
9. stronger
10. basic
11. substa nce that increases $[H^+]$ in solution
12. acid
13. acidic
14. increasi ng
15. a mphoteric

Problems:

1. a. $HCHO_2(aq) + H_2O(l) \rightarrow H_3O^+(aq) + CHO_2^-(aq)$
   b. $HCHO_2/ CHO_2^-$ and H_2O/H_3O^+
   c. 2.20
   d. 11.80

e. 2.9%
f. i. 2.63
ii. 7.1%
g. The preparation of a dilute stock solution as that outlined in the original problem is not useful – it is more useful to use a concentrated stock solution to prepare all dilute solutions.

2. $4.3x10^{-10}$
3. $2.1x10^{-8}$ M
4. 4.4×10^{-5} g
5. pH=5.14
6. 3.44
7. a. $NaHCO_3(aq) + HC_2H_3O_2(aq) \rightarrow NaC_2H_3O_2(aq) + CO_2(g) + H_2O(l)$
   b. $NaHCO_3/H_2CO_3$ and $HC_2H_3O_2/ C_2H_3O_2^-$
   c. Yes, a reaction will occur because there is no product present when these solutions are combined. The value of Q is therefore zero and less than the value of K for the reaction.
8. a. The solution will be basic since the base ionization constant is greater than the acid ionization constant.
   b. Assuming a 1.0 M solution – pH=5.13 & percent ionization = 0.00075%
   c. pH=9.62 and percent ionization = 0.0041%
   d. We can see that the percent ionization of the base is higher than the percent ionization of the acid – this is consistent with the answer given in part a.
   e. $NaHCO_3(aq) + HC_2H_3O_2(aq) \rightarrow NaC_2H_3O_2(aq) + CO_2(g) + H_2O(l)$
   When carbonic acid forms, it decomposes into carbon dioxide and water. The formation of carbon dioxide is observed as fizzing.
9. a. $[OH^-]=1.2x10^{-4}M$ and $[H_3O^+]=8.1x10^{-11}M$
   b. $6.05x10^{-7}$
   c. The base is moderately weak since the base ionization constant is a number to the power of -7.
   d. 0.49%
   e. Yes, the answer in part d is consistent with the answer in part c since the percent ionization is so small.
10. a. $H_2C_6H_6O_6$, $HC_6H_6O_6^-$, $C_6H_6O_6^{2-}$, H_3O^+, H_2O, OH^-
    b. 2.86
11. 21 mL
12. a. $Cr^{3+}(aq) + 7H_2O(l) \rightarrow Cr(H_2O)_5OH^-(aq) + H_3O^+(aq)$
    b. 2.10
13. The oxygen atom in HOCN makes the hydrogen atom more positive and, therefore, more acidic.
14. Ethene can serve as a Lewis acid because the pi orbitals involved in the double bond can be used to accept a lone pair upon rearrangement.

Concept Questions:

1. Water dissociates to form hydronium ions and hydroxide ions which means that it's an acid and a base according to the Arrhenius definition. Water accepts a proton and donates a proton in accordance with the Bronsted-Lowry definition. The transferred proton accepts a lone pair from the water molecule in accordance with the Lewis definition.
2. The percent ionization is larger because the reaction goes further forward, but since there is a smaller amount of the acid, fewer hydronium ions will be produced.
3. 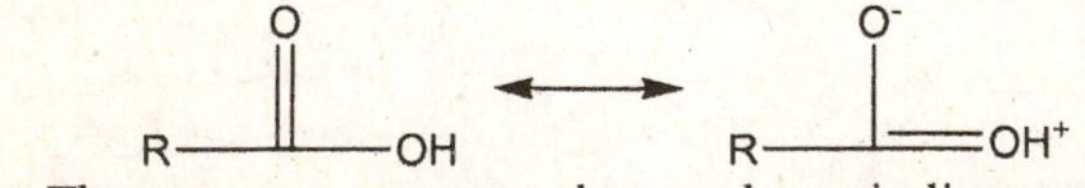

   The resonant structure shown above indicates that the proton is relatively easy to remove (as in an acid).
4. H_3PO_4 is the strongest acid and PO_4^{3-} is the weakest acid. Since the strongest acid has the weakest conjugate base and the weakest acid has the strongest conjugate base, PO_4^{3-} will have the largest K_b value.
5. CO_2 will form carbonic acid when dissolved in water, resulting in an acidic solution.
6. Any other acid will, when dissolved in water, produce hydronium ions. In water, therefore, the other acids don't exist.

7. a. All of these acids dissociate completely to produce hydronium ions and therefore they will all have the same strength.
   b. Using a different solvent will allow the acids to be compared with one another.
   c. HI>HBr>HCl
      $HClO_4$>HCl
      $HClO_4$>H_2SO_4
8. Water contributes 0.0000001 M H^+ - the precision of this number is in the ten millionth place. In the concentration 0.0010 M, the precision is in the ten thousandth place. Since the rules of significant figures say that in addition, the answer has the same precision as the least precise term in the problem, the addition of 0.0000001 M and 0.0010 M will give 0.0010 M.
9. The numbers associated with the hydronium ion concentration for weak acids are very small and must be reported as a factor of ten. The pH scale allows the numbers to be reported in a much simpler form.
10. Amines have a lone pair that can accept a proton.
11. Borax is a base: $Na_2[B_4O_5(OH)_4]\cdot 8H_2O$.

Chapter 16:

Fill in the Blank:

1. pH; added acid or base
2. Lewis acid; Lewis bases
3. its conjugate base
4. solubility product constant.
5. Selective precipitation
6. equivalence
7. increases
8. pK_a
9. molar solubility
10. Henderson -Hasselbach
11. a pH of the $pK_a \pm 1$.
12. a mphoteric
13. polyprotic acid
14. s maller
15. buf fering capacity
16. greater than
17. endpoint
18. a factor of ten
19. for mation constant
20. decreases

Problems:

1. a. 33.1 g $NaC_2H_3O_2$
   b. 4.61
   c. 4.74
   d. Because the molar mass of sodium acetate is greater than the molar mass of acetic acid, adding equal masses of the two will result in a solution that has a greater quantity of acid than base.
   e. 4.75
   f. 14L
2. a. 25.0 mL of HCl
   b. The titration will not have a buffer region because only strong acids and bases will ever be present.
   c. pH=13.16
   d. pH=12.55
   e. pH=7.00
   f. pH=1.39
3. a. As the equivalence point, the pH will be dominated by the conjugate acid of hydrazine and the solution will therefore be acidic with a pH<7.
   b. 4.05mL
   c. 10.75
   d. 10.61

e. 4.46
f. 1.094
g. 0.864
h. 0.749

4. a. 9.246
   b. The pH at half equivalence is approximately equal to the pK_a of the acid (or pK_b of the base).
5. a. pH=2.88
   b. pH=2.48 so ΔpH=-0.40
   c. pH=2.15 so ΔpH=-0.73
   d. The pH in part c is much lower because more acid has been added to the solution. In part c, the effect is essentially to dilute the initial acid solution.
   e. 7.98
   f. Phenol red is the best choice because its pKa is very close to the pH at the equivalence point.
   g.

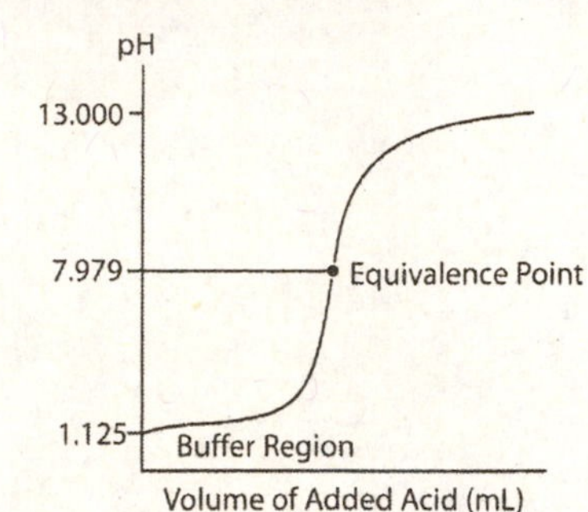

6. a. 6.6×10^{-5} M
   b. 9.8 mg
   c. No, $MgCl_2$ is soluble and the hydronium ion can combine with the carbonate ion to increase the solubility of the magnesium ion.
   d. Yes, $Mg(OH)_2$ is a sparingly soluble salt with a solubility that is two orders of magnitude less than the solubility of magnesium carbonate.
7. Fe^{2+} then Ca^{2+} then Mg^{2+}. The iron carbonate salt has the smallest solubility product constant (it is least soluble) and the magnesium carbonate salt has the largest solubility product constant (it is the most soluble).
8. a. 7.31×10^{-7}M
   b. 1.2×10^{-7} M
   c. 3.57×10^{-11} M
   d. 0.125 M
9. Add HCl to precipitate out $PbCl_2$. Add NaOH to precipitate out Fe^{2+}, Cr^{3+}, and Mg^{2+}. Add $(NH_4)_2HPO_4$ to precipitate out Ca^{2+}. Finally perform a flame test to identify the presence of Na^+.

Concept Questions:

1. They are added in such small amounts that their contribution to the acidity or basicity of the solution is negligible.
2. HInd will be dominant before the equivalence point and Ind^- will be dominant after the equivalence point.
3. The pH at half the equivalence point will be equal to the pK_a of the acid so the pH will be 2.39. The $[OH^-]$ at the equivalence will be equal to the square root of the K_b of the conjugate base multiplied by the concentration of the conjugate base initially present. For a 1M solution, the pOH at the equivalence point is then equal to the negative log of the square root of the K_b value, 5.81 and the pH is 8.19.
4. a. The result is the same whether we allow it to happen in one step or in multiple steps. Splitting it up just allows the mathematical manipulations to be simpler.
   b. No, in actuality the reaction occurs in both directions all at the same time.
   c. The math is more complicated and the calculations will need to be repeated multiple times if we try to mimic reality.

5. The region above the equivalence is not a buffer since it is not representative of a weak acid and conjugate base pair which are needed for a buffer. It looks flat because the pH is dominated by the strong base – as more base is added, the concentration of the acid doesn't really change.

Chapter 17:

Fill in the Blank:

1. free energy
2. decreases
3. non-spontaneous process
4. zero
5. Entropy
6. zero
7. smaller than 1
8. state function
9. maximum work available
10. entha lpy and entropy.
11. increases
12. s maller; the size of the molecule (molar mass)
13. m ultiplied by the same number
14. Thermod ynamics; kinetics
15. negat ive; positive
16. adding tabulated values
17. negat ive
18. distance bet ween standard and equilibrium conditions

Problems:

1. f. $BF_3(l)$ will have the highest entropy. Liquids have larger entropy values than solids and the larger the molar mass the higher the entropy value.
2. $\Delta S>0$ (two aqueous species from one solid species), $\Delta H>0$ (adding heat will increase product formation), and $\Delta G>0$ (this is a sparingly soluble salt – $K<1$).
3. a. ΔS_{vap}(methane) = -21.3 J/°C mol
   ΔS_{vap}(hexane) = -47.0 J/°C mol
   b. intermolecular forces between hexane molecules are greater than the intermolecular forces between methane molecules; as a result, the liquid phase of hexane is more ordered than the liquid phase of methane. When the phase change occurs, a larger change in order occurs with the hexane.
4. a. $2H_2(g) + O_2(g) \rightarrow 2H_2O(l)$ ΔH= -571.6 kJ/mol ΔS=-326.6 J/mol K
   $C(graphite) + O_2(g) \rightarrow CO_2(g)$ ΔH= -393.5 kJ/mol ΔS= 2.9 J/mol K
   $CH_4(g) + 2O_2(g) \rightarrow CO_2(g) + 2H_2O(l)$ ΔH= -890.5 kJ/mol ΔS= -242.9 J/mol K
   b. $2H_2(g) + O_2(g) \rightarrow 2H_2O(l)$ ΔG= -474.2 kJ/mol
   $C(graphite) + O_2(g) \rightarrow CO_2(g)$ ΔG= -394.4 kJ/mol
   $CH_4(g) + 2O_2(g) \rightarrow CO_2(g) + 2H_2O(l)$ ΔG= -818.1 kJ/mol
   c. $2H_2(g) + O_2(g) \rightarrow 2H_2O(l)$ ΔG= -474.2 kJ/mol
   $C(graphite) + O_2(g) \rightarrow CO_2(g)$ ΔG= -394.4 kJ/mol
   $CH_4(g) + 2O_2(g) \rightarrow CO_2(g) + 2H_2O(l)$ ΔG= -818.1 kJ/mol
   d. ΔG= -50.5 kJ/mol
   e. $7.x10^8$
   f. The reaction will be spontaneous at all temperatures.
5. a. $2O(g) \rightarrow O_2(g)$ $\Delta S < 0$
   $H_2O(l) \rightarrow H_2O(g)$ $\Delta S > 0$
   $2H(g) + O(g) \rightarrow H_2O(g)$ $\Delta S < 0$
   $C(graphite) + 2O(g) \rightarrow CO_2(g)$ $\Delta S < 0$
   $C(graphite) + O_2(g) \rightarrow CO_2(g)$ $\Delta S < 0$
   $C(graphite) + 2H_2(g) \rightarrow CH_4(g)$ $\Delta S < 0$
   $2H(g) \rightarrow H_2(g)$ $\Delta S < 0$
   b. $2O(g) \rightarrow O_2(g)$ $\Delta S°$ = -117.0 J/mol K $\Delta H°$ = -498.4 kJ/mol

$H_2O(l) \rightarrow H_2O(g)$	$\Delta S° = 118.8$ J/mol K	$\Delta H° = 44.0$ kJ/mol
$2H(g) + O(g) \rightarrow H_2O(g)$	$\Delta S° = -201.7$ J/mol K	$\Delta H° = -927.0$ kJ/mol
$C(graphite) + 2O(g) \rightarrow CO_2(g)$	$\Delta S° = -114.1$ J/mol K	$\Delta H° = -891.9$ kJ/mol
$C(graphite) + O_2(g) \rightarrow CO_2(g)$	$\Delta S° = 2.9$ J/mol K	$\Delta H° = -393.5$ kJ/mol
$C(graphite) + 2H_2(g) \rightarrow CH_4(g)$	$\Delta S° = -80.8$ J/mol K	$\Delta H° = -74.6$ kJ/mol
$2H(g) \rightarrow H_2(g)$	$\Delta S° = -98.7$ J/mol K	$\Delta H° = -436.0$ kJ/mol

c.

$2O(g) \rightarrow O_2(g)$	$\Delta G° = -463.5$ kJ/mol
$H_2O(l) \rightarrow H_2O(g)$	$\Delta G° = 8.6$ kJ/mol
$2H(g) + O(g) \rightarrow H_2O(g)$	$\Delta G° = -866.9$ kJ/mol
$C(graphite) + 2O(g) \rightarrow CO_2(g)$	$\Delta G° = -857.9$ kJ/mol
$C(graphite) + O_2(g) \rightarrow CO_2(g)$	$\Delta G° = -394.4$ kJ/mol
$C(graphite) + 2H_2(g) \rightarrow CH_4(g)$	$\Delta G° = -50.5$ kJ/mol
$2H(g) \rightarrow H_2(g)$	$\Delta G° = -406.6$ kJ/mol

d.

$2O(g) \rightarrow O_2(g)$	$\Delta G° = -463.4$ kJ/mol
$H_2O(l) \rightarrow H_2O(g)$	$\Delta G° = 8.5$ kJ/mol
$2H(g) + O(g) \rightarrow H_2O(g)$	$\Delta G° = -866.9$ kJ/mol
$C(graphite) + 2O(g) \rightarrow CO_2(g)$	$\Delta G° = -857.8$ kJ/mol
$C(graphite) + O_2(g) \rightarrow CO_2(g)$	$\Delta G° = -394.4$ kJ/mol
$C(graphite) + 2H_2(g) \rightarrow CH_4(g)$	$\Delta G° = -50.5$ kJ/mol
$2H(g) \rightarrow H_2(g)$	$\Delta G° = -406.6$ kJ/mol

e. -801.1 kJ/mol

f. K is huge and will cause an overflow error on most calculators.

g. The reaction will be spontaneous at all temperatures.

6. A reaction will be spontaneous at all temperatures if the entropy change is positive and the enthalpy change is negative.

7. a. $\Delta G° = -174.1$ kJ/mol
   b. 1.02 atm
   c. 5.02%
   d. At temperatures above 314 K.
   e. At low temperatures, the reaction becomes more spontaneous and the equilibrium constant value increases. At high pressures, the reaction will also favor product formation with no change in the equilibrium constant value.

8. a. $\Delta G° = -199.5$ kJ/mol
   b. The reaction is product favored.
   c. 9.297×10^{34}
   d. $\Delta S° = -4.4$ kJ/mol
   e. The entropy change is almost zero reflecting the fact that there are the same number of gas particles on either side of the reaction and that the number of particles per molecule remains constant.
   f. The reaction is enthalpically driven since the free energy change is such a large negative number and the entropy change is such a small negative number.

9. a. 30.%
   b. Isobutane is more stable at room temperature.

Concept Questions:

1. The entropy of the universe increases even though the entropy of the system decreases.
2. At absolute zero a system has no motion or no entropy which is not possible.
3. The entropy change is negative for the condensation and the reaction is exothermic so it will be favorable only at low temperatures.
4. The reaction is carried out more slowly and more reversibly which allows more energy to be extracted – this is in accordance with the concept that the maximum available work is equal to the free energy change of the reaction.
5. This means that neither the product nor the reactant is favored at equilibrium.

6. The change is relatively small because of the hydrogen bonding network which gives a complex system. Methanol also hydrogen bonds and therefore also has a small entropy change of vaporization.
7. The higher the entropy, the more the molecular complexity and the more places for energy to be dispersed when heat is applied. When heat is stored as internal energy, the temperature change is small – this is a large heat capacity.

Chapter 18:

Fill in the Blank:

1. Nernst equation
2. electrolytic
3. force associated with electron motion
4. increase; loss
5. mass; charge
6. coulombs
7. votaic (or galvanic)
8. anode
9. electrical potential; volts
10. dry -cell
11. salt bridg e
12. Over -voltage
13. positive
14. oxidizin g; reduced
15. w ill
16. positive
17. fue l cell
18. zero
19. anode ; cathode
20. concentratio n
21. platinu m
22. w ater

Problems:

1. a. $4Al(s) + 3O_2(g) \rightarrow 2Al_2O_3(s)$
   b. $23H_2O(l) + 3NO_3^-(aq) + 8Al(s) \rightarrow 3NH_3(aq) + 8Al(OH)_4^-(s) + 5H^+(aq)$
   c. $3H_2O(l) + 3Cl_2(g) \rightarrow 5Cl^-(aq) + ClO_3^-(aq) + 6H^+(aq) + 4e^-$
   d. $2HNO_3(aq) + 3H_3AsO_3(aq) \rightarrow 2NO(g) + 3H_3AsO_4(aq) + H_2O(l)$
   e. $2NO_2(g) + 7H_2(g) \rightarrow 2NH_3(g) + 4H_2O(l)$
   f. $3Cu(s) + 8HNO_3(aq) \rightarrow 3Cu(NO_3)_2(aq) + 2NO(g) + 4H_2O(l)$
   g. $2Cr(OH)_3(s) + ClO_3^-(aq) + 4OH^-(aq) \rightarrow 2CrO_4^{2-}(aq) + Cl^-(aq) + 5H_2O(l)$
   h. $2Au^{3+}(aq) + 6I^-(aq) \rightarrow 2Au(s) + 3I_2(s)$
   i. $FeO(s) + CO(g) \rightarrow Fe(s) + CO_2(g)$
   j. $TiO_2(s) + 2Cu(s) + 4H^+(aq) \rightarrow 2Cu^{2+}(aq) + Ti(s) + 2H_2O(l)$
2. a. The cell does not have a salt bridge to maintain electrical neutrality. Without a salt bridge, current will not flow.
   b.

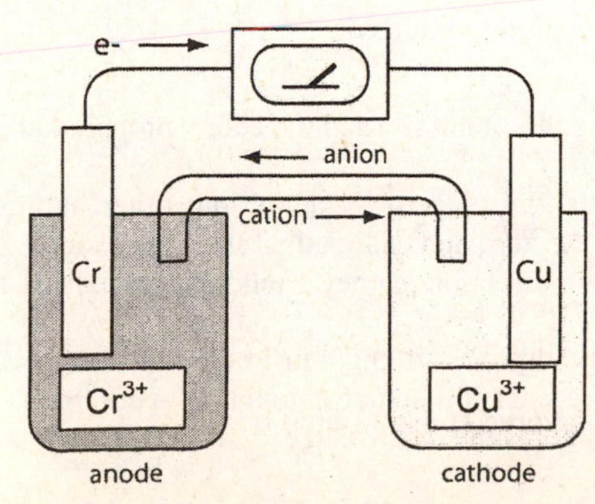

c. Oxidation: $Cr(s) \rightarrow Cr^{3+}(aq) + 3e^-$
Reduction: $Cu^{2+}(aq) + 2e^- \rightarrow Cu(s)$
d. $2Cr(s) + 3Cu^{2+}(aq) \rightarrow 3Cr^{3+}(aq) + 2Cu(s)$ $E° = 1.07V$
e. $Cr(s) \mid Cr^{2+}(0.010\ M) \parallel Cu^{3+}\ (0.20\ M) \mid Cu(s)$
f. $\Delta G° = -619$ kJ
g. $K = 10^{108}$
h. No, the potential that is read will not be the same as the standard cell potential because the concentrations of the reagents are not standard.
i. $E = 1.08V$

3. a. $[Pb^{2+}] = 50.4$ M
b. $[Pb^{2+}] = 0.02$ M
c. The concentration of lead(II) nitrate could be increased or the concentration of nickel(II) nitrate could be decreased.

4. a. 92,800 min.
b. 2.4 A
c. 0.64 A
d. It will take the same amount of time in a more dilute solution. The concentration of the solution is not a factor in calculating the current.

5. a. Yes, the reaction occurs spontaneously since the voltage is a positive number.
b. Oxidation: $Ti(s) + 2H_2O(l) \rightarrow TiO_2(s) + 4H^+(aq) + 4e^-$
Reduction: $Cu^{2+}(aq) + 2e^- \rightarrow Cu(s)$
c. $Ti(s), TiO_2(s) \mid H^+(aq) \parallel Cu^{2+}(aq) \mid Cu(s)$
d. $E° = 1.2110$ V
e. $E = 1.291$ V
f. A cell potential of zero will be measured when the reaction reaches equilibrium.

6. a. Anode: $2Br^-(aq) \rightarrow Br_2(g) + 2e^-$
Cathode: $Na^+(aq) + e^- \rightarrow Na(s)$
The reactions that occur at the anode and cathode are the reactions that have a smaller absolute value of the potential. The closer a negative potential is to zero, the easier it is to carry out that reaction.
b. The reactions will be different because the electrolysis of water is easier to carry out than the oxidation of bromine or reduction of sodium.

7. a. $C_3H_5O_2^- + 2H_2O \rightarrow C_2H_3O_2^- + CO_2 + 6H^+ + 6e^-$
b. $E° = 0.190$ V
$\Delta G° = -110{,}000$ J
c. $\Delta G° = 61800$ J
d. The conversion of lactate to pyruvate is spontaneous while the conversion of NAD^+ to NADH is not spontaneous.
e. When these reactions are carried out together, NAD^+ can be produced.
f. 5.56 mol of NAD^+ will be consumed when 1 mole of lactate is converted to acetate and CO_2.

Concept Questions:

1. a. B < D < A < C
b. B will be the cathode (reduced) and C will be the anode (oxidized).
2. $K_c > 1$, $\Delta G° < 0$, and $E°_{cell} > 0$.
3. a. The standard cell potential equals the measured cell potential when the cell is prepared at standard conditions.
b. The standard hydrogen electrode has a cell potential of zero by definition; other half-cells will have a reduction potential that is in reference to the zero and cannot, therefore, be exactly zero.
4. The electrons in a battery are traveling in the opposite direction as they would spontaneously travel in an electrolytic cell.
5. A negative cell potential represents the minimum amount of work that must be put into a galvanic cell in order for it to run while the actual voltage represents the actual amount of work required. The difference between the minimum and actual amounts gives the overvoltage.

Chapter 19:

Fill in the Blank:

1. nitrogen-14
2. Alpha emission
3. Fusion
4. cyclotron
5. fission,
6. parent; daughter.
7. strong force
8. neutrons; beta decay
9. beta emission
10. nucl ide
11. Geiger -Muller counter
12. m ass defect; energy; mass
13. positron

Problems:

1. a. ${}^{210}_{84}Po \rightarrow {}^{0}_{-1}e + {}^{210}_{85}At$
   b. ${}^{222}_{88}Ra \rightarrow {}^{4}_{2}He + {}^{218}_{86}Rn$
   c. ${}^{234}_{90}Th \rightarrow {}^{0}_{-1}e + {}^{234}_{91}Pa$
   d. ${}^{138}_{57}La \rightarrow {}^{0}_{1}e + {}^{138}_{56}Ba$
   e. ${}^{185}_{79}Au \rightarrow {}^{4}_{2}He + {}^{181}_{77}Ir$
   f. ${}^{40}_{19}K + {}^{0}_{-1}e \rightarrow {}^{40}_{18}Ar$
2. a. Beta decay
   b. Beta decay
   c. Sequential decay
   d. Stable
   e. Positron emission
   f. Positron emission
   g. Beta decay
3. a. ${}^{19}_{10}Ne \rightarrow {}^{0}_{1}e + {}^{19}_{9}F$
   b. 0, 4.18×10^{23} F atoms
   c. 1 half life – 17.22 s.
   d. 27.3 s
4. a. 15500 yr
   b. The age of the wood will be the same. The rate of decay determines the age, not the quantity of wood.
5. a. 0.340407 amu
   8.570 MeV/nucleon
   b. 0.664429 amu
   8.717 MeV/nucleon
   c. 0.958434 amu
   8.584 MeV/nucleon
   d. 1.44296 amu
   8.146 MeV/nucleon
6. a. 0.78 MeV
   b. 3.7095×10^{8} m/s
7. 120 J

Concept Questions:

1. Alpha decay is the least dangerous because it has the smallest penetrating ability even though it is the most energetic type of decay. Gamma radiation emission is the most harmful because it has the largest penetrating ability.
2. Magic numbers must be related to a physically relevant phenomenon such as the energetics of the subatomic particles in the nucleus.

3. There is no collision that needs to occur between the radioactive element and another element.

Chapter 20:

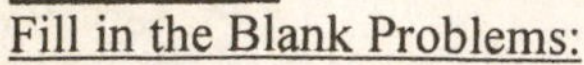

Fill in the Blank Problems:

1. Ethers
2. the most hydrogen atoms
3. structural formula
4. carbonyl
5. isotomers
6. enantiomers; a clockwise or counterclockwise direction
7. *para*-dichlorobenzene
8. racemic mixture
9. double bond
10. cis -trans isomers.
11. Stereoiso mers
12. h ydroxyl
13. oxidation -reduction reactions
14. saturated
15. phen yl
16. an addition pol ymer
17. di mer

Problems:

1.

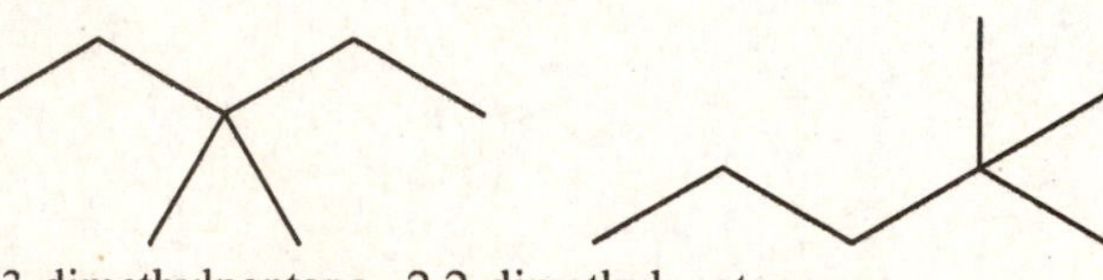

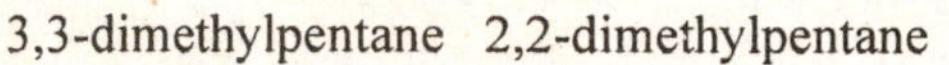

3,3-dimethylpentane 2,2-dimethylpentane

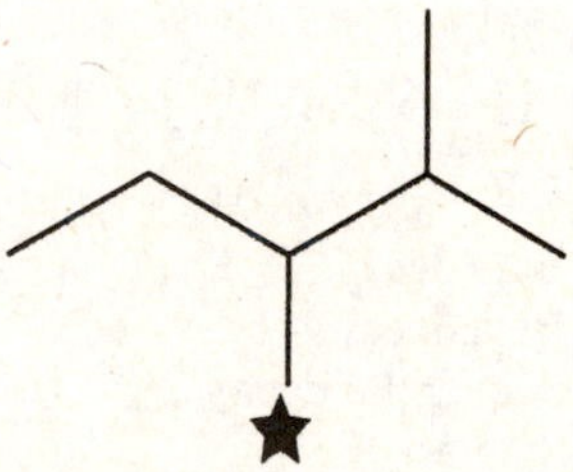

2,3-dimethylpentane

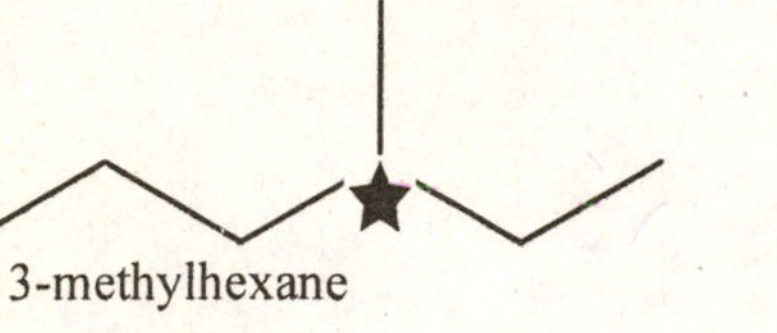

3-methylhexane

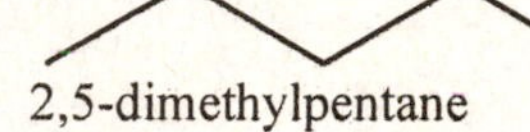

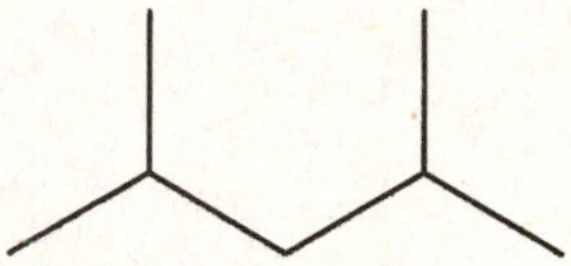

2,5-dimethylpentane

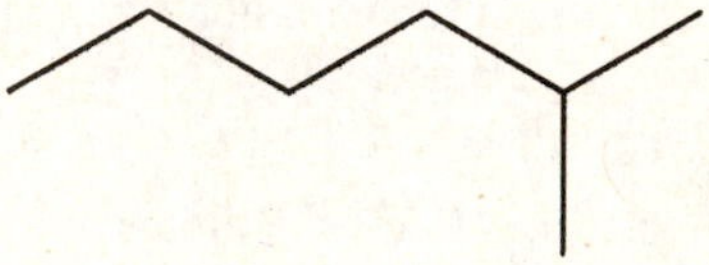

2-methylhexane

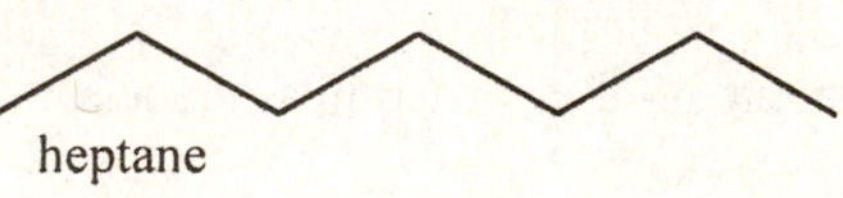

heptane

2.

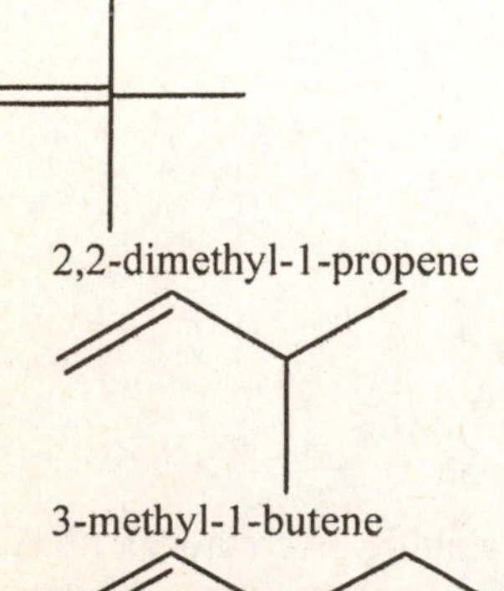

2,2-dimethyl-1-propene

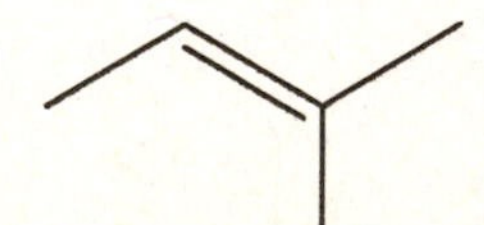

2-methyl-2-butene

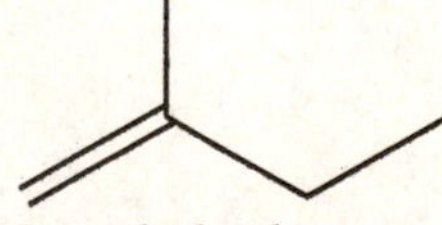

2-methyl-1-butene

3-methyl-1-butene

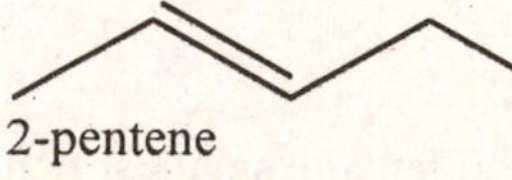

2-pentene

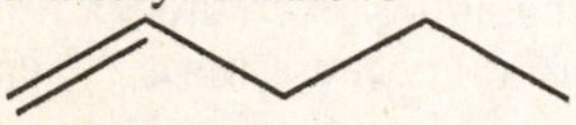

1-pentene

3. a. Ether: ethylpropyl ether

b. Amine: ethyldimethylamine
c. Ester: ethyl ethanoate
d. Benzene: 1,2-difluorobenzene
e. Alcohol: isopropanol or 2-propanol
f. Carboxylic acid: 2-3-dimethylpentanoic acid
g. Carboxylic Acid: 2-propylpentanoic acid
h. Alkyne: 4-methyl-2-hexyne
i. Carboxylic acid: 2,3-dimethyl-4-phenylbutanoic acid
j. Aldehyde: 3-ethylpentanal
k. Alkene: 2,5,6-trimethyl-2-heptene
l. Alcohol: 3-isopropyl-4-methyl-1-pentanol

5. a. F_2 reacts with benzene using iron(III) chloride as a catalyst

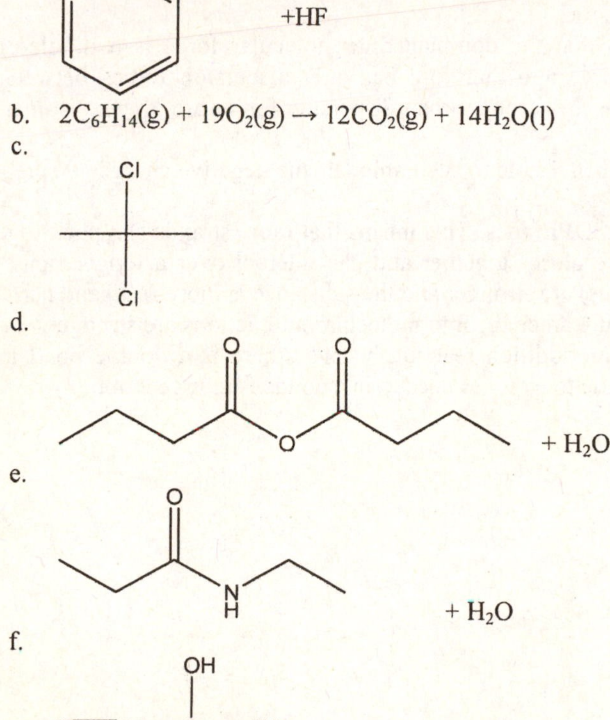

g.

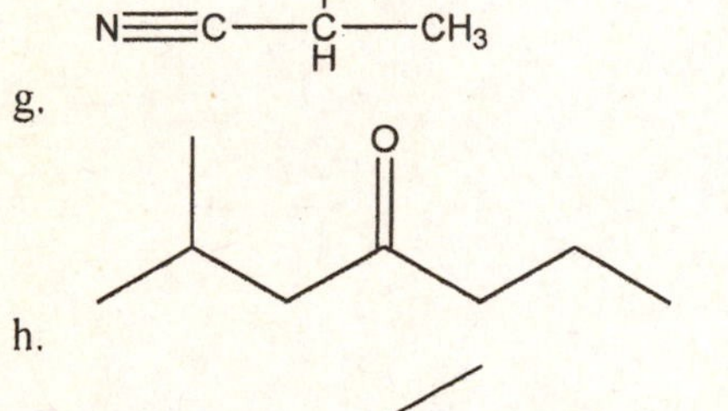

h.

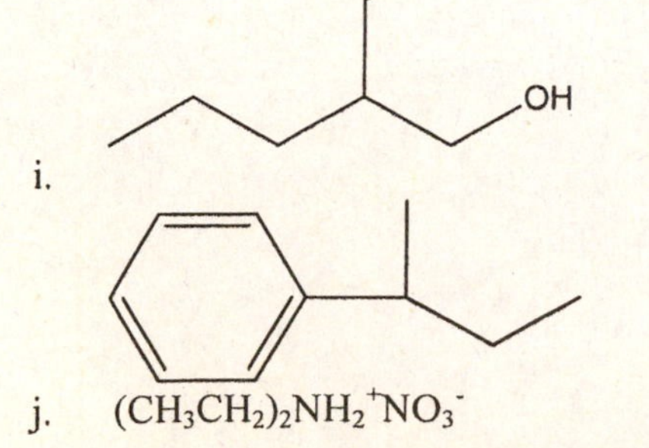

j. $(CH_3CH_2)_2NH_2^+NO_3^-$

6. a.

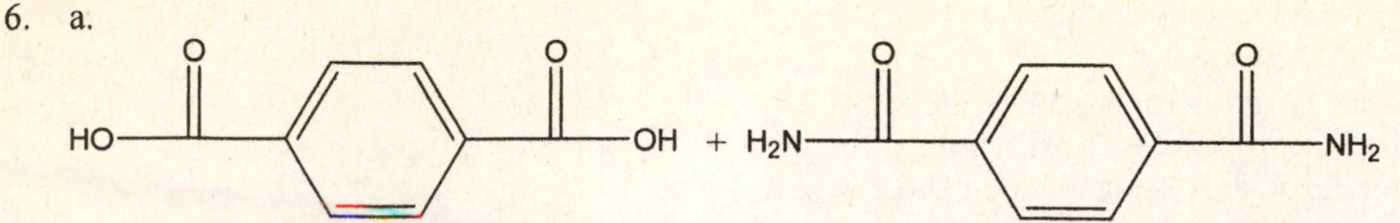

b. Kevlar is formed in a condensation reaction with the elimination of a water molecule as each monomer adds to the polymer.

c. Hydrogen bonds between the oxygen atoms and the hydrogen of the NH group will hold the strands together.

d. At high temperatures, the strands move more freely and the hydrogen bonding network is disrupted.

e. Polymers do not have specific molar masses associated with them since the strand length can vary widely.

Concept Questions:

1. The cis isomer has a dipole moment so that the dominant intermolecular force is a dipole-dipole interaction. The trans isomer has no dipole and therefore has only dispersion forces between the molecules. Since dipole-dipole interactions are stronger than dispersion forces are, the cis isomer will have a higher boiling point.
2. The more substituted carbon atom will be better able to accommodate the negative charge by spreading it out over a larger area.
3. a. HDPE does not have branches while LDPE does. This means that more strands can pack together in the HDPE form; as the strands are closer together and can interact over a longer range, the intermolecular forces (dispersion forces) are stronger and the substance is more rigid and hard.

   b. The HDPE melts at a higher temperature since the intermolecular interactions are stronger.

   c. Polyethylene is probably formed in an addition reaction because there is a double bond in the molecule and there is no water molecule to expel as needed in a condensation reaction.

Chapter 21:

Fill in the Blank Problems:

1. zwitterions
2. sugars; phosphate
3. secondary
4. aldose
5. chromosomes
6. glycosidic
7. polysaccharides
8. glycerol; three fatty acids.
9. amino acids; peptide
10. Primar y
11. six
12. messenger RNA; ribosome
13. fatt y acid; 3 to 19
14. compl ementary; hydrogen bonding interactions
15. phospholipid ; phosphate
16. codon ; amino acid
17. quaternar y
18. un saturated fatty acid

Problems:

1. d>a>c>b
2.

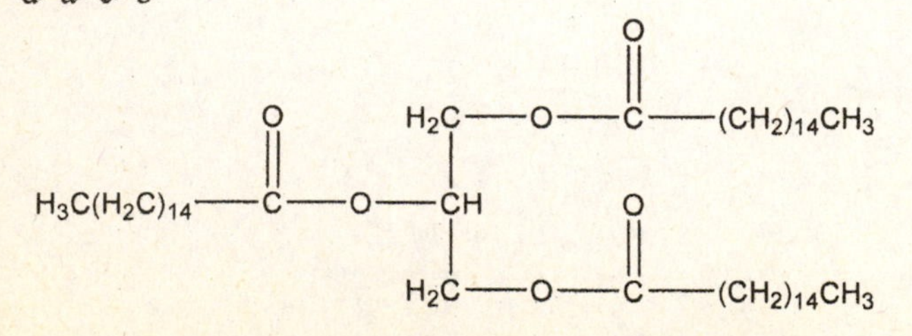

3.

4.

5. a.

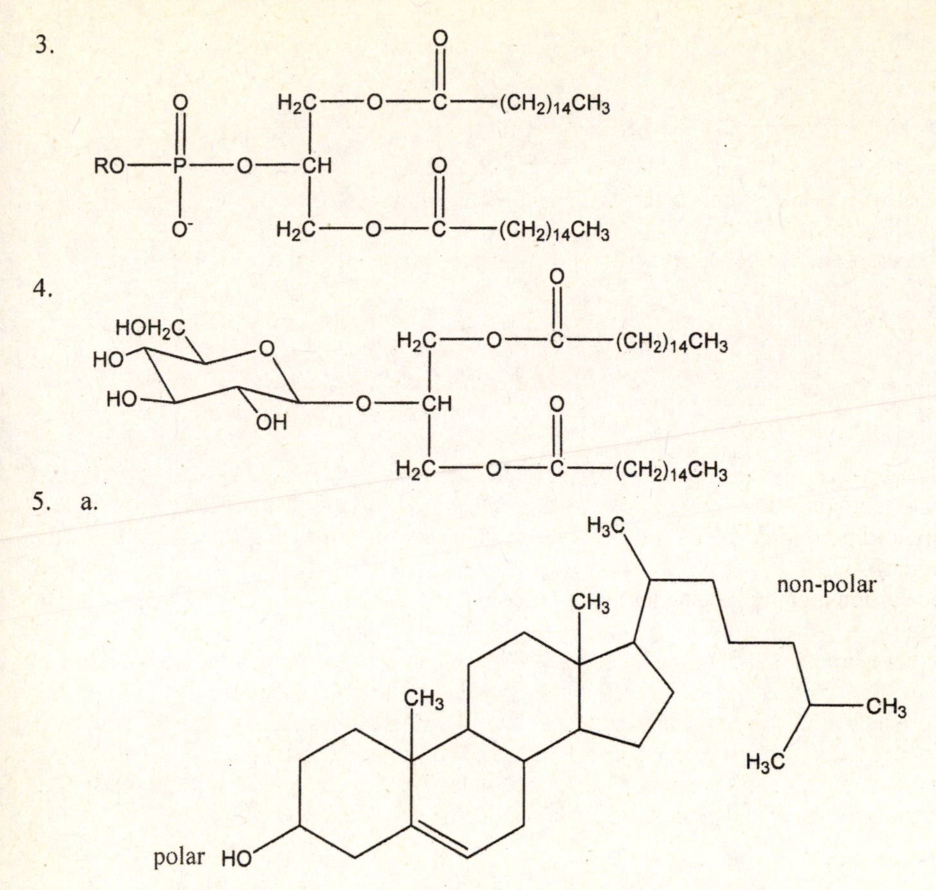

b. Oxidation to make the alcohol and dehydrogenation to form the double bonds.

6. a.

b.

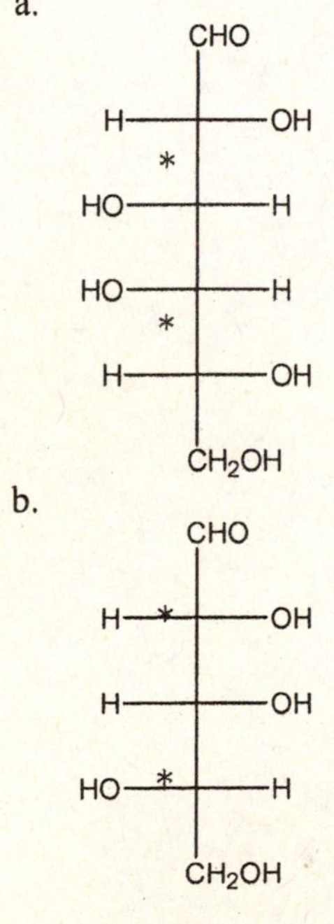

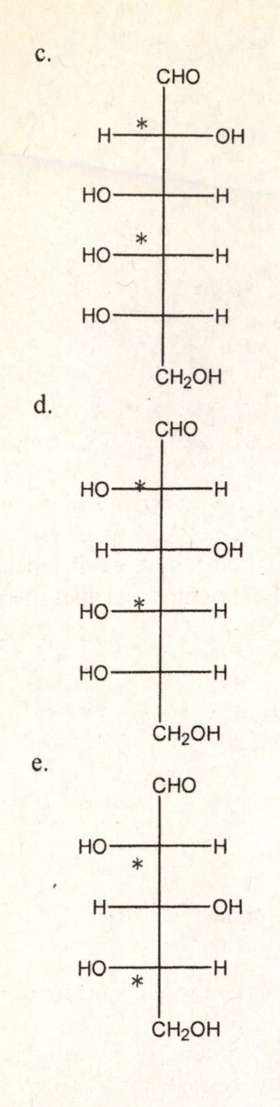

7. All of them are aldoses since they are all aldehydes.
   1. aldohexose
   2. aldopentose
   3. aldohexose
   4. aldohexose
   5. aldopentose
8. Chiral centers are indicated with a star.
9.

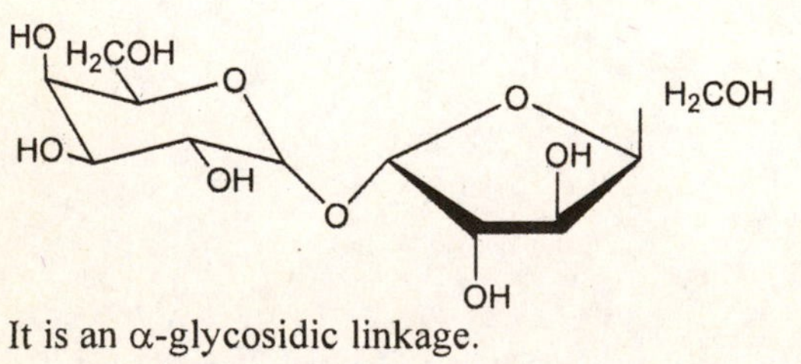

It is an α-glycosidic linkage.

10.

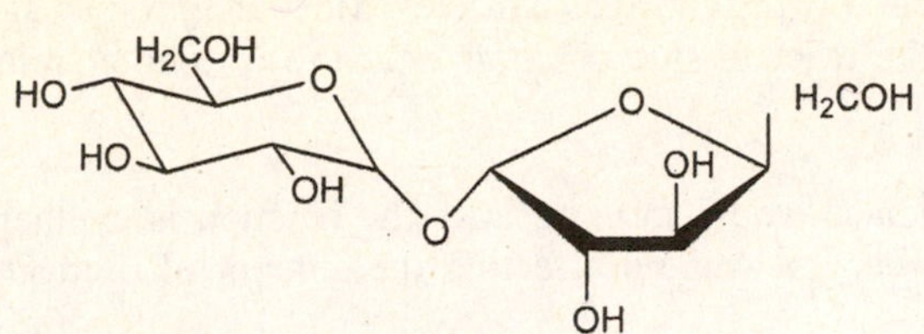

It is an α-glycosidic linkage.

11. Each codon codes for one amino acid and each protein is comprised of hundreds of amino acids so thousands of codons and amino acids will be needed.

Concept Questions:

1. Longer chains will have more extensive intermolecular interactions and will therefore pack more tightly. As the degree of unsaturation increases and the hydrocarbon chains become more rigid, the density will decrease.
2. The more branching, the less dense the substance will be. Cellulose is the least branched and glycogen is the most branched – this means that cellulose is the most dense and glycogen is the least dense.
3. a. Primary structure is the amino acid sequence. Secondary structure is the formation of repeating patterns. Tertiary structure are large scale bends and folds. Quaternary structure describes the way individual units fit together.
   b. Quaternary structure is not observed in proteins with a single unit.
4.

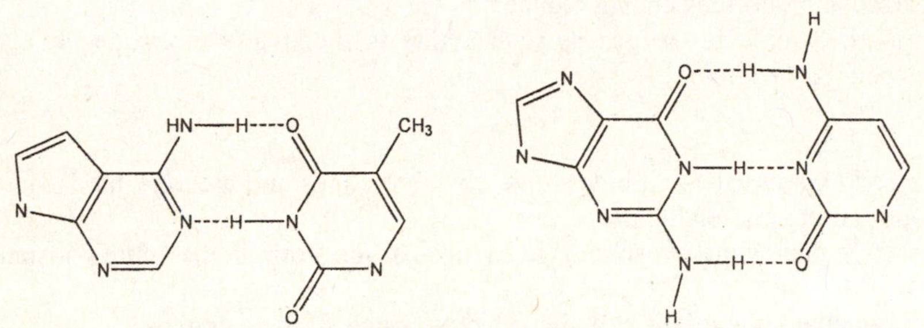

The bases have specific pairings: the A and T have two hydrogen bonds while the G and C have three hydrogen bonds.

5. If the interactions were too strong, then it would require too much energy to access the information stored in DNA. If the interactions were too weak, then the DNA would not be protected enough from damage.
6. Proteins often behave in ways that are dictated by their shapes while DNA does not. As a result, proteins need to have very specific shapes which result in structural complexity.

Chapter 22:

Fill in the Blank Problems:

1. fluorine; weak bond energy and small size.
2. sulfuric acid
3. charcoal
4. orthosilicates
5. Phyllosilicates
6. $B_nH_n^{2=}$
7. Silicates
8. graphite; diamond
9. nitrogen.
10. SiO_2
11. iodine
12. S_8

Problems:

1. a. $3H_2(g) + N_2(g) \rightarrow 2NH_3(g)$

b. High pressures promote the side of the reaction with fewer gas particles. Since the product side of this equation has fewer particles (2) than the reactant side (4), high pressures promote ammonia formation.
c. $\Delta H° = -91.8$ kJ/mol
d. If high temperatures were used, fewer products would form because the reaction is exothermic. Using a catalyst increases the rate of the reaction without affecting the amount of product that forms.

2. This is possible when there are resonance structures. This molecule must have double bonds and resonance structures.
3. a. $4C(s) + S_8(s) \rightarrow 4CS_2(s)$
   b. Reactants: C = 0 and S = 0.
   Products: C = +4 and S = -2
   c. Carbon is being oxidized and sulfur is being reduced.
   d. S=C=S
   e. No other possibilities exist unless we introduce a charge or use a molecule that does not agree with the octet rule. Carbon will not have an expanded octet since it has no d orbitals and sulfur is not a central atom since it is more electronegative than carbon.
4. Silver is oxidized in this reaction.
5. a. $O_2(g) + 2F_2(g) \rightarrow 2OF_2(g)$
   b. Reactants: O = 0 and F = 0
   Products: O = +2 and F = -1
   c. Oxygen is being oxidized and fluorine is being reduced.
   d. No, bromine will not likely oxidize oxygen since bromine is much less electronegative than oxygen as well as fluorine.

Concept Questions:

1. Ionic carbides have ionic bonding; covalent carbides have covalent bonds and are very hard, metallic carbides have metallic bonds and have metallic properties.
2. Red phosphorus is less reactive than white phosphorus is because of the strain in the white phosphorus structure.
3. The empty p orbital of the boron atom overlaps with the filled p orbitals of the halogens.
4. The metal atoms lend the positive charge necessary to balance the negative charge of the silicate. It is possible for a solid to form without the metal ions if there are ions of metalloids present.
5. Nonmetals are more electronegative than metals are and therefore attract electrons when bonding with metals. The gain of electrons by nonmetals is reduction.
6. Nitrogen and oxygen gas are very stable (low energy). In order to form a nitrogen oxide, the stable oxygen and nitrogen bonds must break in favor of weaker nitrogen-oxygen bonds.
7.

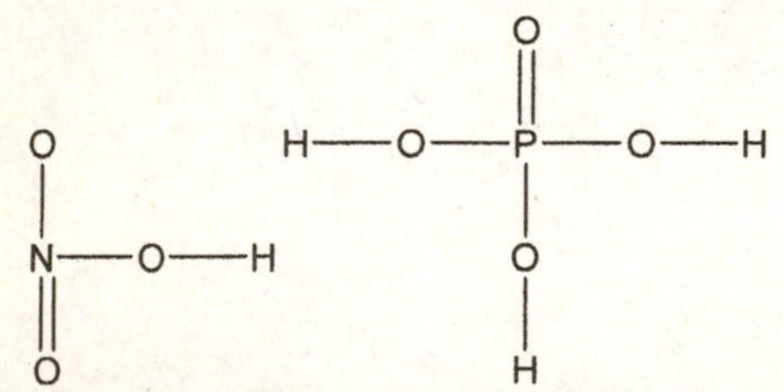

Phosphorus has accessible d orbitals and can therefore bond with an expanded octet. Nitrogen is much smaller and will not expand its valence shell.
8. Halides are very electronegative and therefore tend to be reduced when they interact with other elements.

Chapter 23:

Fill in the Blank Problems:

1. leaching
2. arc-melted
3. metallurgy
4. octahedral; tetrahedral

5. Refining
6. minerals
7. substitutional alloy
8. Powder metallurgy
9. alloy
10. bronze ; brass
11. t wo-phase region
12. Calci nation
13. ferro magnetic
14. gan gue

Problems:

1. a.

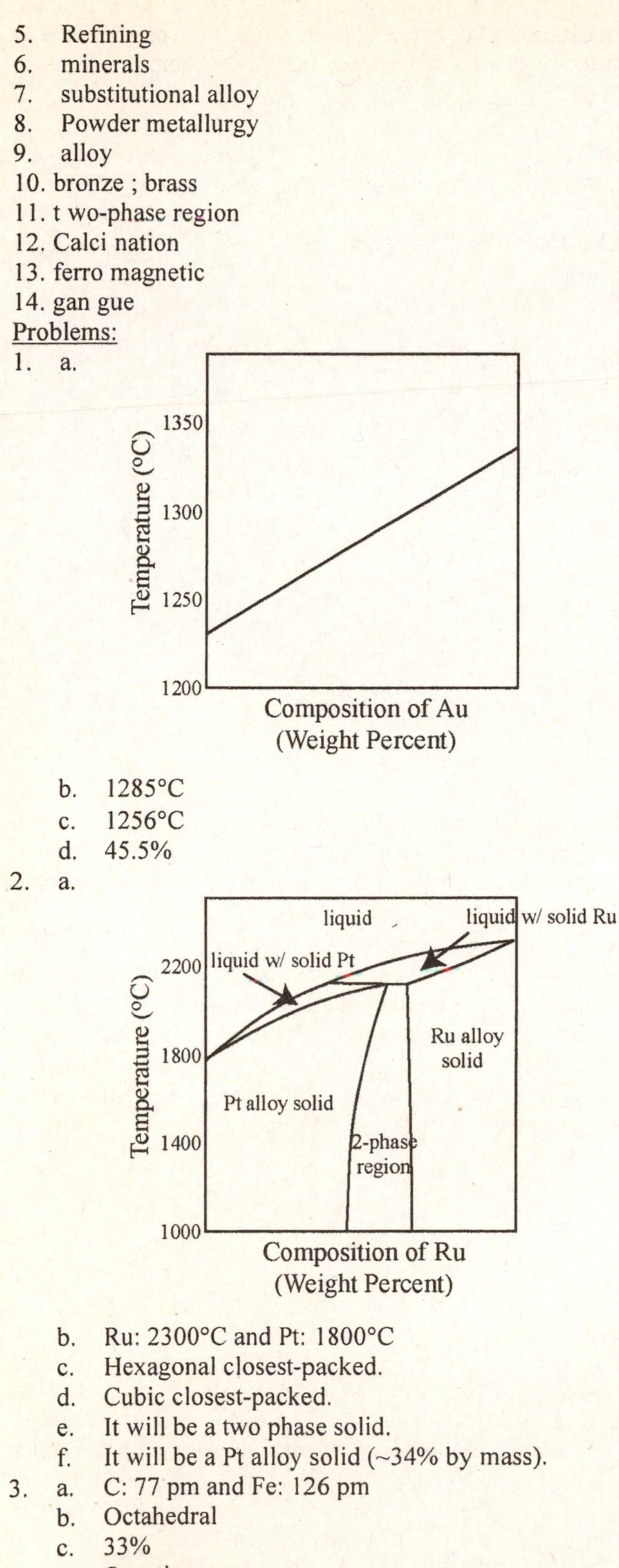

   b. 1285°C
   c. 1256°C
   d. 45.5%

2. a.

   b. Ru: 2300°C and Pt: 1800°C
   c. Hexagonal closest-packed.
   d. Cubic closest-packed.
   e. It will be a two phase solid.
   f. It will be a Pt alloy solid (~34% by mass).

3. a. C: 77 pm and Fe: 126 pm
   b. Octahedral
   c. 33%

Concept Questions:

1. Malleable, good conductors of heat, good conductors of electricity, and they tend to lose electrons. Malleability and conductivity are a result of the metallic bonding with bands formed over the entire metal instead of discrete bonds. The tendency to lose electrons is a result of the fact that metals are not very electronegative.

2. a. They tend to lose electrons or tend to form cations.

   b. $3MnO_2(s) + 4Al(s) \rightarrow Mn(s) + 2Al_2O_3(s)$
      Reactants: Mn = +4, O = -2, Al = 0
      Products: Mn = 0, O = -2, Al = +3
      $TiCl_4(g) + 2Mg(s) \rightarrow 2MgCl_2(l) + Ti(s)$
      Reactants: Ti = +4, Cl = -1, Mg = 0
      Products: Ti = 0, Cl = -1, Mg = +2
      $3FeCr_2O_4(s) + 8Al(s) \rightarrow 4Al_2O_3(s) + Cr(s) + 3Fe(s)$
      Reactants: Fe = +2, Cr = +3, O = -2, Al = 0
      Products: Fe = 0, Cr = 0, O = -2, Al = +3
      $Hg(g) + SO_2(g) \rightarrow HgS(s) + O_2(g)$
      Reactants: Hg = 0, S = +4, O = -2
      Products: Hg = +2, S = -2, O = 0
      $Zn(l) + CO(g) \rightarrow ZnO(s) + C(s)$
      Reactants: Zn = 0, C = +2, O = -2
      Products: Zn = +2, C = 0, O = -2

Chapter 24:

Fill in the Blank Problems:

1. strong-field; weak-field
2. increases
3. ligands
4. Coordination
5. spectrochemical series
6. oxidation state; coordination number
7. lanthanide contraction
8. Geometric; cis-trans and fac-mer.
9. Low-spin
10. polyden tate ligands or chelates.
11. z^2 and x^2-y^2; xy, xz, and yz
12. Le wis acid; a Lewis base.

Problems:

1. a. Potassium hexacyanoferrate(II)
   b. Tetraamminetetrachloroplatinum(IV)
   c. Hexacarbonylchromium(0)
   d. Triamminetriaquachromium(III) chloride
   e. Tetraamminedichlorocobalt(III) chloride
   f. Amminetrichloroplatinate(II) ion
   g. Trisethylenediamminenickel(II) chloride
   h. Tetraamminebromochlorocobalt(II)
2. a. $[Co(NH_3)_5Br]SO_4$
   b. $[Co(NH_3)_5SO_4]^+$
   c. $[CoF_6]^{3-}$
   d. $[Cu(H_2O)_2(CN)_4]Cl_2$
   e. $[Co(en)_3]Cl_3$
   f. $[Cr(NH_3)_5Cl]SO_4$
   g. $Pt(NH_3)_2Cl_2$
3. a. Coordination isomer
   b. Linkage isomer
   c. Coordination isomer

4. a.

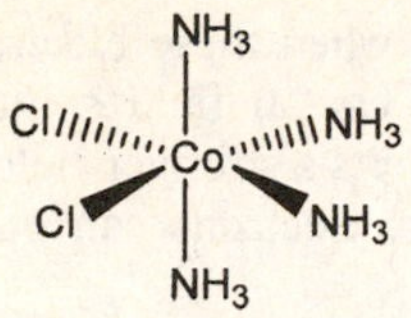

cis 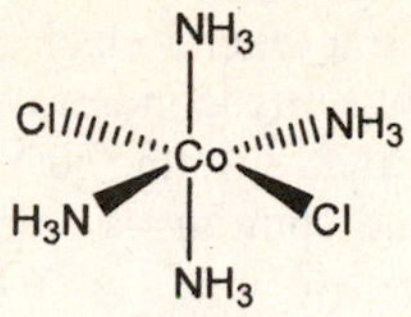 trans

b. No isomers exist
c. No isomers exist
d.

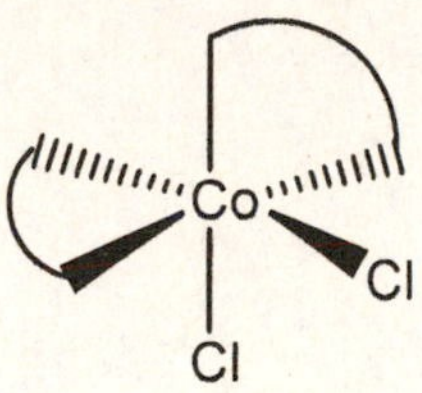

cis 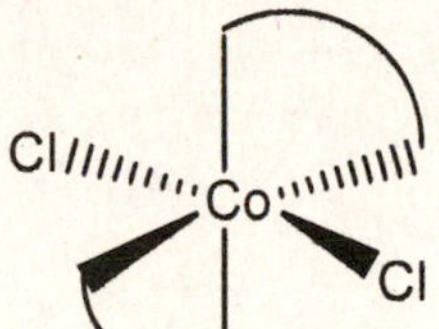 trans

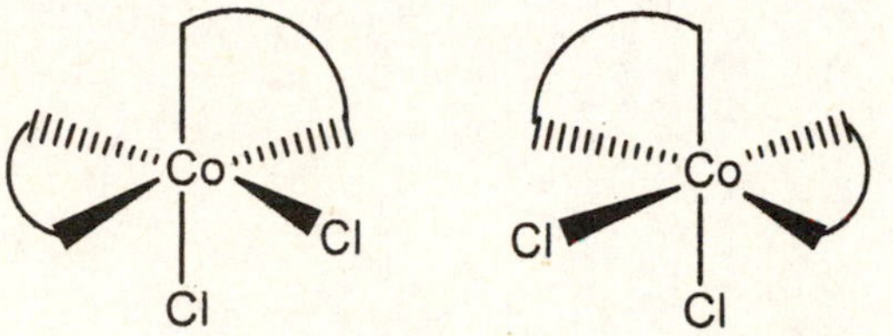

enantiomers

e.

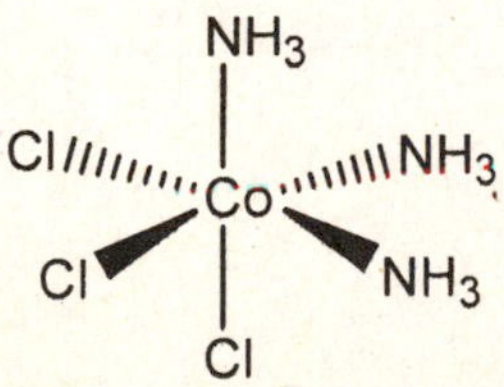

fac 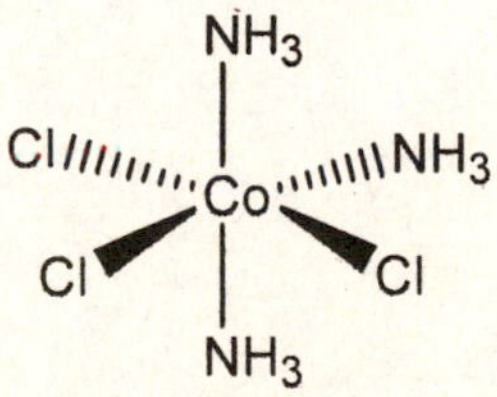 mer

f. No isomers exist.

5. a. 0
b. 0
c. 0
d. 3
e. 0
f. 4

6. a = b = c = e < d < f

7. A green compound absorbs red light which is low in energy. The compounds with the smallest crystal field splitting will absorb red light so d and f will appear green.

Concept Questions:

1. a. Br and SO_4 have different field splitting properties so the transition from the lower to higher energy state will have different energies.
b. Isomers with ligands that have similar crystal field splitting abilities will look the same.
c. The violet compound has a smaller field splitting so it is more likely to be magnetic.
d. SO_4^{2-} must exert a stronger ligand field since the energy difference between the levels is greater.
e. Green light: ~520 nm or $4x10^{-19}$J. Yellow light: ~575 nm or $3x10^{-19}$J. And we see that the green light is higher in energy and so the red complex has a larger crystal field splitting.

2. See Figures 24.13, 24.17, and 24.18. The tetrahedral configuration appears to promote magnetism more because the energy between the levels is always smaller.
3. a. In the original problem, electrical energy is used to excite electrons; when these electrons relax back down, they emit light with an energy that corresponds to the crystal field splitting. The compound that appears green has a larger crystal field splitting so Y exerts a stronger field.
   b. The magnetic compound is MX_6. Since the field splitting is smaller, electrons can occupy the higher energy level.
   c. When illuminated with white light, the compounds will appear as their complementary colors: MY_6 will appear red and MX_6 will appear green.
   d. X donates more electron density to the metal since the CO bond energy changes more in the metal compound with X than the one with Y.